ADVANCES IN ENZYMOLOGY

AND RELATED AREAS OF MOLECULAR BIOLOGY

Volume 77

ADVANCES IN ENZYMOLOGY

AND RELATED AREAS OF MOLECULAR BIOLOGY

Founded by F. F. NORD

Edited by ERIC J. TOONE

DUKE UNIVERSITY, DURHAM, NORTH CAROLINA

VOLUME 77

WILEY

A JOHN WILEY & SONS, INC., PUBLICATION

Library of Congress Cataloging-in-Publication Data:

ISBN 978-0-470-63835-4

Printed in Singapore

eBook ISBN: 978-0-470-92052-7
oBook ISBN: 978-0-470-92054-1
ePub ISBN: 978-0-470-92293-4

10 9 8 7 6 5 4 3 2 1

CONTENTS

CONTRIBUTORS

LEONARD AMARAL, Unit of Mycobacteriology, Instituto de Higiene e Medicina Tropical, Universidade Nova de Lisboa, Lisbon, Portugal

RAVI D. BARABOTE, Department of Plant Sciences, University of California, Davis, CA 95616

KIM E. BETTANEY, Institute for Membrane and Systems Biology, Astbury Centre for Structural Molecular Biology, Faculty of Biological Sciences, University of Leeds, Leeds, LS2 9JT, UK

VICTORIA I. BUNIK, School of Bioinformatics and Bioengineering, and Belozersky Institute of Physico-Chemical Biology, Moscow Lomonosov State University, Moscow, Russian Federation

ARTHUR J. L. COOPER, Department of Biochemistry and Molecular Biology, New York Medical College, Valhalla, NY 10595

NATALIA DUDAREVA, Department of Horticulture and Landscape Architecture, Purdue University, West Lafayette, IN 47907

ANNA FÀBREGA, Departamento de Microbiología Clínica, Centro de Diagnóstico Biomédico, Hospital Clínic, Facultad de Medicina, Universidad de Barcelona, IDIBAPS, Barcelona, Villarroel, 170; 08036 Barcelona, Spain

SEAMUS FANNING, Centre for Food Safety, UCD Veterinary Sciences Centre, University College Dublin, Belfield, Dublin 4, Ireland

JOE A. FRALICK, Department of Microbiology and Immunology, Texas Tech University Health Sciences Center, Lubbock, TX 79430

PETER J. F. HENDERSON, Institute for Membrane and Systems Biology, Astbury Centre for Structural Molecular Biology, Faculty of Biological Sciences, University of Leeds, Leeds, LS2 9JT, UK

ALVARO HERNÁNDEZ, Departamento de Biotecnología Microbiana, Centro Nacional de Biotecnología and CIBERESP, Darwin 3, Cantoblanco, 28049 Madrid, Spain

JEFFREY N. KEEN, Institute for Membrane and Systems Biology, Astbury Centre for Structural Molecular Biology, Faculty of Biological Sciences, University of Leeds, Leeds, LS2 9JT, UK

ANNE-BRIT KOLSTO, Laboratory for Microbial Dynamics, Department of Pharmaceutical Biosciences, University of Oslo, P.O. Box 1068 Blindern, 0316 Oslo, Norway

JASMIN K. KROEGER, Laboratory for Microbial Dynamics, Department of Pharmaceutical Biosciences, University of Oslo, P.O. Box 1068 Blindern, 0316 Oslo, Norway

DONG LENG, Institute for Membrane and Systems Biology, Astbury Centre for Structural Molecular Biology, Faculty of Biological Sciences, University of Leeds, Leeds, LS2 9JT, UK

FENG LONG, Molecular, Cellular and Developmental Biology Interdepartmental Graduate Program, Iowa State University, Ames IA 50011

PIKYEE MA, Institute for Membrane and Systems Biology, Astbury Centre for Structural Molecular Biology, Faculty of Biological Sciences, University of Leeds, Leeds, LS2 9JT, UK

JOSÉ LUIS MARTÍNEZ, Departamento de Biotecnología Microbiana, Centro Nacional de Biotecnología and CIBERESP, Darwin 3, Cantoblanco, 28049 Madrid, Spain

INGERID NES, Laboratory for Microbial Dynamics, Department of Pharmaceutical Biosciences, University of Oslo, P.O. Box 1068 Blindern, 0316 Oslo, Norway

HIROSHI NIKAIDO, Department of Molecular and Cell Biology, Barker Hall, University of California, Berkeley, CA 94720-3202

JEAN-MARIE PAGÈS, UMR-MD-1, IFR88, Facultés de Médecine et de Pharmacie, Université de la Méditerranée, Marseille, France

IAN C. PAULSEN, Department of Chemistry and Biomolecular Sciences, Macquarie University, Sydney, Australia, 2041

JOHN T. PINTO, Department of Biochemistry and Molecular Biology, New York Medical College, Valhalla, NY 10595

QINGHU REN, Department of Chemistry and Biomolecular Sciences, Macquarie University, Sydney, Australia, 2041

IGNASI ROCA, Departamento de Microbiología Clínica, Centro de Diagnóstico Biomédico, Hospital Clínic, Facultad de Medicina, Universidad de Barcelona, IDIBAPS, Barcelona, Villarroel, 170; 08036 Barcelona, Spain

MATHEW D. ROUTH, Molecular, Cellular and Developmental Biology Interdepartmental Graduate Program, Iowa State University, Ames, IA 50011

NICHOLAS G. RUTHERFORD, Institute for Membrane and Systems Biology, Astbury Centre for Structural Molecular Biology, Faculty of Biological Sciences, University of Leeds, Leeds, LS2 9JT, UK

MASSOUD SAIDIJAM, Institute for Membrane and Systems Biology, Astbury Centre for Structural Molecular Biology, Faculty of Biological Sciences, University of Leeds, Leeds, LS2 9JT, UK; currently, Department of Molecular Medicine and Genetics, School of Medicine, Hamedean University of Medical Sciences, Hamedan, Iran

MICHAEL J. SAN FRANCISCO, Department of Biological Sciences, Texas Tech University, and Department of Microbiology and Immunology, Texas Tech University Health Sciences Center, Lubbock, TX 79409

JOHN V. SCHLOSS, Department of Pharmaceutical Sciences, University of New England, Portland, ME 04104

WILLIAM M. SHAFER, Department of Microbiology and Immunology, Emory University School of Medicine, Atlanta, GA 30322, and Laboratories of Microbial Pathogenesis, VA Medical Center, Decatur, GA 30033

RICHARD E. STRAUSS, Department of Biological Sciences, Texas Tech University, Lubbock, TX 79409

CHIH-CHIA SU, Department of Chemistry, Iowa State University, Ames, IA 50011

GERDA SZAKONYI, Institute for Membrane and Systems Biology, Astbury Centre for Structural Molecular Biology, Faculty of Biological Sciences, University of Leeds, Leeds, LS2 9JT, UK; currently, Institute of Pharmaceutical Analysis, Faculty of Pharmacy, University of Szeged, Szeged, Hungary

JOSE THEKKINIATH, Department of Biological Sciences, Texas Tech University, Lubbock, TX 79409

GOVINDSAMY VEDIYAPPAN, Department of Microbiology and Immunology, Texas Tech University Health Sciences Center, Lubbock, TX 79430

JORDI VILA, Departamento de Microbiología Clínica, Centro de Diagnóstico Biomédico, Hospital Clínic, Facultad de Medicina, Universidad de Barcelona, IDIBAPS, Barcelona, Villarroel, 170; 08036 Barcelona, Spain

ALISON WARD, Institute for Membrane and Systems Biology, Astbury Centre for Structural Molecular Biology, Faculty of Biological Sciences, University of Leeds, Leeds, LS2 9JT, UK

ZHIQIANG XU, Institute for Membrane and Systems Biology, Astbury Centre for Structural Molecular Biology, Faculty of Biological Sciences, University of Leeds, Leeds, LS2 9JT, UK

EDWARD W. YU, Molecular, Cellular and Developmental Biology Interdepartmental Graduate Program, Department of Chemistry, Department of Physics and Astronomy, and Department of Biochemistry, Biophysics and Molecular Biology, Iowa State University, Ames, IA 50011

YARAMAH ZALUCKI, Department of Microbiology and Immunology, Emory University School of Medicine, Atlanta, GA 30322

QIJING ZHANG, Department of Veterinary Microbiology and Preventive Medicine, College of Veterinary Medicine, Iowa State University, Ames, IA 50011

PREFACE

Bacterial infection remains a leading cause of death in both the Western and developing world. Although the discovery of penicillin heralded the beginning of the age of antibiotics, the phenomenon of resistance emerged almost immediately. Since that time it has become clear that the effective pharmaceutical treatment of bacterial infection will never be complete, but rather, will continue as a complex fight on the ever-shifting battlefield of bacterial evolution and gene transfer. Bacterial drug resistance is most often the result of one of four broad mechanisms: metabolic inactivation of the active species, mutation of the drug target, the emergence of alternate metabolic pathways that obviate the effect of antibiotic action, and removal of active drug agent by active transport. Because of the extraordinary importance of antibiotic resistance in modern medicine, this volume represents the first of a series on the biochemical basis of antibiotic resistance.

Although it had long been imagined that the complex cell wall of gram-negative organisms was responsible for the resistance to antibiotic treatment so commonly observed, it became apparent through the 1980s that the resistance was the result of a complex series of multidrug efflux transporters. Indeed, nearly 40 such transporters have been identified to date, and at least 20 are proposed to act as multidrug efflux pumps. Efflux pumps are divided broadly into five families, based on sequence homology: the ATP-binding cassette family, which harnesses the energy of direct ATP hydrolysis to move solutes against a gradient, and the major facilitator superfamily, small multidrug resistance family, resistance–nodulation–cell division superfamily, and multiple antibiotic and toxin extrusion protein family, all of which use ion gradients as energy sources.

Among the various classes of bacterial efflux pumps the resistance–nodulation–division (RND) family is widely considered as the most important to the broad phenomenon of bacterial resistance, and we have focused on this group here. Nikaido and co-workers performed some of the earliest work toward discovering the role of this group of proteins in bacterial resistance and in elucidating their mechanism of action. In a review of the RND family of efflux pumps, he provides both a unique historical perspective on the discovery of the family, and a review of biochemical, biophysical, and genetic studies on both the activity and inhibition of this extraordinarily important group of proteins. The activity of bacterial efflux pumps operates under extraordinary multilevel control, and Amaral and co-workers report on the nature and mechanism of that control, inducers of RND efflux pump activation, and inhibitors of bacterial efflux pumps. Finally, Yu and co-workers review the extensive body of knowledge that has accumulated on the structure of tripartite RND transporters vital to the pathogenicity of several important human pathogens, including *Escherichia coli, Campylobacter,* and *Neisseria,* and describe the use of this structural information for the elucidation of mechanism and the development of inhibitors.

The largest and structurally most diverse group of secondary membrane transport proteins is the major facilitator superfamily (MFS) group. Despite the group's size—some 15,000 members, representing 25% of all known prokaryotic membrane transporters—relatively little is known about the fundamental biochemistry and structural biology of the group. This lack of knowledge is largely the result of the composition of MFS proteins; as transmembrane proteins they are hydrophobic, poorly behaved in solution, and difficult to purify. In the fourth chapter, Henderson and co-workers review the strategies relevant to the expression and purification of members of the MFS class of efflux pumps.

The quinolone antibiotics occupy a special place in the history of medicinal chemistry, as being among the first antibiotics developed against malaria. Later, quinolone antibiotics were widely used for the treatment of urinary tract infections, and several later-generation drugs, including ciprofloxacin, remain in use today for the treatment of a variety of bacterial infections. The utility of almost all quinolone antibiotics is greatly limited by the phenomenon of multidrug resistance, due in large part to the activity of efflux pumps. In the fifth chapter, Vila and co-workers consider the role of various efflux pumps in quinolone resistance in important human pathogens, including gram-positive and gram-negative bacteria, in Enterobacteriaceae and in other species of clinical relevance, including *Haemophilus influenzae, Campylobacter,* and *Mycobacterium.*

Most transporter proteins do not function to expel drugs, but rather, work to import various vital solutes into cells. Indeed, nearly 650 families of transporter proteins have been identified and catalogued, grouped roughly into 10 or so superfamilies. In addition to the massive complexity afforded by the sheer number of transporters, the study of transmembrane transporters is further hampered by the structural complexity of multimeric transmembrane proteins and the difficulties associated with culturing many pathogens. As a result of this complexity, much of our knowledge of efflux-pump-attributable multidrug resistance arises from genomic studies of pathogens and pathogenic flora. In the sixth chapter, San Francisco and co-workers report on the genomics of efflux pumps in both bacteria and fungi, and how such approaches offer novel avenues for the development of new therapeutic agents.

A final contribution to Volume 77 comes from Cooper and co-workers. This final review is unrelated to bacterial efflux pumps and multidrug resistance, but rather, considers the mechanism of paracatalyltic reactions involving various electrophiles, including oxygen, induced by carbanion-generating enzymes. In such reactions a carbanionic reductant is inter-cepted by an electrophilic species, leading to off-pathway products and, in many instances, to modification of enzyme side chains. The ubiquitous presence of oxygen as an electrophilic species renders this process of special importance in many biological settings, and the deleterious effect of various enzyme-bound reactive oxygen species contribute to various pathophysiological conditions. Cooper and co-workers consider the en-zymes susceptible to such paracatalytic reactions and consider the role of the intermediates and lesions in cancer and in neurodegenerative disease.

ERIC J. TOONE

STRUCTURE AND MECHANISM OF RND-TYPE MULTIDRUG EFFLUX PUMPS

By HIROSHI NIKAIDO, *Department of Molecular and Cell Biology, Barker Hall, University of California, Berkeley, California*

CONTENTS

I. Introduction

It has long been known that gram-negative bacteria are usually much more resistant than gram-positive bacteria to the actions of antibiotics and

Advances in Enzymology and Related Areas of Molecular Biology, Volume 77
Edited by Eric J. Toone   Copyright © 2011 John Wiley & Sons, Inc.

chemotherapeutic agents. The cells of the former group are surrounded by an extra membrane layer, the outer membrane (OM), and it was suspected that this additional membrane layer acted as a general barrier for the influx of agents. Indeed, the nonspecific diffusion channels of OM, the porins, limit the influx of small hydrophilic agents, because their channels are quite narrow (7 by 11 Å in the *Escherichia coli* OmpF porin) (1, 2). In addition, the lipid bilayer domain of the OM is unusual in its extreme asymmetry by having the outer leaflet composed nearly exclusively of lipopolysaccharides containing only saturated fatty acid chains (3), and the very low fluidity of this outer leaflet decreases the spontaneous permeation rates of hydrophobic probe molecules by nearly two orders of magnitude compared with conventional phospholipid bilayers containing many unsaturated fatty acid residues (4). Nevertheless, when the permeability coefficients of β-lactams across *E. coli* OM were measured, even the slowest-penetrating compounds were found to equilibrate across the outer membrane usually within a minute, that is, in a time span that is much shorter than the generation time of *E. coli* (5), in part because the surface-to-volume ratio is very high in these small bacterial cells. It was thus clear that the OM barrier alone cannot generate significant levels of resistance, and another major process must work synergistically with this barrier (5). With β-lactams, the ubiquitous periplasmic β-lactamases can fulfill this role. However, with other classes of antibiotics, the nature of this second process remained obscure until the constitutive multidrug efflux pumps were found to be widespread in gram-negative bacteria and to act synergistically with the OM barrier (6, 7).

Gram-negative bacteria usually contain multidrug efflux pumps belonging to several families, such as ABC (ATP-binding cassette), SMR (small multidrug resistance), MFS (major facilitator superfamily), MATE (multiple antibiotic and toxin extrusion), and RND (resistance–nodulation–division) (8–10). Among these, most are pumps located in the cytoplasmic membrane and must pump out drugs rapidly into the periplasm because the drugs can penetrate back into cytosol frequently by spontaneous diffusion. Only the RND pumps (and a few exceptional pumps belonging to other families) exist in a tripartite form traversing both the OM and the inner membrane, in a manner first suggested by Wandersman and co-workers for a protein-secreting apparatus of gram-negative bacteria (11). This complex involves, in addition to the RND pump protein located in the inner membrane, an outer membrane channel such as TolC of *E. coli* (belonging to OMF [outer membrane factor] family (12)), and a periplasmic adaptor protein (belonging to the MFP [membrane fusion protein] family (13)). As shown in Figure 1, this construction allows the bacteria to pump out drug molecules directly into the external medium. This is a huge advantage for bacteria, because the drug in the medium has to cross the low-permeability OM in order to reenter the cells, in contrast to the drug molecules in the periplasm, which can easily penetrate the high-permeability inner membrane. Thus, the tripartite RND pumps are expected to produce drug resistance very effectively by working synergistically with the outer membrane barrier (6, 7).

Wild-type strains of most gram-negative bacteria are resistant to most lipophilic antibiotics (for *E. coli* they include penicillin G, oxacillin, cloxacillin, nafcillin, macrolides, novobiocin, linezolid, and fusidic acid), and this "intrinsic resistance" was often thought to be caused by the exclusion of drugs by the outer membrane barrier. Indeed, breaching the outer membrane barrier does sensitize *E. coli* cells to the drugs just mentioned (14). However, inactivation of the major and constitutively expressed RND pump AcrB also

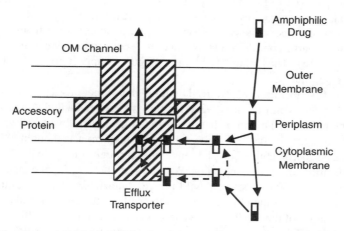

Figure 1.   Early schematic view of the tripartite pump complex. Note that amphiphilic drugs (empty and solid rectangles represent hydrophobic and hydrophilic parts of the molecule) are hypothesized to be captured from the periplasm–plasma membrane interface or possibly from the cytosol (or the cytosol–membrane interface). For the latter process, two possible pathways are envisaged (dashed arrows): Either the substrate is flipped over to the outer surface of the membrane first and then follows the regular periplasmic capture pathway, or it follows a different capture pathway from the cytosol. For the AcrAB–TolC system, the efflux transporter corresponds to AcrB, the accessory protein to AcrA, and the OM channel to TolC. [Modified from (7).]

makes the bacteria nearly completely susceptible to these agents [the minimal inhibitory concentration (MIC) of a lipophilic penicillin, cloxacillin, goes down from $512\,\mu g/mL$ in the wild type to only $2\,\mu g/mL$ (15)] even in the presence of the intact outer membrane barrier. Thus, the characteristic intrinsic resistance of gram-negative bacteria owes as much to the RND pumps as to the outer membrane barrier.

Since some tripartite pumps give resistance to drugs that cannot penetrate the inner membrane, such as the dianionic compound carbenicillin, they were proposed to be able to capture drug molecules from the periplasm (6, 7). Finally, in order to produce significant levels of drug resistance, the drugs that were pumped out into the periplasm by the simple pumps located in the inner membrane need to be captured and then extruded across the OM into the medium by tripartite RND pumps (16, 17). For example, in *Pseudomonas aeruginosa*, the expression of a simple tetracycline pump TetA produces only a very modest degree of resistance ($MIC = 32\,\mu g/mL$) in the absence of the main tripartite pump MexB–MexA–OprM, but MIC is raised up to $512\,\mu g/mL$ if the tripartite pump is, in addition, expressed at the normal, constitutive level. That the high resistance is not due to the pumping of tetracycline by the tripartite pump alone is seen from the fact that the tripartite pump, in the absence of TetA, produces an MIC of only $4\,\mu g/mL$ (16). Thus, these two types of pumps appear to act in a truly synergistic manner.

The first example of an RND pump that functions as a multidrug exporter was identified in 1993 (18); this was AcrB (called AcrE in that paper) in *E. coli*. The immediately upstream gene, *acrA*, was known to be involved in acriflavine resistance, but the mechanism was thought to be the decreased permeability of the envelope through the global chemical modification of its components. We showed that acriflavine penetrates rapidly across OM in both the wild-type and *acrAB*-inactivated strain, yet accumulates at a much higher level in the cells of the latter. We also noted the similarity of the AcrB sequence to CzcA, a divalent cation efflux pump, and proposed that AcrB functions by actively pumping out acriflavine, not by strengthening the passive permeability barrier of OM. The extremely wide substrate specificity of this pump was indicated by the fact that inactivation of the *acrAB* genes made *E. coli* hypersusceptible not only to several dyes but also to detergents (such as SDS, Triton X-100, and bile salts) and to a wide range of antibiotics, including macrolides, β-lactams, tetracycline, chloramphenicol, fusidic acid, and novobiocin (but not aminoglycosides) (Figure 2).

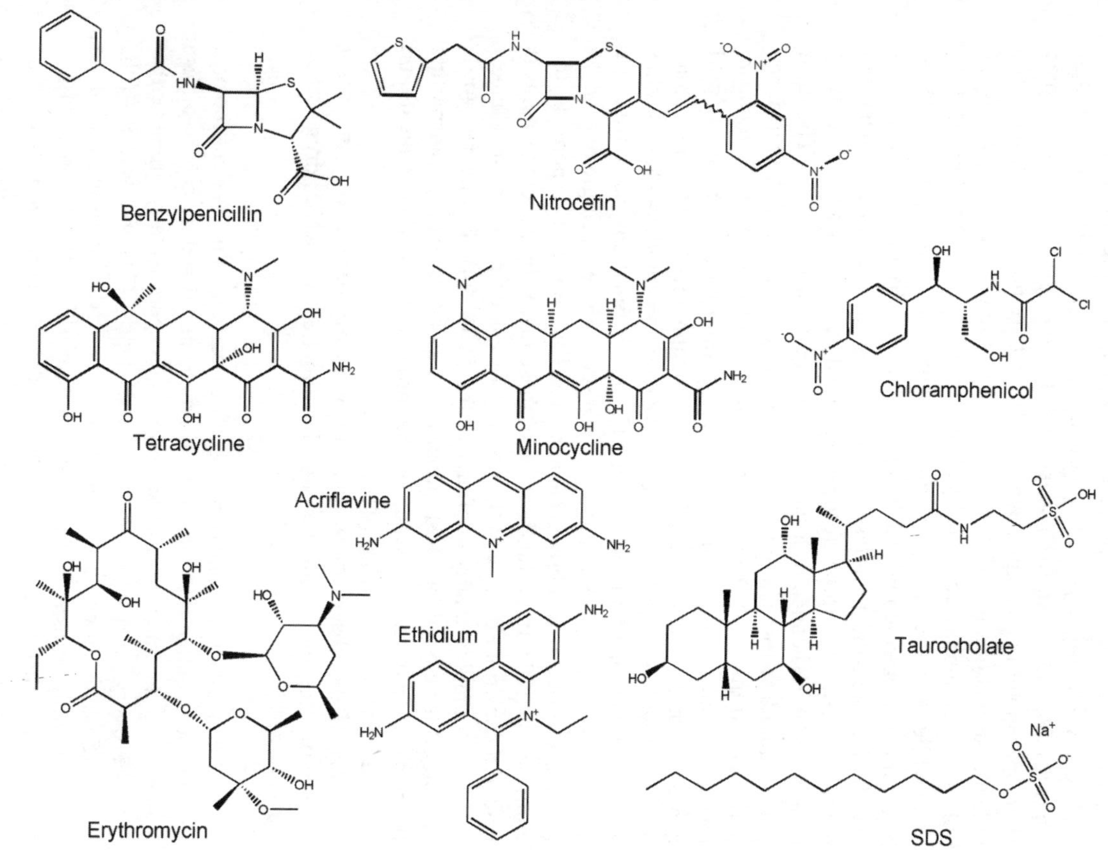

Figure 2. Examples of substrates for the AcrB pump.

Chloramphenicol

Taurocholate

SDS

Nitrocefin

Minocycline

Acriflavine

Ethidium

Benzylpenicillin

Tetracycline

Erythromycin

5

At about the same time, Poole and co-workers (19) found that
P. aeruginosa mutants capable of growth in the presence of an iron
chelator, α,α'-dipyridyl, overproduced an outer membrane protein
(OprM, called OprK in the paper), and that the *oprM* gene was a part
of an operon with upstream genes *mexAB* that are homologous to *acrAB*.
Indeed, the inactivation of these genes made the organism hypersusceptible to chloramphenicol, tetracycline, ciprofloxacin, and streptonigrin.
Meanwhile, our studies on clinical isolates of *P. aeruginosa* whose
β-lactam resistance could not be explained fully by the combination of
OM barrier and the periplasmic β-lactamase showed that unidentified
multidrug efflux pump(s) are overexpressed in these strains, leading to
simultaneous resistance to a large number of agents (20, 21). The most
likely candidate for this pump was found to be the MexAB–OprM, the
major constitutive RND multidrug efflux transporter of *P. aeruginosa*,
through collaboration with the Poole laboratory (22). MexAB–OprM
extrudes not only fluoroquinolones, chloramphenicol, and tetracycline,
but also novobiocin and a number of β-lactam antibiotics. These studies
also gave the first indication that the pump functions as a tripartite
complex, as *mexA*, *mexB*, and *oprM* genes comprised a single operon; in
E. coli, the outer membrane channel TolC is coded elsewhere on the
chromosome (23). Importantly, each of these three component proteins is
essential for drug efflux, and the absence of even one component makes
the entire complex totally nonfunctional (18, 19, 24). To this day, AcrB
and MexB have been the most intensively studied RND drug transporters
in bacteria and have served as the prototype for biochemical and
structural studies of such pumps.

Some of the observations made in this early period were important
in formulating our concepts on the function of RND drug efflux
transporters.

1. Some β-lactam compounds, such as carbenicillin and ceftriaxone,
   which contain multiple charged groups and therefore cannot diffuse
   across the inner membrane, as verified experimentally, are nevertheless very good substrates of MexAB–OprM (21) and AcrAB– TolC
   (15). This observation indicates that the RND pumps are capable of
   capturing substrates from the periplasm (6, 7, 21). The periplasmic
   capture also fits with the observation that the same pump can catalyze
   the efflux of substrates with diverse charges: uncharged, anionic,

cationic, or zwitterionic (7). Export of these substrates from cytosol will generate different effects on membrane potential, which may be difficult to deal with.

2. Although the pumps often handle a very wide range of substrates, they all seem to have a significant lipophilic portion (7, 15). This observation led to a hypothesis that many substrates become partially partitioned, from the periplasm into the outer leaflet of the inner membrane (see Figure 1), before being captured by the pump (7). This association with the membrane surface would involve an energetic cost in terms of decreased entropy, but a similar phenomenon is well known to occur with the interaction of amphiphilic anesthetics with lipid bilayers (25). An alternative interpretation of the requirement of lipophilic domain in the ligands, however, might be that hydrophobic interaction plays a predominant role in the binding of substrates at the binding pocket of the pump (see Section IV).

Finally, it should be mentioned that the tripartite architecture is not completely limited to RND pumps. In *E. coli*, an MFS efflux pump EmrB is known to form a tripartite structure together with the periplasmic adaptor EmrA and the outer membrane channel TolC (26), and to pump out weakly acidic or largely uncharged substrates, such as the proton uncoupler carbonyl cyanide *m*-chlorophenylhydrazone (CCCP), nalidixic acid, and thiolactomycin (27). An ABC efflux pump, MacB, pumps out various macrolides with the periplasmic adaptor MacA, as well as the OM channel TolC (28). Although MacA has been crystallized (29) and an elegant proteoliposome reconstitution of MacB has been achieved (30), the totality of our knowledge on these non-RND systems cannot be compared with that on the RND systems, and we concentrate on RND systems in this chapter.

Recent reviews deal with various aspects of bacterial multidrug efflux; in addition to the comprehensive reviews cited earlier (9, 10), there are reviews emphasizing different aspects, such as multidrug resistance (31, 32), structure (33–35), and mechanisms (36). We do not discuss the regulation of pump production [which was reviewed recently (37)] or the clinical microbiology of the pumps, as they are outside the scope in this chapter. We also do not discuss in detail the outer membrane channels nor the periplasmic adaptor proteins [reviewed recently by Zgurskaya and associates (38)].

II. Biochemical and Genetic Studies

A. RANGE OF SUBSTRATES

Comparison of MIC values between the wild-type and $\Delta acrAB$ strains (7, 18, 24) showed that AcrB of *E. coli* can handle a very wide range of compounds. These include cationic dyes such as acriflavine, crystal violet, ethidium bromide, and rhodamine 6G; and antibiotics such as penicillins, cephalosporins, fluoroquinolones, macrolides, chloramphenicol, tetracyclines, novobiocin, fusidic acid, oxazolidinones, and rifampicin; and detergents such as Triton X-100, sodium dodecyl sulfate, and bile acids. The range even extends to simple solvents such as pentane and cyclohexane (39, 40) (Figure 2). It was obvious that there is no structural similarity between most of these compounds. Furthermore, the charge states of the ligands are diverse, some nonionic, some anionic, some cationic, and others containing multiple ionizable groups (see Figure 2). However, there was one common feature: that the substrates were either relatively lipophilic or at least contained lipophilic domains (7). This led to the hypothesis that the substrates first interact with the membrane lipid bilayer (7), and become captured either from within the bilayer or from the bilayer–aqueous interface (7) (Figure 1). However, alternative interpretations are possible, as mentioned earlier.

The capture of substrates from the periplasm or a location that is in rapid equilibrium with periplasm was described in Section I. In this connection it is interesting that the general inhibitor of *P. aeruginosa* and *E. coli* RND pumps, MC-207,110 (phenylalanyl-arginyl-β-naphthylamide (41), which apparently is a favored substrate of the pumps, affected the carbenicillin MIC only to a small extent (41). Thus, the addition of this inhibitor decreased the MICs of drugs such as erythromycin and levofloxacin to almost exactly the same degree as the genetic inactivation of the MexB–MexA–OprM pump. In contrast, the inhibitor produced only a fourfold decrease in carbenicillin MIC, although the genetic inactivation decreased it 512-fold. A similar discrepancy, a strong decrease in MIC by genetic inactivation vs. little decrease in the presence of the inhibitor, was also found for tetracycline and ethidium bromide (41). Since carbenicillin must be captured exclusively from periplasm, these results were sometimes interpreted to mean that this particular inhibitor acts more effectively for drugs that are predominantly captured in the cytosol. Currently, we are more inclined to assume that AcrB captures all drugs exclusively from the periplasm (see Section IV). If so, these results are perhaps caused by the binding of various substrates to different subdomains of the large binding pocket of the transporter.

As described in Section I, the drug efflux function of AcrB requires the simultaneous presence of the periplasmic adaptor protein AcrA and the outer membrane channel TolC. Chemical cross-linking showed that the adaptor protein AcrA can be cross-linked to the RND transporter AcrB in intact cells (42), and more recently, the entire trimeric complex was isolated from gently lysed *E. coli* cells without cross-linking (43). Interestingly, the association of the component proteins was stabilized by the presence of substrates (43).

When the crystallographic structures of the transporters AcrB (44) and, most recently, MexB (45), the adaptor proteins MexA (46, 47) and AcrA (48), and the outer membrane channel TolC (49) and OprM (50) became available (Figure 3), it became possible to speculate on the details of interaction among these components. Space does not permit a discussion of the details of these structures, but some of the prominent features will be listed.

1. From the amino acid sequences, AcrB (18) and other RND pumps (51) were expected to contain 12 transmembrane helices, with two extremely large periplasmic domains between helices 1 and 2 and between helices 7 and 8. Indeed, one can estimate that the size of the periplasmic domain would almost surpass that of the transmembrane domain, a prediction borne out with the x-ray crystallographic structures (44, 45). What the amino acid sequence did not tell us was that these pumps existed as trimers and that the tip of the large periplasmic domain, often called the *TolC-docking domain*, would fit the internal end of the TolC channel (44).

2. The adaptor proteins had a very elongated shape, as predicted from physicochemical studies (52). This shape was created by a very long α-helical hairpin structure on top of another elongated domain, the "lipoyl" domain, which usually carries bound lipoic acid in various enzymes. Finally, this structure was followed by a β-barrel domain. [The structure of the fourth domain at the opposite end of the molecule from the α-hairpin, called the *membrane-proximal domain*, was solved recently (53) and is discussed below.] All these

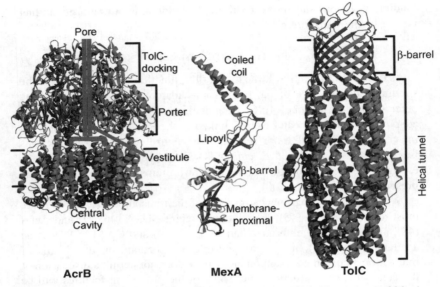

Figure 3.   Crystallographic structures of MexA, AcrB, and TolC. The structure of MexA (monomer), rather than that of AcrA, is shown here because the coordinates for the entire protein sequence have been published (53). The pictures were drawn by using Pymol (http://www.pymol.org), based on PDB files 1IWG (AcrB), 2V4D (MexA), and 2VDE (TolC). The approximate positions of membrane bilayers are shown by horizontal black lines. In the trimer structures of AcrB and TolC, each protomer is shown in cyan, mauve, and blue, and the MexA structure is colored in a rainbow scheme from the blue of the N-terminus to the red of the C-terminus. In the periplasmic domain of AcrB, the boundaries of the "TolC-docking" and "Porter" subdomains are indicated. The large central cavity (red dotted lines) is connected to the periplasm through vestibules, one of whose opening is shown as a red ellipse between the protomers colored in green and cyan. The apparently closed central pore that connects the central cavity to the top of TolC docking domain is indicated by red vertical lines. In MexA, the four domains are indicated. Although the AcrA and MexA are modified in the cell by the addition of a fatty acyl chain at the N-terminus plus the addition of glyceride (containing two additional fatty acyl chains), this is not shown in the figure. The β-barrel domain and the helical tunnel domain of TolC are indicated. (See insert for color representation.)

separate domains create a flexible structure, again as predicted by site-directed spin labeling studies (54).

3. The structure of TolC (49) was perhaps the most spectacular. It is a trimer containing the OM-traversing channel of a β-barrel, with each protomer contributing four transmembrane β-strands of the

12-stranded barrel. Even more interestingly, the channel in the β-barrel is continued by a long helical tunnel, approximately 130 Å long, made of long α-helical hairpins and extending deep into the periplasm.

These structures obviously suggest that the tip of the periplasmic domain of AcrB, or the TolC-docking domain, contacts the end of the TolC periplasmic tunnel. This contact was also confirmed by introducing cysteine residues to the "top" of AcrB and the "bottom" of TolC channel, and observing the formation of a disulfide linkage (55). Although this end of the TolC tunnel is closed in the first structure published, there are now structures in partially open states (56).

Modeling efforts for the trimeric complex were also helped by studies that involved alteration of the structure of component proteins. For example, we found that the adaptor protein of another *E. coli* tripartite system, YhiU (also called MdtE), could not replace AcrA in the formation of tripartite system containing AcrB, and tested chimeric AcrA molecules containing different parts of YhiU (57). Within the 398-residue sequence of AcrA, residues 1–290 could be substituted by the YhiU sequence, but substitution of residues 290–357 produced a protein inactive in forming a functional tripartite complex with AcrB, suggesting that the C-terminal portion of the adaptor proteins was important in interaction with the RND pump proteins. Study of component interaction in vitro by the use of titration calorimetry also gave results (58) consistent with the chimeric AcrA data. Thus, both the full-length AcrA and its C-terminal half (residues 172–397) are associated with AcrB, but the N-terminal half (residues 24–172) or the central fragment (residues 45–315) showed no evidence of interaction with AcrB. In this connection, it must be mentioned again that until very recently the published x-ray crystallographic structures of the adaptor proteins were incomplete, corresponding roughly to the central fragment just mentioned, thus lacking the most important C-terminal part that interacts with the RND transporter. In another approach it was noted that AcrA can function with MexB and TolC, but the full spectrum of drug resistance found in the MexAB–OprM system was not attained (59). Random mutagenesis of AcrA resulted in the isolation of mutants that work better with MexB, and some of them were located in the β-barrel domain.

As regards the interaction between AcrA and TolC, Lobedanz and co-workers (60) were able to carry out a fine resolution analysis by cross-linking Cys residues of AcrA, introduced by site-directed mutagenesis, with amino groups on the surface of TolC, establishing how only one side of the α-helical-coiled coil hairpin in AcrA interacts tightly with the intramolecular groove in the TolC helical bundle.

Hypothetical models of the tripartite assembly were proposed in the past. In one model (46), trimers of MexB and OprM are surrounded by either 12 or six MexA proteins. In another (61), nine AcrA protomers circle the trimers of AcrB and TolC. However, there were some major problems in building these models.

1. The oligomeric state of the adaptor proteins is uncertain. Cross-linking studies showed that overexpressed AcrA existed as dimers and trimers in the cell (42), whereas AcrA lacking the N-terminal lipid modification behaved as monomers when isolated (52). Adaptor proteins have been crystallized as giant complexes containing more than 10 protomers (46–48). These are obviously crystallographic artifacts that do not tell us anything about the physiological state of their self-association.

2. As mentioned above, the structure of the adaptor proteins used in modeling lacked the extreme N-terminus and a large C-terminal domain, the portion known to be essential for the interaction between the adaptor and the RND pump protein.

3. The stoichiometric ratio between the adaptor and the pump protein (or TolC–OprM) within the tripartite complex was not known.

A giant stride in this area was recently made by the Koronakis group (53). They have refined the data obtained previously for the MexA crystal (47) and succeeded in obtaining the structure of the hitherto missing domain containing the N-terminus and the large C-terminal segment. This is now known to form the fourth domain or membrane-proximal domain of the adaptor, containing an α/β structure with a characteristic concave surface (Figure 3). By using this MexA structure as a template, they built the complete structure of AcrA. Finally, by applying the method used earlier to study the interaction between TolC and AcrA (60) [i.e., site-directed insertion of Cys followed by cross-linking], they were able to determine precisely where defined portions of AcrA fit into which surface of AcrB. Interestingly, in this cross-linking study in intact cells, AcrA seems to interact with only one contiguous stretch of AcrB within the periplasmic domain, thus suggesting a 1 : 1 stoichiometric ratio between

these two proteins. Based on these observations, a data-driven model of the
AcrAB complex could be built, which is shown as Figure 4.

Another significant advance in the study of component interaction was
the in vitro examination of association between MexA and OprM (62).

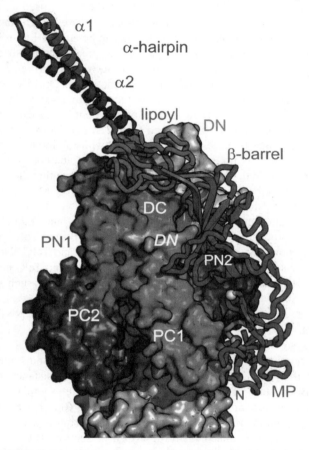

Figure 4.   Model of an AcrAB complex based on the recently refined AcrA structure and the
cross-linking data. The periplasmic domain of an AcrB monomer is shown in a space-filling
model, with each subdomain (PC, periplasmic, C-terminal; PN, periplasmic, N-terminal; DC,
docking, C-terminal; and DN, docking, N-terminal) colored differently. The AcrA is shown as
a green ribbon model, and the domains [α-hairpin, lipoyl, β-barrel, and MP (membrane-
proximal)] are identified in green letters. [From (53), with permission from the authors.] (See
insert for color representation.)

Such studies are difficult because both are membrane proteins [although some adaptor proteins appear to work in intact cells without the lipid modification at the N-terminus (52)], and the adaptor inserted into one membrane must interact with the OM channel inserted into another membrane (see Figure 1). Reffay and co-workers (62) used an ingenious system of L_3 phase (sometimes called "sponge phase") surfactants, where the distance between neighboring bilayers can be adjusted. In this system, the lateral diffusion rates of MexA become greatly diminished when it forms a complex with the OprM trimer, a much larger protein, in another bilayer. The MexA immobilization as a function of OprM concentration showed that the MexA monomer to OprM trimer stoichiometry was strongly dependent on pH [a finding reminiscent of an earlier study of AcrA conformation (54)] and ranged from 2 at pH 7.5 to 6 around pH 6. The former value is more or less consistent with the model of Symmons et al. (53), but it is not yet clear how the second adaptor protein can be accommodated in the final tripartite complex. We note that these ratios are consistent with the abundance of MexA, MexB, and OprM protomers in the cell reported by the Nakae group (63): namely, 2 : 1 : 1. However, in *E. coli*, AcrA appears much more abundant, Tikhonova and Zgurskaya (43) reporting an abundance of 10 : 1 : 3 for AcrA, AcrB, and TolC protomers; this ratio, however, may reflect the fact that AcrA and TolC also function for other systems: AcrA, for example, for AcrD (64) and TolC not only for all other RND systems but also for some non-RND systems such as EmrAB and MacAB.

One observation that may be pertinent here is that some RND systems apparently contain a pair of adaptor proteins. For example, the TriABC–OpmH system of *P. aeruginosa*, which pumps out triclosan, contains two adaptor protein genes (*triA* and *triB*) within the operon, and studies clearly showed that both genes are needed for triclosan resistance (65). The ZrpADBC system of *Serratia* sp. appears to be yet another example of such a system containing a pair of adaptor proteins (66). If we assume that the pump proteins in these systems function as trimers according to the functionally rotating mechanism (described in Section III.C), each protomer of the pump is likely to contact a pair of the adaptor proteins, giving rise to a 2 : 1 : 1 ratio between the adaptor, pump, and OM channel proteins. However, an alternative arrangement of 1 : 1 : 1 stoichiometry is still possible, producing a situation in which different adaptor–pump complexes would behave similarly to the different pump protomers in the MdtBC-type complexes (see Section IV.E).

C.   KINETIC BEHAVIOR OF THE AcrB PUMP

To understand the behavior of the pump in intact cells, it is obviously essential to know its kinetic constants. Hints on the binding affinity of various substrates were obtained by using them as competitive inhibitors in the reconstitution assay of AcrB (67) (see below). In this assay, which depended on the export of fluorescent phospholipids by the purified AcrB protein, taurocholate, a conjugated bile salt, inhibited the reaction most strongly (presumably by competing as a substrate), the 50% inhibition occurring at 15 μM. In contrast, antibiotics were less efficient inhibitors, and cloxacillin and erythromycin caused 50% inhibition only around 100 μM. These results are important in showing that the conjugated bile salts are probably the preferred, and most likely the natural, substrates of the AcrB pump of *E. coli*, an organism that must live in an environment full of these membrane-disrupting detergents. In the same study (67), proton movement accompanying the pumping of substrates was also measured. Although a single concentration of substrates (0.2 mM) was used, again the conjugated bile salts were most effective in increasing the proton flux. However, these assays do not measure substrate binding per se, and we do not know the true affinity of each substrate to the pump. (We should also mention that running similar assays in intact cells is difficult because both the substrate and the competitive inhibitor must diffuse across the outer membrane before binding to the pump.)

There was a recent attempt to measure the binding of substrates to purified AcrB protein by the use of fluorescence polarization (68). Because binding of fluorescent substrates to a large AcrB trimer is expected to slow the tumbling of these molecules, this method offers advantages. The apparent dissociation constants reported ranged from 5.5 μM for rhodamine 6G to 74 μM for ciprofloxacin. However, the method does not give reliable information on binding stoichiometry, and because the authors did not use, as controls, mutant AcrB proteins that are altered in substrate binding, it is unclear whether the binding occurred to the true binding sites or to some extraneous pockets that bind lipophilic molecules in a nonspecific manner.

The efflux of fluorescent dyes from cells after their energization was followed in several studies. Typically, cells are preloaded with dyes that become fluorescent only within the membrane or the cytosol. The preloading requires deenergization of the RND pump by a proton conductor such as CCCP (carbonyl cyanide *m*-chlorophenylhydrazone), and reenergization of the cells is done by adding compounds such as glucose or formate. If everything goes perfectly, the time-dependent decrease in fluorescence intensity should follow the integrated form of the Michaelis–Menten equation, which should give us the kinetic constants K_m and V_{max}. In the first paper reporting the use of one of these probes (69), the results were not convincing because the efflux rates from *P. aeruginosa* strains either producing or not producing the MexB–MexA–OrpM efflux complex showed little difference, presumably because deenergization with cyanide could not be reversed rapidly. However, efflux assays were performed successfully since then as a qualitative assay, for example with NPN (*N*-phenylnaphthylamine) and *E. coli* (41) and with DASPEI (2-[4-dimethylamino]styryl-*N*-ethylpyridinium iodide) and *E. coli* (70). No effort to use the efflux curve for quantitative analysis of the transport process was reported. We have tried to do this quantitative analysis with a fluorescent dye, Nile Red, using *E. coli* (70a). However, the system is extremely complex, because CCCP used in the dye-loading period may still persist in the cells, at least in the beginning, and the full membrane energization with carbon sources will take a few seconds, thus making analysis of the early phase of the efflux curve quite difficult.

An early study that attempted to determine the kinetic constants of efflux was that of Narita et al. (63). Here the authors calculated the rate of ethidium efflux from the difference in the ethidium influx rates between efflux-incompetent and efflux-competent cells of *P. aeruginosa*, and reported a V_{max} value of 100 nmol/s per 3×10^8 cells, said to correspond to a turnover number of $500 \, s^{-1}$. However, the authors did not show how these numbers were arrived at. In fact, inspection of the data shows that the efflux rate plotted against intracellular ethidium concentration was essentially linear, and little sign of saturation can be observed. Thus, these conclusions are not convincing.

Recently, we had the first success with the estimation of kinetic constants of the AcrB pump (71). With attempts using intact cells, the difficulty is always with estimation of the substrate concentration within the periplasm, where it is presumably captured by the pump. We solved this problem by using β-lactams as substrates. Since β-lactams are hydrolyzed by the periplasmic β-lactamase, and since we know the K_m and V_{max} values of the enzyme, we can calculate the periplasmic concentration of β-lactams, C_p, from the Michaelis–Menten equation if we determine spectrophotometrically the β-lactam hydrolysis rate by intact cells. Once the periplasmic concentration C_p is known, the rate at which the β-lactams cross the outer membrane by simple diffusion through porin channels (V_{in}) can be calculated from the permeability coefficient of the OM, because it is proportional to the concentration difference of the β-lactam between the external medium and the periplasm. The efflux rate V_e is then the difference between the influx rate V_{in} and the hydrolysis rate V_h, or $V_{in}-V_h$. Initially, this approach was applied successfully to the influx, efflux, and hydrolysis of a cephalosporin, nitrocefin, and we found that the efflux occurs with the Michaelis–Menten saturation kinetics, with an apparent K_m value of about 5 μM (Figure 5A). There was little sign of positive cooperativity, a result that is interesting in view of the functional rotating mechanism of the pump discussed later in this chapter. A turnover number of about $10\,s^{-1}$ was calculated. The low turnover number was as predicted in early studies, on the ground that the tripartite pumps may have to deal only with the small number of drug molecules that have trickled through the effective outer membrane barrier (7).

Interestingly, when the assay was extended to cephalosporins that were clinically useful, such as cephalothin, cefamandole, and cephaloridine, kinetics with strong positive cooperativity was observed (Figure 5B). Although it is not clear why such kinetics were not observed with nitrocefin, we note that nitrocefin has a much lower K_m value than that of these compounds. Interestingly, compounds showing strong cooperativity also have high V_{max} values, so the ratio of V_{max} to $K_{0.5}$ (the substrate concentration giving a half-maximal rate, K_m in the case of noncooperative kinetics) did not change much for the four compounds tested (range: 0.5 to $1.9 \times 10^{-2}\,cm^3\,mg^{-1}\,s^{-1}$), although the V_{max} for cephaloridine was 80 times higher than that for nitrocefin. The results on cephaloridine merit close attention for several reasons.

1. We have shown that cephaloridine cannot diffuse across the cytoplasmic membrane (15, 21). Thus, these results represent the kinetics of drug export initiated by the periplasmic capture of the drug.
2. The criterion for the evaluation of the range of substrates for an efflux pump was usually the comparison of MICs of the drug in wild-type cells and pump gene deletion mutants. By this criterion, the presence or absence of AcrAB made little difference to cephaloridine MICs (15, 70), so this compound was thought to be either a nonsubstrate or at best a marginal substrate for the pump. However, detailed kinetic analysis showed that cephaloridine may be pumped out very rapidly, with the turnover number approaching $1000\,s^{-1}$ (71), although the

A. Nitrocefin

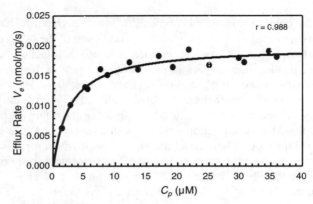

B. Cephaloridine

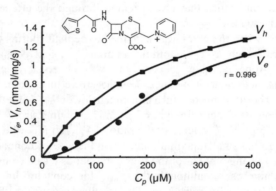

C. Cefazolin

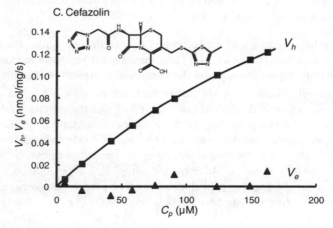

$K_{0.5}$ value was very high (288 μM). Thus, the MIC comparison can be a poor and sometimes a misleading indicator of the pump function.

In this case, the AcrAB function made little difference to the MIC values, because the bacteria are killed at β-lactam concentrations usually below 5 μM (72, 73), and in this concentration range, the efflux is insignificant because of the high $K_{0.5}$ value and especially because of the strongly sigmoidal kinetics. The high turnover number for some substrates explain the observation that AcrAB can pump out simple solvents (39, 40), as these compounds must traverse the OM bilayer rapidly because they are small and uncharged and must be pumped out very rapidly by the efflux pump in order to produce resistance.

Interestingly, one cephalosporin, cefazolin, showed no sign of getting pumped out by the AcrAB pump (Figure 5C). This compound has very hydrophilic substituents (a tetrazole and a thiadiazole) at the 7- and 3-positions, respectively, and presumably the binding of the substrates to the AcrB binding site requires the presence of lipophilic moiety, preferably an aromatic ring, in at least one of these positions [see the article by Nikaido et al. (15)]. Kinetic constants for the AcrAB-mediated efflux of penicillins were also determined by a similar approach (73a).

Recently, binding of 24-, 25-, or 27-hydroxysterols to Niemann–Pick C1 protein (NPC1), a member of the RND family involved in cholesterol trafficking in vertebrate cells (74), was reported (75). The binding took place despite the presence of a high concentration (1%) of the detergent Nonidet P40, and the half-saturation was reached with about 0.1 μM of 25-hydroxysterol. 7-, 19-, or 20-Hydroxysterol did not bind to this protein. In comparison with bacterial RND pumps, the NPC1 protein has an N-terminal extension consisting of an extramembranous domain of about 240 residues and an additional transmembrane segment. This extramembranous domain is secreted from cultured cells as a soluble dimer, which bound 25-hydroxysterol with a strong affinity ($K_d = 10$ nM) (76). Since this domain is uniquely present in NPC1, it is unclear whether these beautiful

◄ ——————————————————————————————

Figure 5.   Kinetics of AcrAB–TolC-catalyzed efflux of nitrocefin (A), cephaloridine (B), and cefazolin (C). The rates of efflux from the periplasm (V_e) are plotted against the periplasmic drug concentration (C_p) for each drug. Nitrocefin efflux follows a classical Michaelis–Menten kinetics, but cephaloridine efflux shows a sigmoidal kinetics. The rates of β-lactamase-catalyzed hydrolysis (V_h) for cephaloridine and cefazolin are shown for comparison. [From (71).]

results would help us in our effort to understand the ligand-binding process in efflux transporters (see Section IV.B).

D. GENETIC STUDIES OF AcrB AND ITS RELATIVES

To identify domains responsible for pump functions, large segments of genes were replaced with sequences coming from another homolog, producing chimeric genes. For example, Elkins and Nikaido (77) used *E. coli* AcrB, which shows strong resistance to ciprofloxacin, novobiocin, fusidic acid, erythromycin, and taurocholate, and AcrD, which generates no resistance to these agents except a modest one against taurocholate. Replacing both of the large periplasmic loops of AcrD with those from AcrB created a transporter pumping out all of the tested AcrB substrates efficiently. Replacing loops of AcrB with those from AcrD created a transporter that behaved like AcrD in terms of its substrate specificity. Finally, replacing the transmembrane regions of AcrD with corresponding sequences from AcrB did not alter the substrate specificity of the transporter. These results show that the substrate specificity is determined nearly entirely by the periplasmic domain (and that the critical binding of substrates presumably occurs here), and the transmembrane domain probably plays a role only in energy coupling. That the exchange of transmembrane domain does not alter the substrate specificity was also confirmed later by using *P. aeruginosa* MexB and MexY (78).

Tikhonova and co-workers (79) used AcrB and *P. aeruginosa* MexB. The results of chimeric exchange suggested that the substrate specificity was determined largely by the second external loop (called PC2 in AcrB) of these transporters, containing residues 612–849, a conclusion consistent with the location of the substrate-binding pocket identified in the asymmetric trimer crystal of AcrB (see Section IV.A). [This study was recently refined by the use of site-directed mutagenesis (80), which pinpointed the difference in macrolide extrusion activity to residue 616, which is Gly in AcrB and Asn in MexB]. They also showed that replacing the region coding for the transmembrane helices 8 to 12 of AcrB with the MexB sequence produced a somewhat MexB-like pattern for a few drugs (a lower efflux of ethidium bromide and lincomycin, and a higher efflux of cinnoxacin); currently, it is difficult to explain these results in terms of structure, although the elongation of helix 8 into the periplasmic domain appears crucial in the conformational change that leads to drug extrusion (Section III.C). Finally, this group showed that the N-terminal periplasmic loop

(PN1 and PN2) and the first part of the C-terminal loop (PC1) essentially determine the interaction of the RND pumps with their cognate periplasmic adaptor proteins, a conclusion that is in complete agreement with the recent study based on the crystal structures (see Figure 4).

Use of point mutations could in principle allow us to produce more detailed knowledge on the binding and export of substrates. In an approach pioneered by Lomovskaya's laboratory, random mutants of *P. aeruginosa* MexD that acquired the ability to extrude carbenicillin, which is not handled by the wild-type MexD, were isolated (81). All mutants mapped to the periplasmic domain, none to the transmembrane domain, further supporting the conclusion mentioned above. Among the residues identified, Phe608 (corresponding to Phe610 of AcrB) is a part of the substrate-binding pocket more recently identified (see below). Some others [Gln34, Gln89 (corresponding to Glu 89 of AcrB), Asn673(Thr676)] are on the walls of the large lateral cleft of the periplasmic domain, which is discussed below (Section IV.B) as a potential site of periplasmic entry for substrates. These residues are shown as space-filling models in the AcrB structure (Figure 6).

Bohnert et al. (82) used an *E. coli* strain that lacked both AcrB and AcrF and instead overproduced another RND pump, YhiV. A spontaneous mutant obtained after repeated exposure to levofloxacin was shown to owe its increased resistance to the change of an aliphatic Val610 residue of the YhiV into an aromatic Phe residue. This mutant pump has an altered specificity. It produces a stronger resistance to relatively small aromatic compounds such as levofloxacin, linezolid, and tetracycline, but the resistance to nonaromatic, bulky macrolides becomes weaker than in the parent protein. This residue, which corresponds to Val612 of AcrB, is seen to be a part of the substrate-binding pocket in the asymmetric AcrB trimer structure (discussed in Section IV.A).

Middlemiss and Poole (83) used a different approach and carried out an in vitro random mutagenesis of the *mexB* gene from *P. aeruginosa*. This is a more comprehensive approach, but is expected to generate more "noise." Indeed, the group of mutants that decreased the level of resistance to most drugs included the presumed proton relay mutants in Asp407 and Asp408 of the transmembrane domain (see Section III.B), or Gly220 mutant in the "peg" that is inserted into the periplasmic domain of the neighboring protomer. The mutants that are significantly altered in the substrate specificity, in contrast, often contained alterations in the periplasmic domain, as expected. Among these, alterations of Ala618 and Arg716

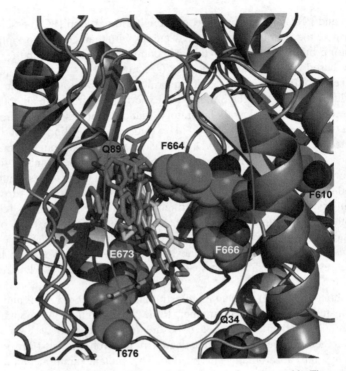

Figure 6.   Large cleft of the AcrB periplasmic domain, seen from the outside. The opening of the cleft is indicated approximately by the large red ellipse. AcrB residues corresponding to those MexD residues affecting the substrate specificity (81) (Phe610, Gln34, Glu89, and Thr676) are shown in space-filling models, as well as those that were shown to affect drug pumping by site-directed mutagenesis (88) (Phe664, Phe666, and Glu673). The binding of drugs to the Asn109Ala mutant AcrB (88), discussed in Section III.A, is shown by stick models (ciprofloxacin, yellow; nafcillin, green; ethidium, magenta; rhodamine 6G, salmon; and Phe-Arg-β-naphthylamide, orange). (See insert for color representation.)

(corresponding to Ala618 and Arg717 of AcrB) occur on the opposing walls of the large lateral cleft. However, there were several mutants in the transmembrane domain, and their phenotypes remain to be explained.

Point mutations can also be introduced into the transmembrane domain by site-directed mutagenesis. In 1999, inspection of amino acid sequence of an RND-family toxic cation efflux pump CzcA of *Ralstonia* sp. and other RND pumps showed that there are several conserved charged residues in the transmembrane domain, including Asp402, Asp408, and Glu415 (84). Changing these residues into nonacidic residues abolished the cation pumping activity. This was followed up a few years later by the site-directed mutagenesis study of MexB (85). In MexB, the Asp402 and Asp408 of CzcA are replaced by a consecutive pair of aspartate residues, Asp407 and Asp408 (corresponding to Asp407 and Asp408 in AcrB) on transmembrane segment (TMS) 4, and both of them were found to be essential for the pump function. This study also identified an essential Lys939 (corresponding to Lys940 of AcrB) located in TMS 10 and suggested that these three residues produce a proton-relay network. Finally, we have identified Thr978 of AcrB on TMS 11 as another, functionally essential component of this network (86) (see Figure 10 below). (This study also identified Arg971, located close to the cytosolic end of TMS11, as an essential residue (86), although it is far away from the network of other charged residues.) These results emphasize the role of the transmembrane domain in proton translocation, which presumably provide energy for drug pumping by the periplasmic domain.

Site-directed mutagenesis was also applied to residues assumed to be involved in the binding or movement of substrates. When we reported on the crystal structures of AcrB with ligands in the central cavity of the transmembrane domain (87), we were aware of the potential problem that the ligand might be binding to any hydrophobic pocket of the protein, which has nothing to do with the normal pathway of the exported ligand. Thus, we changed residues that appeared to be involved in the binding through site-directed mutagenesis, and we obtained an apparently assuring result that conversion of Phe386 to alanine nearly totally abolished the resistance to ethidium, rhodamine 6G, and dequalinium (87). However, this result was obtained by using a very high copy number vector of the pUC series, and we are now aware that this approach may possibly produce misleading results by making the cell membrane leaky through strong overproduction of intrinsic membrane proteins. Indeed, Phe386Ala mutation hardly affected susceptibility of E. coli to most drugs when tested with expression by the vector pSPORT1 (88). Nevertheless, Hearn and co-workers (89), using a Pseudomonas fluorescens homolog of AcrB, EmhB, expressed from a medium copy number plasmid, found that decreased efflux of dequalinium was produced by the same change (Phe386Ala) as well as the change of Asn99 to alanine. Furthermore, changing Asp101, which is located at the "ceiling" of the central cavity, into alanine decreased the efflux of most drugs as well as of polycyclic hydrocarbons.

With AcrB mutant Asn101Ala, described later in Section III.A, the efflux activity was decreased by changing into alanine the residues Glu673, Phe664, or Phe666, all located in the right (looking into the center of molecule with the periplasmic domain up) wall or the bottom of the large periplasmic cleft, close to its entrance (89) (Figure 6). With the Cu^+ and Ag^+ RND-type efflux pump CusA of *E. coli*, changing into leucine the methionine residues 573, 623, and 672, all close to the entrance of the periplasmic large cleft in the AcrB model, strongly reduced its activity (90). Together with the binding of substrates to this cleft observed in the Asn101Ala mutant AcrB (89), described below, these results favor the assumption that the periplasmic uptake of substrates by these pumps involves the transient binding to, and then entry from, the large periplasmic cleft.

Dastidar et al. (91) introduced cysteine residues to various positions of *Haemophilus influenzae* AcrB, chosen on the basis of their effects on substrate specificity in MexD (81) (described above), and measured, in intact cells, their accessibility to a relatively hydrophilic probe, fluorescein maleimide in the presence and absence of efflux substrates such as erythromycin, novobiocin, cloxacillin, ethidium bromide, Triton X-100, and the inhibitor Phe-Arg-β-naphthylamide. Most substrates, except ethidium, inhibited strongly the modification of cysteine at position 288, which corresponds to Gly290 of AcrB that is close to the bottom of the substrate-binding pocket in the asymmetric trimer structure (described in Section IV.A). Interestingly, the modification of cysteine at position 601, corresponding to Phe617 of AcrB, one of the residues lining the substrate-binding pocket, is inhibited by Triton X-100, but not by other substrates. These data suggest the large size (and/or the flexibility) of the binding pocket that would accommodate different ligands in different ways (see also Section IV.B).

Murakami et al. (70) used cysteine scanning mutagenesis to examine the role of the narrow central "pore" of the trimer, which appears to be closed in the crystal structure of AcrB. Although alterations of the residues facing the center of the pore, including Asp101 just mentioned, were found to decrease the extrusion of drugs, the interpretation is complicated by the fact that the pore is made by a three-stranded coiled-coil structure composed of all three protomers of AcrB, so that intersubunit disulfide bonds are formed in some cases. Now that the drug-binding site and the likely drug extrusion pathway were discovered within the AcrB protomer (Section III.C), it seems unlikely that the interprotomeric central pore plays a direct role in substrate movement.

E. MEMBRANE VESICLES AND PROTEOLIPOSOME RECONSTITUTIONS

In the biochemical studies of most bacterial transporters, the most useful approaches were often the use of membrane vesicles, either right-side-out or inverted. However, with RND family multidrug efflux pumps, this approach was not successful, although it was pursued vigorously in several laboratories, including our own. The likely reasons for this failure include the hydrophobic nature of most substrates, which allows a spontaneous equilibration of transported substrates across the membrane bilayer. Furthermore, "transport" by RND transporters may not change the transmembrane location of the substrates since the major pathway of substrate capture is from the periplasm. Another possibility may be that the periplasmic adaptor protein, which seems to be necessary to activate the transporter (see below), becomes stripped off during the preparation of the membranes.

The first successful functional reconstitution of an RND pump was achieved by Zgurskaya and Nikaido (67) in 1999. Cells containing *acrAB* genes on a high-copy-number pUC plasmid were grown without induction, and the membrane proteins were solubilized with 5% Triton X-100 overnight in the cold. The native AcrB protein contains four histidine residues among the eight residues at its C-terminus, and thus the protein could be purified by adsorption to a Cu^{2+}-chelate matrix followed by elution with imidazole, always in the presence of Triton X-100. Just before the reconstitution, AcrB was again adsorbed to the Cu^{2+}-chelate matrix, and Triton X-100 was exchanged with octyl-β-D-glucoside by elution with a buffer containing the latter detergent as well as imidazole. Since AcrB formed aggregates in octylglucoside within a few hours, it was necessary to carry out reconstitution into proteoliposomes immediately by the octyl-glucoside dilution method.

Innovative approaches were required for assays of the transport activity of AcrB, because most of the substrates of AcrB are quite hydrophobic (7) and are expected to diffuse spontaneously across the membrane bilayer, as mentioned above. Thus, it was impossible to rely on the quantitation of ligands accumulated within the intravesicular space. Inspired by the report that mammalian P-glycoprotein, which also transports hydrophobic compounds, has activity as a phospholipid flippase (92), we added 7-nitrobenz-2-oxa-1,3-diazol-4-yl (NBD)-labeled fluorescent phospholipid to the phospholipid mixture used in proteoliposome reconstitution. Phospholipids, however, are likely to become reinserted into the original bilayer even when they are expelled from it by the AcrB pump. To minimize the effect of this process, an excess of "acceptor" liposomes, which did not contain AcrB, were used to "trap" the fluorescent phospholipids extruded. Finally, the amount of the NBD-labeled phospholipids remaining in the AcrB-containing "donor" proteoliposomes was estimated by initially quenching the NBD fluorescence through the fluorescence energy transfer to rhodamine-labeled phospholipids added initially (together with NBD-labeled phospholipids) to the proteoliposome reconstitution mixture.

Phospholipid extrusion, as indicated by time-dependent increase in NBD fluorescence, required the addition of Mg^{2+} (Figure 7), which is consistent with the observation that in intact cells, another RND-family multidrug efflux pump MexXY requires the presence of Mg^{2+} in the

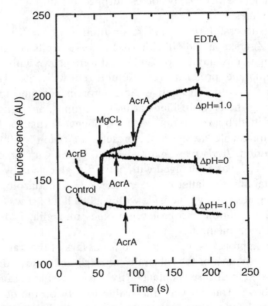

Figure 7. Phospholipid extrusion assay of reconstituted AcrB. When a transmembrane pH gradient was generated by diluting the vesicles (internal pH, 7.0) into a buffer of pH 6.0, donor vesicles not containing any AcrB (bottom trace) do not show much change in fluorescence, nor do AcrB-containing vesicles in the absence of pH gradient (middle trace). However, when AcrB-containing vesicles are exposed to pH gradient, there is an expected gradual increase in NBD fluorescence in the presence of Mg^{2+}. Note that the rate of fluorescence increase is strongly accelerated if AcrA (15 µg/mL) is also added. [From (67).]

medium for its function (93). Remarkably, strong pumping also required the addition of lipid-free AcrA. This result was interpreted as the result of AcrA protein connecting the donor and acceptor vesicles (67). However, knowledge of the structure of AcrA (see Figures 3 and 4) makes this hypothesis unlikely, because AcrA can connect to the second membrane only through TolC, which was not present in our system. Rather, AcrA could have activated the pumping activity of AcrB directly, in view of the fact that AcrD, a close homolog of AcrB, is activated in a reconstituted system by AcrA (64) (see below).

To show more directly that AcrB functions as a proton/drug antiporter, the system was energized by using the valinomycin-induced efflux of K^+, which was converted into an interior-acidic proton gradient in the presence of KCl. By measuring the intravesicular pH with a fluorescent, membrane-impermeable pH indicator pyranine, we confirmed that proton efflux occurred accompanying the pH-gradient-driven operation of the pump in the presence of substrates (67) (Figure 8). When the number of protons moved per vesicle is calculated, it appears that the AcrB pump is functioning extremely slowly, with a turnover rate of less than one per minute. We note that the assay was carried out without the addition of Mg^{2+} and AcrA, which were needed for the activation of AcrB according to the fluorescent phospholipid extrusion assay. Possibly an even more important factor could have been the location of the substrates, which were added to the extravesicular space that corresponded to cytosol as the acidified vesicle interior corresponded to the periplasm. Thus, it seems likely that we were measuring the slow spontaneous influx of substrates into the intravesicular space, followed by their capture by the periplasmic domain of AcrB.

The affinity of substrates for AcrB was estimated in two ways. First, the substrates were added as potential competitors in the fluorescent phospholipid efflux assay. This showed that conjugated bile acids, such as taurocholate, were the most effective inhibitors of phospholipid efflux. Second, the potential substrates were added in the proton efflux assay, and again taurocholate was the most effective in producing proton efflux among the compounds tested (Figure 8). These results fit the notion that the natural substrate for the AcrB pump are the bile salts, most of which exist in the conjugated form in the intestinal tract of higher animals.

Soon afterward, the successful reconstitution of another RND pump, the CzcA protein of *Ralstonia*, was reported (83). CzcA is not a multidrug pump, but catalyzes the export of toxic divalent cations such as Zn^{2+}, Co^{2+}, and Cd^{2+}. The PCR-amplified gene was inserted into a commercial

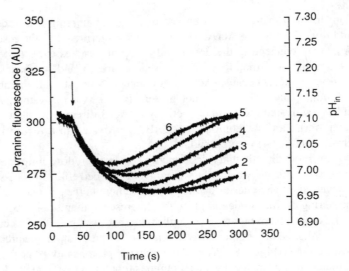

Figure 8.   Proton flux assay of reconstituted AcrB. AcrB-containing vesicles were made in the presence of 1 mM pyranine, a fluorescent pH indicator, and the extravesicular pyranine was removed by gel filtration. The interior-acidic pH gradient was generated by diluting the vesicles containing 0.1 M KCl into 0.1 M NaCl and then by adding valinomycin (arrow). The acidic internal pH was maintained for several minutes in the absence of drug (curve 1). When erythromycin (curve 2), chloramphenicol (curve 3), cloxacillin (curve 4), glycocholate (curve 5), or taurocholate (curve 6) were present at 0.2 mM, the internal pH returned to its initial value more rapidly, presumably due to the proton efflux accompanying the influx of drugs by the AcrB antiporter. [From (67).]

vector pASK-IBA3, which supplies a C-terminal, eight-residue tag. The protein was purified by affinity chromatography with a modified streptavidin matrix that binds the C-terminal tag (94). Solubilization and purification were performed in the presence of phospholipids (95). The purified CzcA was mixed with Triton X-100-treated liposomes, and the proteoliposomes were formed by removal of the detergent with Bio-Beads. A pH gradient was imposed across the membrane by diluting proteoliposomes containing 100 mM Tris-Cl at pH 5.0 into 100 mM Tris-Cl at pH 7.0. Since the substrates do not diffuse spontaneously across the membrane bilayer, the pumping activity can be assessed by traditional methods, and indeed, vesicles were shown to accumulate divalent cations such as Zn^{2+}. Better results were obtained by acidifying the vesicle interior by the spontaneous efflux of NH_3 from proteoliposomes loaded with 0.5 M NH_4Cl. The data with Zn^{2+} were reported to show a sigmoidal kinetics with a Hill coefficient of 2, although only four concentrations of the substrate were used. These data, unfortunately, may not have much physiological relevance because the half-maximal rate was achieved at an absurdly high concentration of 6.6 mM, where micromolar (or even lower) values of $K_{0.5}$ were expected. Furthermore, the experiment measured the movement of cations from outside (corresponding to cytosol) to intravesicular space (corresponding to periplasm), whereas the real pump is likely to export substrates from periplasm to the external medium. Indeed, we now know that a similar cation pump, CusAB system captures ions such as Cu^+ from the periplasmic binding protein CusF before their efflux (96, 97).

Reconstitution of another multidrug efflux RND pump AcrD was achieved in 2005 (64). AcrD was of special interest to us, as some of its substrates, aminoglycosides, are very hydrophilic and are not expected to cross membrane bilayers spontaneously. AcrD, in a form with a C-terminal hexahistidine tag, was expressed from a high-copy-number plasmid with isopropyl-β-D-thiogalactoside (IPTG) induction, and was purified in dodecylmaltoside. The AcrD preparations in dodecylmaltoside were used directly for proteoliposome reconstitution by octylglucoside dilution method, without the prior removal of dodecylmaltoside. Routine assays relied on the proton efux determined with the intravesicular pH indicator pyranine, as described above. It was soon discovered that no acceleration of proton flux occurred, with aminoglycosides (70 μM) either outside or inside vesicles, unless AcrA (with the lipidation site removed) was added at the time of reconstitution, and thus to the interior of the vesicles. It was known that AcrA was required for the activity of AcrD in intact cells (77), but this was expected because AcrA (or one of its homologs) would be needed for the construction of a tripartite efflux machinery. In the proteoliposomes containing AcrD, a tripartite structure is not needed for efflux, and the result strongly suggested that AcrA and its homologs of the MFP family had another important function in directly activating the pumping function of the cognate RND efflux transporter. A similar situation was found more recently with the MFP MacA for the activation of an ABC efflux transporter MacB (30).

Results obtained with streptomycin are shown in Figure 9. When streptomycin was added to the external medium, the drug is occupying a more alkaline compartment that is similar to the cytosol in intact cells. Under such a setup, we could not detect stimulation of proton efflux (curve 2); however, a strong stimulation was observed when streptomycin

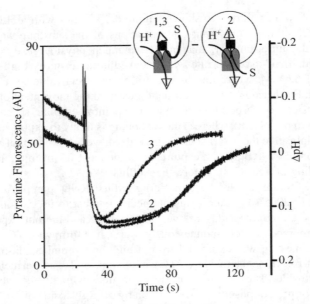

Figure 9. Proton flux assay of reconstituted AcrD with streptomycin. AcrD-containing proteoliposomes were made in KCl, and an interior-acidic pH gradient was generated by diluting them into NaCl and by adding valinomycin at 30 s. The insets show the experimental setup. The gray object denotes AcrD, with its periplasmic domain shown as ellipses. The black rectangle denotes AcrA, and the substrate (streptomycin, 70 μM) is shown as "S". Curve 1, AcrD proteoliposomes containing Mg^{2+} and AcrA inside, but with no streptomycin. Curve 2, the same proteoliposome as in curve 1, to which streptomycin was added to the external medium. Curve 3, AcrD-proteoliposomes made on the same day that contained Mg^{2+}, AcrA, and streptomycin in the intravesicular space. [From (64).]

was added to the more acidic intravesicular space, corresponding to the periplasm (curve 3). These results are important, as they were the first results showing that at least one drug was captured *exclusively* from the periplasm. Adding kanamycin, tobramycin, gentamicin, or amikacin to the interior of the vesicles similarly produced a strong increase in the efflux of protons (64). However, these aminoglycosides seemed to produce some stimulation even when they were added to the outside medium, and we then concluded that these drugs may be captured both from the periplasm and from the cytosol. However, examination of the data (64) shows that the stimulation tended to be marginal or small when the drugs were added to the outside medium, and I now believe that data were overinterpreted at that time and that the capture occurs only from the periplasm with all aminoglycosides. We tested the intravesicular accumulation of $[^3H]$gentamicin by adding it to the external medium (64). Although the vesicles seemed to accumulate the drug slowly, the background accumulation was also significant, and the calculated turnover number of AcrD was in the range of $0.01\,s^{-1}$. This is insignificant in comparison to the now determined turnover number of AcrB for cephalosporins, between 10 and $1000\,s^{-1}$ (Section II.C), and it fits with the notion that there is little evidence for the capture of aminoglycosides from inside the cytosol.

III. Structure of AcrB

A. SYMMETRIC STRUCTURES OF AcrB

Crystallography supplied essential pieces of knowledge on the mechanisms of the RND pump complex. The crystallographic structures of the outer membrane channels and of the periplasmic adaptor proteins were described briefly in Section II.B. Again, as mentioned earlier, Murakami et al. solved the structure of AcrB as a homotrimer with a threefold rotational symmetry in 2002 (44) (Figure 3), the first crystallographic structure for a proton-motive-force-driven transporter solved.

We mentioned earlier that AcrB probably captures its substrates from the periplasm, and the prediction was made in the early days that drugs may first partition into the membrane–periplasm interface (Figure 1). Indeed, the AcrB trimer structure showed that there was a small opening (vestibule) between subunits at the bottom of the periplasmic domain, close to the external surface of the membrane bilayer (44), and this led to the top of the large central cavity in the transmembrane domain (Figure 3). It was thus hypothesized that the drugs diffused through the vestibule into the central cavity, where it was captured. Indeed, co-crystallization of AcrB with various drugs (rhodamine 6G, ethidium, dequalinium, and ciprofloxacin) produced crystals with drug molecules within this central cavity (98). However, our conclusions were less than watertight. First, the resolution was poor (3.5 to 3.8 Å), and thus the identification of the drug molecules could not be made with absolute confidence, although drugs with poly-aromatic structures were used precisely for the ease of detection in the crystal structure. Second, the wall of the central cavity, where these drugs were thought to be bound, appeared quite open and totally unlike the traditional ligand-binding sites. To explain drug binding to such sites, we had to propose the existence of "composite" binding sites with the involvement of head groups of phospholipids that were assumed to be present in the central cavity (87). To confirm that the bound drugs were on their correct path to export, we applied site-directed mutagenesis to change Phe386, which seemed to be close to most substrates, into Ala, and found that the efflux activity was nearly abolished (87). However, as described already (Section II.D), this result is likely to be an artifact resulting from the use of high-copy-number vectors. Thus, our earlier conclusion that the drugs bind to the walls of the central cavity is not totally convincing. Nevertheless, the AcrB–YajC co-crystal obtained accidentally from *E. coli* grown in ampicillin-containing medium (99) was reported to contain two ampicillin molecules per AcrB protomer at this position near the ceiling of the central cavity, although the resolution was again poor (3.5 Å) and it is difficult to imagine that these ampicillin molecules survived attack by the high levels of plasmid-coded β-lactamase during production and purification of the protein.

In one symmetric structure of AcrB, the revealed N-terminal structure was shown to narrow the cytosolic opening of the central cavity somewhat (100). However, it is unclear whether this has any bearing on the function of the pump, as the role of the central cavity in substrate binding is uncertain, as described above.

We obtained another set of drug co-crystallization data with a site-directed mutant of AcrB, Asn109Ala (88). In the simple hypothesis we had at that time for drug extrusion, the drug molecules are captured near the ceiling of the central cavity, enter the central pore bordered by three helices (corresponding roughly to Asp99 through Leu117) from the three protomers, and pass through this pore to reach the top of the periplasmic domain (see Figure 3). However, the pore in the crystal structure appears closed. Since the side chain of Asn109 appears to extend deeply into the center of the pore, we naively thought that changing it to alanine might weaken the interaction of helices from the three subunits and "loosen up" the pore. The mutant AcrB functions almost normally as judged from the resistance levels to several antimicrobial agents. Interestingly, when the mutant protein was crystallized in the presence of several ligands [ethidium, rhodamine 6G, ciprofloxacin, nafcillin, and the inhibitor Phe-Arg-β-naphthylamide (41)], ligands were found to be bound not only to the ceiling of the central cavity [as seen earlier (98)] but also to the periplasmic domain, near the entrance of the large, deep, external cleft (Figure 6). Again, there was the problem that the ligands could have become bound to any lipophilic pocket, which may not be on the proper pathway for export. Changing residues that appeared to be interacting with the ligands into alanine through site-directed mutagenesis produced a drastic decrease in resistance to most drugs in the case of Glu673, and more modest, yet significant, decreases with Phe666 and Phe664. Although we cannot exclude the possibility that they may inactivate the transporter by a mechanism unrelated to ligand binding, the more recent asymmetric AcrB structures (discussed in Section III.C) are not inconsistent with the hypothesis that this is the portal for the periplasmic entry of substrates. The putative pathway for the drugs to the binding site is closed in the "access protomer" (see below), which resembles the protomers in the symmetric trimer in conformation; thus, the drug molecules cannot go further and may be forced to interact with the entrance of this pathway in the symmetric structure. Furthermore, there are random and site-directed mutation data that suggest the importance of residues within the cleft (Section II.D)(see also Figure 6) as well as the recent data using Cys mutants (see below).

Finally, it should be noted that the symmetric AcrB crystals soaked in deoxycholate contained deoxycholate molecules within the external cleft of the periplasmic domain (101), in a position essentially identical to the various substrates mentioned above.

B.   ATTEMPTS TO DETERMINE THE STRUCTURE OF AcrB DURING THE DRUG EXPORT CYCLE

Because of the uncertainties involved in the modeling of ligand electron density and in the possibility of nonspecific interaction between the ligand and the protein surface, we tried a different approach. We crystallized AcrB mutants in which one of the transmembrane domain residues putatively involved in proton translocation, Asp407, Asp408, Lys940 (corresponding to Lys939 of MexB), or Thr978 (see Section II.D), was changed to Ala (86, 102). In the crystal structure (44), the side chains of these residues are very close and appear to interact with each other either by salt bridge or hydrogen bonding (Figure 10). We assumed that during the normal function cycle of AcrB, a proton antiporter, one of these residues will be altered in its protonation state, most probably by the protonation of Asp, and this may disrupt the tight interactions among these residues and eventually, cause conformational changes of the entire protein, which would be coupled to ligand export. If so, our mutant proteins would mimic this transient, protonated intermediate state during the process of drug extrusion, and

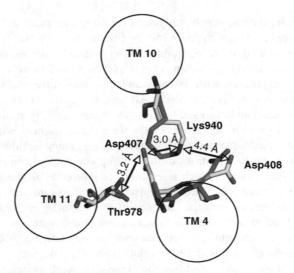

Figure 10.   Putative salt bridge/H-bonding network in the transmembrane domain of the wild-type AcrB protomer (yellow stick model) and the Asp407Ala mutant (green stick model), based on PDB files 1IWG and 2HQC. The view is along the line perpendicular to the membrane plane, from the periplasmic side. Note the prominent flipping of the Lys940 side chain, which is also seen with other mutants of the network (not shown). [From (102).] (See insert for color representation.)

allow us to see this elusive structure that cannot be deduced from the ground-state structures of the wild-type protein. When the crystal structures (without ligands) of these mutant proteins were examined (102), we found the following.

1. The tight interaction between Asp407, Asp408, Lys940, and Thr978 indeed became disrupted, and the Lys940 side chain, which interacted strongly with both of the carboxyl groups of Asp407 and Asp408 in the wild-type AcrB, went through a strong flipping motion and now faced away from these residues (Figure 10). Remarkably, the alterations were quite similar regardless of which residue was changed to Ala (102).

2. An extensive conformational alterations were found in the transmembrane domain (102). Again, it was reassuring that the conformational alterations were similar in all four mutants, confirming that they were

due to the disruption of the salt bridge/hydrogen-bonding network among these four residues rather than to specific effects of individual mutations. Most of the changes involved the shortening or melting of helices, except that helix 8, one of the helices connecting the trans-membrane domain to the periplasmic domain, became extended into the periplasmic domain (102).

3. When the structure of the periplasmic domain was examined, however, there were only very small changes (102), a disappointing result because we expected major conformational alterations in the periplasmic domain during substrate binding and extrusion (see above). The reason for this failure was suggested from the studies describing the asymmetric trimer structure of AcrB, described below.

C. SOLUTION OF THE ASYMMETRIC AcrB TRIMER STRUCTURE

At about the same time as our study of the proton-relay network mutants, several papers appeared describing the asymmetric trimer structure of AcrB (103–105) (Figure 11). In contrast to the symmetric structures seen before, in these structures each protomer, especially its periplasmic domain, takes a unique conformation (Figure 11). All groups identified one protomer in which a pocket is opened up in the periplasmic domain, a likely substrate-binding site, and in the study by the Murakami group (103), this pocket was indeed found with a bound substrate molecule (minocycline or doxorubicin). The interpretation of electron density as the bound substrate was strengthened by using minocycline that was labeled with a heavy atom, bromine. This protomer is called the *binding protomer* by the Murakami group or T (for "tight") by others. Other protomers are called the *access* (or L for "loose") and *extrusion* (or O for "open") *protomers*. In the extrusion protomer, which shows the largest conformational alteration, the binding pocket becomes collapsed, and there appears to be an opening of the pathway from the binding pocket to the top (central funnel) of the periplasmic domain. Inspection of the structures of the transmembrane domains shows that the tight association of Asp407–Asp408–Lys940–Thr978 becomes disrupted in the extrusion protomer, probably as a consequence of the protonation of Asp407 and/or Asp408; indeed, the arrangement of these side chains here is nearly identical to that found in our mutant proteins (102) (Figure 10, green sticks). This is consistent with our hypothesis that the disruption of the

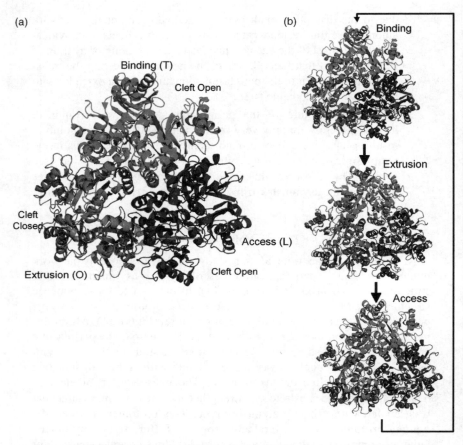

Figure 11. Periplasmic domain of the asymmetric crystal structure of AcrB (A) and the functionally rotating mechanism proposed (B), from studies of Murakami et al. (103), drawn using Pymol from PDB file 2DRD. The view is from the top (periplasm), and the transmembrane domains were removed for clarity. The three protomers, each with a distinctive conformation, are shown in different colors. In part A, the binding protomer (green) contains the bound ligand minocycline (in a stick model). When the protonation of the Asp407, Asp408, or both changes the alignment of proton relay residues shown in Figure 10, the conformational alteration is transmitted to the periplasmic domain, which is changed into an extrusion protomer, which extrudes the drug molecule outward. After reprotonation of Asp residue(s) in the transmembrane domain, the initial conformation of the Access protomer is restored. Thus, each protomer was proposed to go through cyclic changes of conformation (part B, where each protomer retains its initial color scheme) in a functionally rotating manner (103–105). Note that the large external cleft of the periplasmic domain becomes closed in the extrusion protomer. (See insert for color representation.)

network among the four residues mentioned alters conformation of the periplasmic domain. However, the asymmetric structure also shows that extensive conformational changes in the periplasmic domain require complementary accommodation by the neighboring protomers (103–105). This was not possible in our construct, in which all protomers corresponded to the mutant protein.

As mentioned already, the extrusion protomer is the subunit where the conformation of the periplasmic domain is altered most strongly from the conformation seen in the symmetric trimer. Presumably, this conformational alteration is driven by the extension of helix 8 toward the periplasm. This change, in turn, is caused by disruption of the salt bridge/hydrogen-bonding network among Asp407, Asp408, Lys940, and Thr978, as seen in the structures of both the wild-type protein in the asymmetric model (103–105) and the structure of the proton-relay network mutants (102).

The asymmetric structure was important because the structure of one single trimer suggests three stages in the drug extrusion process. Thus, the protomers appear to go through a cyclic conformational change, from the open, access conformation (which would allow the access of substrates), through the ligand-bound, binding confirmation, and finally, the extrusion conformer, which propels the substrates out and whose transmembrane domain shows signs of disrupted network among proton-translocating residues (103–105) (as shown in Figure 11B). Thus, this mechanism is called the *functionally rotating mechanism*; it is reminiscent of the mechanism of F_1-ATPase, although there is nothing that corresponds to the physical rotation of its γ-subunit. The "power stroke" of this entire process occurs during the change of binding into extrusion conformation, which involves the largest movement of various sections in the periplasmic domain. This large conformational alteration is initiated apparently by the protonation of one or both of the Asp residues (Asp407 and Asp408) in the midst of the transmembrane domain, which leads to a significant elongation of TM helix 8 into the periplasmic domain and all the subsequent conformational changes in the periplasmic domain.

Recently, the reactivity of the carboxyl groups in AcrB was assessed by labeling with a hydrophobic reagent, dicyclohexylcarbodiimide, followed by CNBr cleavage of the protein and analysis of the fragments by mass spectrometry (106). It was found that Asp408 was especially reactive to carbodiimide, with a high pK_a value of 7.4, implicating the protonation/deprotonation of at least this Asp residue as a key step in the proton

translocation pathway, and eventually, in generating conformational alterations in AcrB that leads to drug extrusion.

IV.   Mechanism of Drug Efflux

A.   MAIN SUBSTRATE-BINDING SITE

The possible route of substrate binding and extrusion is suggested by the
structure of the asymmetric trimer (103–105). In the binding protomer, the
periplasmic domain contains an expanded binding pocket containing
several aromatic and hydrophobic residues [Phe136, 178, 610, 615, 617,
and 628; Val139 and 612; Ile277 and 626; and Tyr327 (104)]. This pocket
was indeed found to contain the drugs minocycline and doxorubicin (103),
and its role is also supported by the mutation data with the AcrB homolog
YhiV (82) previously described as well as the more recent data on AcrB
(107). In the latter study, each of the Phe residues in the binding site was
changed to Ala by site-directed mutagenesis. Among them, Phe610Ala
mutant became strongly more susceptible to almost all drugs, and Phe178A
and Phe615Ala had a somewhat less extensive effect. Changing Phe628
and Phe136, and especially Phe617, decreased the AcrB activity only to a
limited range of substrates, perhaps understandably, as these side chains
occupy a peripheral position in the binding pocket.

The significance of this binding site is also suggested by the observation
that the pocket becomes collapsed in the extrusion protomer. Finally, its
physiological significance is supported further by a comparison of its
surface in AcrB and in the homology-modeled AcrD (36); although the
surface is entirely hydrophobic in AcrB, in the aminoglycoside pump AcrD
it contains many oxygen atoms that may function in the binding of the
hydrophilic basic substrates.

B.   PATHWAYS FOR DRUGS: A CURRENT HYPOTHESIS

How do the substrates reach the binding pocket? All three groups that
studied the asymmetric trimer structure agree that there is an open pathway in
the binding protomer between the binding pocket and the external large cleft
of the periplasmic domain of the binding protomer (102–104) (see Figure 12).
Although this cleft was earlier thought by some investigators to be the site of
interaction between AcrA and AcrB, in the most up-to-date model of
AcrA–AcrB interaction this cleft remains fully open (53) (Figure 4). [The
entry of the substrate from the cleft was advocated by Lomovskaya and

(A)   (B)

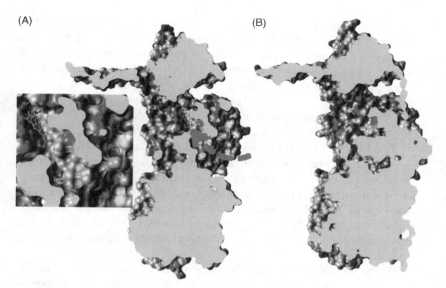

Figure 12.   Cutout view of the binding protomer with the bound minocycline in a green stick representation (A) and the extruding protomer (B), both from PDB file 2DRD, drawn using the UCSF Chimera package (133). The wide passageway from the external cleft (dashed green arrow) seen in A becomes closed in B. The passageway in A was also enlarged, so that the extremely hydrophobic surface in this area can be seen more easily. [Modified from (36), with permission from Elsevier.] (See insert for color representation.)

Totrov (108) before the appearance of the asymmetric crystal structures.] We have seen that the possibility of substrate entry from the lateral cleft is also supported by the results of random mutagenesis data (e.g., (81) and (83), described in Section II.D) and the results of site-directed mutagenesis (e.g., (88) and (89) in the same section) (see Figure 6). Finally, co-crystallization studies showed the presence of bound substrates at the entrance of this large cleft (88), as described in Section III.B (Figure 6).

We have recently examined experimentally the path(s) of the drugs within the periplasmic domain of AcrB (108a). We first changed each of the 48 residues that may line the drug path into Cys, and tested if the Cys residue was covalently modified by fluorescein maleimide. This reagent did not label sites obviously outside the path, even when they were surrounded by hydrophobic residues; thus the labeling appears to occur only when the reagent behaves as the AcrB substrate, and becomes concentrated in the drug pathway. Some of the residues thus identified were also shown to block the efflux of Nile Red when convalently modified by bulky sub-stituents, further confirming their location in the substrate path. In this manner, we showed that, in addition to the residues lining the binding pocket identified in the asymmetric crystal (103), the residues at the entrance of the large lateral cleft (Figure 6) were also in the substrate path, in support of the idea that the substrates mainly enter the AcrB periplasmic domain from the lateral cleft and then reach the binding pocket.

The Murakami group found, in addition, the second pathway, which connects the binding site with the vestibule, the passageway between protomers, located close to the external surface of the inner membrane. A similar pathway between the binding site and the vestibule area could also be seen in the Darpin-containing crystal, which so far has shown the highest resolution (105). Thus, there is a possibility that the substrate may go through the initial part of the vestibule and then go into the binding pocket.

In the extrusion protomer, because of the conformational changes probably initiated by the protonation of Asp408 in the transmembrane domain, the binding pocket becomes narrower and the entrance pathway(s) become closed, and at the same time a new pathway from the binding pocket to the funnel-like structure at the top becomes open (Figure 12). This would favor the extrusion of drug molecules into the TolC channel.

Thus, the basic principles of the drug pumping mechanism, based on the functionally rotating model, are now reasonably well understood. However, some details are still at the stage of speculation. For example, for the drug extrusion to occur with a reasonable efficiency, the drug molecule must remain in the binding pocket for some time. How would this be achieved when both the drug entry pathway and the pocket are surrounded primarily by hydrophobic side chains (Figure 12)? Furthermore, how does the AcrB pump accommodate such a wide range of substrates, some of which are shown in Figure 2? The answer to these questions is not immediately apparent from the publications describing asymmetric crystal structures (103–105), because the only drug molecules found in co-crystals are minocycline and doxorubicin, despite trials with many dozens of drugs by the Murakami group. Because of this situation, we have tried a computer-simulated docking approach (108b). With the binding protomer of minocycline-bound AcrB (103), the binding of about two dozen known AcrB substrates was estimated. Minocycline was shown to bind to the apo-protein in a location and conformation found in the crystal structure, although this was not the complex predicted to be the most stable. This result validates the use of the computational approach, at the same time suggesting that the energy scoring function of the program may not yet be perfect. Currently, there is no way to include the movement of the backbone of this rather flexible protein in simulation. Thus, the details of the binding interactions of drugs predicted give us only broad hints. Nevertheless, examination of the possible binding modes of many substrates was instructive, in a way somewhat similar to the usefulness of comparison of many homologous sequences. Many substrates (including minocycline, tetracycline, and nitrocefin) prefer to bind, at least in this modeling exercise, with their hydrophobic domain bound to the narrow groove of the binding site, with their hydrophilic group(s) exposed either at the "top" and/or "bottom" minipockets within the site. These subdomains are much more hydrophilic than the central groove. This is very similar to the binding of minocycline (103) (for the structure, see Figure 2), which is bound to the groove mainly by using its hydrophobic middle of the molecule and has its one polar domain (containing the acidic OH, an amino group, and an amide group) on top and its other polar group (an amino group) at the bottom. Murakami et al. (103) argue that the polar groups on top interact with an Asn274 side chain, and the amino group at the bottom interacts with Gln176. However, none of these groups are at hydrogen-bonding distance, and are at best 4 to 5 Å away. Thus, it seems best not to think of these interactions as being specific. Rather, the top and bottom of a minocycline seem to fit into more hydrophilic, larger minipockets which may even be filled with water.

Interestingly, some ligands are predicted to bind nearly exclusively to the bottom pocket and its neighboring hydrophobic patch, completely avoiding the central groove where the bulk of minocycline binds. This group includes the inhibitor NMP (naphthylmethylpiperazine (109)) as well as chloramphenicol. These observations are consistent with the finding that the efflux of nitrocefin, a groove-binder, is inhibited by the simultaneous presence of another groove-binder, minocycline, but not at all by the presence of a cave-binder, chloramphenicol (102b).

In this light, we can imagine the movement of substrates as follows. Initially, substrates are attracted to the hydrophobic surfaces of the pathways, either from the periplasm through the large external cleft of AcrB periplasmic domain or possibly from near the external surface of the plasma membrane through the vestibule. However, the protein–substrate complex is not in its most stable conformation here, because the hydrophilic part of the substrate is not stabilized optimally. The final stabilization is achieved at the binding site, because of the presence of two features here: one, interface(s) for hydrophobic interactions, and the other, rather large hydrophilic pockets next to it. The substrate range of AcrB is likely to be determined essentially by this structure of the binding site. The substrates bound here is then ready for extrusion through the conformational change of the binding protomer into the extrusion protomer.

C. BIOCHEMICAL STUDIES OF CONFORMATIONAL ALTERATIONS

Although the structure of the asymmetric AcrB trimer strongly suggests the mechanism of ligand extrusion coupled to proton influx, this mechanism remains a hypothesis. We set out to prove this hypothesis by carrying out biochemical studies of the pump. The proposed cyclic change involves the opening and closing of the large external cleft in the periplasmic domain (Figure 11). Thus, if the opening of the cleft is prevented, we can predict that the cyclic conformational changes will be prevented and that the pump will cease to function. We introduced pairs of cysteine residues at various positions on the opposing walls of the cleft (110). Although single cysteine mutations produced little inhibition of transport, proteins containing double mutations (Asp566Cys and Thr678Cys; Phe666Cys and Thr678Cys; Phe666Cys and Gln830Cys) on the opposing walls of the cleft were strongly compromised in function. This is likely to be due to the forced closing of the cleft by disulfide bonds formed between these residues, but we cannot rule out the possibility that the trimeric assembly failed because of the premature formation of disulfide bonds between protomers. Indeed, with cysteine pairs at other positions, there were data suggesting this interpretation. To avoid this problem, we expressed mutated AcrB in a host strain with a defective DsbA, a periplasmic disulfide oxidoreductase that plays a major role in the formation of disulfide bonds in this compartment (111, 112). Indeed, in this strain, the AcrB protein containing Phe666Cys and Thr678Cys (or Gln830Cys) largely retained its transport activity, presumably because disulfide bond formation did not occur in the absence of DsbA enzyme activity. We tried to see, by using these mutants, if the cross-linking and inactivation of AcrB can be observed in real time by using a fast-acting disulfide cross-linking agent based on the methanethiosulfonate groups (113). Indeed, addition of such cross-linkers to cells that are pumping out a fluorescent dye, ethidium, was shown to stop the function of the pump instantaneously (110), thus providing biochemical support for the rotating mechanism of drug transport.

A similar approach was also pursued by the group of K. M. Pos (114). They introduced pairs of cysteines at a wide range of positions, importantly including those that were expected to become closer in the access or the binding protomers, in addition to those in the extrusion protomer. The double cysteine mutant of Ser132 and Ala294, which are expected to become closer only in the access protomer, inactivated the pump as estimated by the decreased resistance to a number of drugs, and the pumping activity was increased to a detectable extent upon the addition of disulfide-reducing agent dithiothreitol to intact cells (114).

D.   USE OF COVALENTLY LINKED TRIMERS

The functional rotation mechanism of AcrB drug pumping process (103–105) predicts that if one of the three protomers is defective in the proton relay network, the pumping action by the entire trimer should come to a halt. This hypothesis could not be tested by the disulfide cross-linking experiment described above, because all protomers had the same Cys mutations. We devised a way to test this hypothesis by creating a giant gene in which three *acrB* sequences were connected together through a linker sequence (115). When the sequence of the cytosolic, horizontal helix (Met496 through Arg540) connecting the two halves of the trans-membrane domain was used as the linker, the linked trimer was expressed well as a single giant protein and produced drug resistance levels sometimes even higher than that produced by the monomeric *acrB* gene. When only one of the three-component *acrB* sequences in the giant gene was changed to include mutations in the proton-relay network, the entire trimer became inactive. When only one of the three *acrB* sequences was made to contain the Phe666Cys and Gln830Cys, cross-linking of these two cysteine residues by a fast-acting methanethiosulfonate cross-linker instantaneously inactivated the entire trimer, regardless of the position of the altered protomer in the giant sequence (Figure 13). These results strongly support the notion that the trimer acts by a functionally rotating mechanism.

E.   RND SYSTEMS APPARENTLY REQUIRING PAIRED TRANSPORTERS

The RND transporters we have discussed so far, AcrB, MexB, or AcrD, exist as homotrimers and are now assumed to work by a functionally rotating mechanism. There are, however, RND transporter genes that occur in tandem, such as MdtBC, discovered by our group (116) as well as by

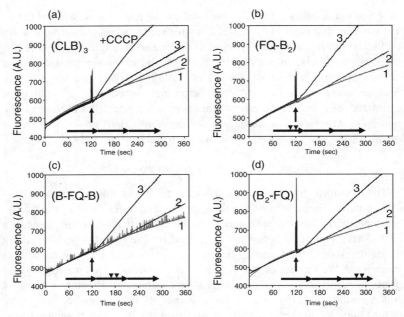

Figure 13. Use of the covalently linked AcrB trimer showing that inactivation of one protomer is sufficient in inactivating the entire trimer. Covalently linked trimers had constructions shown at the bottom of each panel with only one protomer containing Phe666Cys and Gln830Cys mutations (shown as downward facing arrowheads). These residues are located on opposite walls of the cleft, and the linked AcrB trimer remains active, as shown by the slow net entry of the dye ethidium bromide (because of efflux) when it is expressed as a *dsbA* mutant strain, in which the disulfide bond formation is inefficient in the periplasmic space (curve 1). Modification of individual Cys residues by 5-MTS reagent in contrast produced little inactivation (curve 2). Real-time inhibition of pump function through closure of the periplasmic cleft is then achieved by the addition of a fast-acting methanethiosulfonate SH cross-linker MTS-2-MTS (curve 3).

Yamaguchi's group (117). Both groups showed that both of the RND transporters, MdtB and MdtC, had to be present for the full pumping activity. Interestingly, Baranova and Nikaido (116), who relied on the expression of these genes from the chromosome, found that MdtB or MdtC alone (in the presence of MFP MdtA) was completely inactive, whereas Nagakubo et al. (117), who expressed the genes from plasmids, reported that MdtC alone gives some marginal activity.

When the C-terminus of either the B or C protein was labeled with hexahistidine tag, metal affinity purification produces a complex containing both MdtB and C, confirming that these proteins produce a heteromulti-meric complex (115a). The size of the complex showed that it was a trimer, and the average composition was B_2C. To establish the precise composition of the active trimer, the *mdtB* and *mdtC* genes were connected by linker sequences to produce covalently linked trimers of all possible combina-tions. The results showed that the highest activity was obtained only when the linked trimer contained two copies of MdtB and one copy of MdtC.

Since the heterotrimer contains two different protomers, the functionally rotating mechanism seems unlikely for this pump, at least in its unmodified form. Furthermore, when the proton-translocation pathway residues such as Asp410 (corresponding to Asp407 of AcrB) were changed to Ala in one of the MdtB sequences, the complex lost its activity, although similar changes in the MdtC sequence (e.g., in Asp401, corresponding to Asp407 of AcrB) resulted only in a small change in activity (H. S. Kim and H. Nikaido, manuscript in preparation). Since we know from AcrB that protonation in this network of amino acids is needed for the extrusion of substrate drugs (see above), we assume that MdtB performs this function by changing from the binding conformation to the extrusion conformation. We do not know at present the role of MdtC. It could simply play a role in mechanically stabilizing the trimeric complex, in facilitating the capture of substrates in the neighboring MdtB unit, or in actually capturing the substrates by itself. Regardless of the precise roles of protomers, it is clear that the operating principle here is somewhat different from that of the homotrimers. Interest-ingly, MdtC is a basic protein with the calculated pI of 8.4, in stark contrast to other RND transporters that are acidic (AcrB, 5.4; AcrD, 5.6; MdtB, 5.2). We also do not know if this heterotrimeric structure gives advantages, if any, in terms of pumping activity. Such RND pumps requiring a paired set of transporters are also found in related bacteria, such as SdeAB of *Serratia marcescens* (118), whose substrate range is limited as in MdtBC (bile acids and novobiocin). However, SmeJK of *Stenotrophomonas maltophilia* is reported to have a wider range of substrates (119).

V. AcrAB–TolC and Drug Resistance

We have so far outlined our current knowledge on the mechanism of the AcrAB–TolC complex, beginning with the capture of substrates either from the periplasm or from the periplasm–inner membrane interface, followed by the conformational change in the AcrB periplasmic domain initiated by the flow of proton(s) through the AcrB transmembrane domain, finally resulting in the extrusion of drugs into the external medium through the long channel of TolC. However, gram-negative RND-type efflux pumps have attracted attention because they are thought to play a large role in drug resistance, a major public health problem, which is becoming more and more acute because this type of multidrug efflux greatly diminishes the activity of most of the antimicrobial agents currently available (31). Thus, this review would not be complete without discussing the effect of AcrAB–TolC on the drug resistance. Furthermore, the mechanisms and kinetics of the pump we have discussed so far cannot be confirmed unless they explain, quantitatively, the level of resistance finally achieved in the living bacterial cell.

The analysis is the simplest with β-lactams, as their target is in the periplasm and as most of them are relatively hydrophilic, so that we can neglect their diffusion across the inner membrane. We have attempted this analysis at the time when multidrug efflux pumps were not yet known (73). In that study we considered the influx across the outer membrane, which follows Fick's first law of diffusion, and the drug degradation by the periplasmic β-lactamase, which follows Michaelis–Menten kinetics. By solving these two simultaneous equations for the external drug concentration (C_O) that would just produce the periplasmic concentration that would kill the cell (C_{inh}), we could predict MIC values for β-lactams. Thus,

$$\text{MIC} = C_O = C_{inh} + \frac{V_{max}}{PA} \frac{C_{inh}}{(C_{inh} + K_m)} \tag{1}$$

where V_{max} and K_m are the usual kinetic constants of the β-lactamase, P is the permeability coefficient of the β-lactam across the outer membrane and A is the area of the outer membrane per unit weight of bacteria.

This procedure was applied to four strains of E. coli, producing widely different levels of the chromosomally coded AmpC β-lactamase (V_{max} of 0.013 to 16.6 nmol mg^{-1} min^{-1} with cephaloridine), and the MIC values predicted were very close to the values observed for all of the cephalosporins tested. However, with some penicillins, there was a significant difference, which was most pronounced with carbenicillin, whose MIC values were usually eight times higher than the values predicted. We can now understand these results at least qualitatively. With most cephalosporins, at very low periplasmic concentrations where the cells become killed [C_{inh} is usually less than 5 μM (73)], the efflux is insignificant, especially because of its sigmoidal kinetics (71); thus, exclusion of efflux from consideration produces only small errors. However, with carbenicillin, efflux is important, as seen from the strongly decreased MIC values in *acrAB*-inactivated strain (15, 72), whereas the chromosomal β-lactamase of *E. coli* works poorly with this compound (72); hence, the prediction without consideration of efflux underestimates MIC values greatly.

We can also carry out a more quantitative analysis. The strain used by Nagano and Nikaido (71) for the analysis of nitrocefin was a rather artificial construct, containing a mutant porin with a wider channel, to facilitate the analysis. Nevertheless, with this strain, when both the AcrAB efflux pump and the AmpC β-lactamase were deleted, the nitrocefin MIC dropped from 38 μM to 2.2 μM. We can calculate the MIC value expected in the parent strain as follows.

When the periplasmic concentration reaches C_{inh}, the rate of efflux V_e is $V^e_{max} C_{inh}/(K^e_m + C_{inh})$, and the rate of hydrolysis is $V^h_{max} C_{inh}/(K^h_m + C_{inh})$, both from the Michaelis–Menten relationship, where the superscripts *e* and *h* indicate constants related to efflux and hydrolysis, respectively. Then, since $V_{in} = PA(C_O - C_{inh})$, C_O or MIC can be calculated as $C_{inh} + (V_e + V_h)/P/A$. As was noted earlier in the analysis without the consideration of efflux (73), the constant that is most difficult to estimate is C_{inh}. It has to be smaller than 2.2 μM, the MIC of the $\Delta acrAB$ $\Delta ampC$ mutant, but how much smaller? As there may be other efflux pumps that slowly pump out nitrocefin, we assume arbitrarily that C_{inh} is one-half of this concentration. Then the calculations give us the predicted MIC of 72 μM, or 36 μg/mL. Although this is twice the observed MIC of 18 μg/mL, the discrepancy is not too large, considering that MIC was determined under somewhat different conditions in comparison with the kinetic constants (e.g., temperature and medium).

An interesting exercise is to estimate the effect of outer membrane barrier on MIC values. We can see that the more permeable outer membrane would decrease the second term of the formula used above, by increasing *P*, and would make the MIC closer to C_{inh}. Thus, the effect of efflux (or periplasmic hydrolysis) will become minimized. Conversely, in an organism with a more efficient outer membrane barrier, the term $(V_e + V_h)/P/A$ will become larger, and thus efflux (and hydrolysis) will contribute more effectively in raising the MIC values, the situation we find, for example, in *P. aeruginosa* (20–22).

When different strains were compared, it was evident that the MIC values did not have a linear relationship with the amount of efflux pumps produced. Here a relevant study is the rigorous analysis from the group of Frère (120) on the effect of the β-lactamase level and of the target sensitivity to MIC. It was shown that if a high-affinity β-lactamase is overproduced, a 5000-fold difference in the sensitivity of the target (penicillin-binding proteins) will produce only a twofold difference in MIC. We can see in equation (1) that under certain conditions the $C_{inh}/(K_m + C_{inh})$ term approaches 1, so that the MIC will be affected solely by the balance between permeability and the enzyme concentration (i.e., by $V_{max}/P/A$), and not much by the sensitivity of the target, or C_{inh}. These conditions are attained, for example, when the K_m of the enzyme is very low, as with cefotaxime, in comparison with C_{inh}. Thus, the presence of β-lactamase will "mask" the difference in target sensitivity. Of course, if the β-lactamase and efflux were totally absent, there would be a 5000-fold difference in MICs. Although at the time of the work by Frère's group, the kinetic behavior of the AcrB pump was not known, the conclusions may be relevant, as the efflux follows saturation kinetics such as enzymatic hydrolysis.

It must be remembered, however, that the K_m of the AcrB pump is usually much higher than C_{inh} (see Section II.C), and under such conditions the sensitivity of the target will continue to affect the magnitude of MIC. If we neglect the sigmoidal nature of the efflux kinetics (Section II.C), we can calculate the situation with cephalothin as follows. Here the K_m (actually, $K_{0.5}$) of efflux is about 100 μM (71), much larger than C_{inh}, which earlier we estimated to be around 1 μM for cephalosporins. Thus, the $C_{inh}/(K_m + C_{inh})$ term of equation (1) may be approximated as C_{inh}/K_m (i.e., 0.01 in this case). The V_{max} value for cephalothin efflux was about 1 nmol s^{-1} mg^{-1} cells (71). As P for cephalothin is around 10^{-5} cm/s (73) and A is 132 cm^2/mg cells, the $V_{max}/P/A$ term in (1) will be around 1000 μM and the second term of (1) will be close to 10 μM, yielding a predicted MIC of 11 μM (neglecting the periplasmic hydrolysis). If the target becomes twice as resistant, with a C_{inh} value of 2 μM, the MIC predicted will be 22 μM, following the changes in target sensitivity in a nearly proportional manner. If the efflux rate is decreased by one-half, the MIC predicted will be 6 μM, an almost but not completely proportional change. In reality, the sigmoidal nature of the efflux kinetics is likely to produce significant deviations from proportionality in these situations.

Carrying out similar analysis for antibiotics that have targets in the cytoplasm is much more complicated, because we have to consider the additional step of fluxes across the cytoplasmic membrane. In addition, the efflux across the cytoplasmic membrane may be catalyzed by "singleton" pumps located in this membrane, rather than by AcrB, as shown recently by Tal and Schuldiner (121) for ethidium bromide (pumped out into the periplasm by an MFS pump MdfA and an SMR pump EmrE). In this case, the drug efflux from cytosol to external medium would require successive action by at least two pumps, the export into periplasm cata-lyzed by a singleton pump and then the export from periplasm to external medium catalyzed by the AcrB complex. The kinetic behavior of some of the singleton pumps has been elucidated in great detail: for example, that of EmrE (122–124). However, the kinetic constants of AcrB for non-β-lactam substrates are currently unknown, and building mathematical models of transport of these compounds would be very difficult at present.

There are some interesting observations on the effect of the presence of AcrB on MICs of antibiotics with intracytoplasmic targets. Thus, a mutation in the ribosomal protein L22 lowers the affinity of ribosomes to erythromycin (about a fivefold increase in K_d), and makes the E. coli cells resistant to this drug, raising the growth-inhibitory concentration (at 50% growth) from 100 μg/mL to 800 μg/mL (125). In a strain with a deletion of the tolC gene (therefore, in the virtual absence of efflux), this difference largely disappears, the 50% growth-inhibitory concentration increasing from 3.5 μg/mL to only 5 μg/mL by the introduction of the same L22 mutation (125). A similar observation was made in another laboratory (126), which found that the difference in the 50% growth-inhibitory concentrations between the wild type and an L22 mutant strain decreased from tenfold in the efflux-competent background to only 1.5-fold in the ΔacrAB background. These authors considered the possibility that erythromycin might increase the translation of AcrAB proteins, but could not find any evidence for such a phenomenon. Synergy between the efflux and the target mutations was also found in Campylobacter spp. (127). Thus, these observations are the opposite of masking of the target sensitivity by β-lactamase mentioned above.

An attempt has been made to explain this phenomenon of *resistance masking* in efflux-deficient strains by mathematical modeling (128, 129). Since nothing is known about the kinetics of efflux, the efflux pump is assumed to operate at a substrate concentration far below K_m, and the efflux is assumed proportional to intracellular drug concentration. Most important, because the increase in cell volume due to growth was assumed to be the only mechanism that decreases the drug concentration inside, the rate of net influx of drugs through a loosely defined "cell envelope" needed to be set extremely low, on an order of magnitude similar to the growth rate. I remain quite skeptical about the validity of such a modeling approach carried out in the almost total absence of experimental constraints, as nothing is known about the kinetics of OM diffusion of macrolides or about that of active efflux of these compounds. Possibly the exaggeration of the difference in target sensitivity by efflux could be brought about by the sigmoidal kinetics of the pump (71), but we need to know much more before we get into more realistic simulations.

VI. Coda

As a participant who was there at the discovery of *E. coli* AcrB as a multidrug efflux pump (18) 17 years ago, I am impressed and gratified by the amount of progress that scientists have made in this area. At that time, the mere notion that a single transporter is capable of exporting such a wide variety of compounds (see Figure 2) and that it may capture substrates in the periplasm was met with a great deal of skepticism, and the publication of our reports on the *P. aeruginosa* multidrug efflux process (20, 21) was delayed for at least several months because the reviewers simply refused to believe in the existence of such a process.

In contrast, we now know the detailed three-dimensional structures of all of the components of the tripartite pumping complex (Figures 3 and 4), and we understand the mechanism of drug extrusion, at least in its basic outline. It seems likely that drugs are captured from the periplasm perhaps exclusively, because so far, structures do not give any hint as to how the drugs captured in the cytosol may be brought out to the periplasmic binding site. We understand the pathway(s) through which drugs reach the binding site and how the protonation of one or two Asp residues in the transmembrane domain causes conformational changes that are transmitted to the periplasmic domain to produce drug extrusion. We understand that the AcrB protomers within the trimeric complex act by a functionally rotating mechanism, so that all of them have to be functional for any drug molecule to be extruded, a situation that produces sigmoidal kinetics with some substrates. We are even approaching the stage where we can understand how various substrates bind to the various subdomains of the large binding sites (Section IV.B).

However, many questions remain. I hope that I emphasized the need to correlate our biochemical knowledge of pump action to the final effect of the pump in raising MIC (Section V), an area still in its infancy. Indeed, the kinetics of AcrB has been examined only for cephalosporins and peni-cillins, and we totally lack comparable data for all other classes of antimicrobial compounds. We need to know more precisely how various drugs bind to the binding site. The computer docking studies (Section IV.B) can give us only hints, and they remain to be proven by real biochemistry. We also do not have any idea how AcrA stimulates the function of AcrB.

The functions of the adaptor (or MFP), AcrA and its homologs, have been reviewed (38). However, it should be mentioned that the MFP CusB of the Cus system, a toxic cation exporter of *E. coli*, is known to bind the substrate Cu^+ (96), and most excitingly, direct transfer of metal ion from the periplasmic binding protein CusF to CusB was recently demonstrated (97). Thus, it seems quite possible that in this system the substrate ion flows from the periplasmic binding protein to the periplasmic adaptor protein, and then, finally, to the periplasmic domain of an RND transporter, CusA.

There are also unresolved questions on the function of tripartite RND multidrug pumps in intact cells. For example, *P. aeruginosa* strains over-expressing a tripartite pump MexCD–OprJ were often found to have reduced functions of other tripartite pumps, MexAB–OprM and MexXY–OprM, although the latter pumps are expressed at the normal level (130). Another curious observation is that TonB, a protein that functions in energizing the outer membrane siderophore receptors, is needed in *P. aeruginosa* for functioning of the MexAB–OprM pump (131, 132). As far as the authors are aware, there is no explanation of this effect at the molecular level, but this reminds us that a small protein YajC was accidentally co-crystallized with AcrB (99). Although the deletion of gene *yajC* had only a small effect on drug susceptibility, this observation certainly requires further follow-up study, especially because this co-crystal was obtained from cells expressing both YajC and AcrB at normal constitutive levels, without the lopsided, plasmid-driven overexpression of only AcrB.

Acknowledgements

The studies in the author's laboratory are supported by U.S. Public Health Service grant AI 09644. I thank V. Koronakis for supplying the high-resolution version of Figure 4.

References

1. Nikaido, H., and Vaara, M. (1985) Molecular basis of bacterial outer membrane permeability, *Microbiol. Rev. 49*, 1–32.

2. Nikaido, H. (2003) Molecular basis of bacterial outer membrane permeability revisited, *Microbiol. Mol. Biol. Rev. 67*, 593–656.

3. Kamio, K., and Nikaido, H. (1976) Outer membrane of *Salmonella typhimurium*: accessibility of phospholipid head groups to phospholipase C and cyanogen bromide-activated dextran, *Biochemistry 15*, 2561–2570.

4. Plésiat, P., and Nikaido, H. (1992) Outer membranes of gram-negative bacteria are permeable to steroid probes, *Mol. Microbiol. 6*, 1323–1333.

5. Nikaido, H. (1989) Outer membrane barrier as a mechanism of antibiotic resistance, *Antimicrob. Agents Chemother. 33*, 1831–1836.

6. Nikaido, H. (1994) Prevention of drug access to bacterial targets: permeability barriers and active efflux, *Science 264*, 382–388.

7. Nikaido, H. (1996) Multidrug efflux pumps of gram-negative bacteria, *J. Bacteriol. 178*, 5853–5859.

8. Saier, M. H., Jr., Paulsen, I. T., Sliwinski, M. K., Pao, S. S., Skurray, R. A., and Nikaido, H. (1998) Evolutionary origins of multidrug and drug-specific efflux pumps in bacteria, *FASEB J. 12*, 265–274.

9. Li, X.-Z., and Nikaido, H. (2004) Efflux-mediated drug resistance in bacteria, *Drugs 64*, 159–204.

10. Li, X.-Z., and Nikaido, H. (2009) Efflux-mediated drug resistance in bacteria: an update, *Drugs 69*, 1555–1623.

11. Wandersman, C. (1992) Secretion across the bacterial outer membrane, *Trends Genet. 8*, 317–322.

12. Paulsen, I. T., Park, J. H., Choi, P. S., and Saier, M. H., Jr. (1997) A family of gram-negative bacterial outer membrane factors that function in the export of proteins, carbohydrates, drugs and heavy metals from gram-negative bacteria, *FEMS Microbiol. Lett. 156*, 1–8.

13. Dinh, T., Paulsen, I. T., and Saier, Jr., M. H. (1994) A family of extracytoplasmic proteins that allow transport of large molecules across the outer membranes of gram-negative bacteria, *J. Bacteriol. 176*, 3825–3831.

14. Vaara, M. (1993) Antibiotic-supersusceptible mutants of *Escherichia coli* and *Salmonella typhimurium*, *Antimicrob. Agents Chemother. 37*, 2255–2260.

15. Nikaido, H., Basina, M., Nguyen, V., and Rosenberg, E.Y. (1998) Multidrug efflux pump AcrAB of *Salmonella typhimurium* excretes only those β-lactam antibiotics containing lipophilic side chains, *J. Bacteriol. 180*, 4686–4692.

16. Lee, A., Mao, M. S., Warren, M. S., Misty, A., Hoshino, K., Okamura, R., Ishida, H., and Lomovskaya, O. (2000) Interplay between efflux pumps may provide either additive or multiplicative effects on drug resistance., *J. Bacteriol., 182*, 3142–3150.

17. Tal, N., and Schuldiner, S. (2009) A coordinated network of transporters with overlapping specificities provides a robust survival strategy. *Proc. Natl. Acad. Sci., USA* (e-published).

18. Ma, D., Cook, D. N., Alberti, M., Pon, N. G., Nikaido, H., and Hearst, J. E. (1993) Molecular cloning and characterization of *acrA* and *acrE* genes of *Escherichia coli*, *J. Bacteriol. 175*, 6299–6313.

19. Poole, K., Krebes, K., McNally, C., and Neshat, S. (1993) Multiple antibiotic resistance in P*seudomonas aeruginosa*: evidence for involvement of an efflux operon., *J. Bacteriol. 175*, 7363–7372.

20. Li, X. Z., Livermore, D. M., and Nikaido, H. (1994) Role of efflux pump(s) in intrinsic resistance of *Pseudomonas aeruginosa*: resistance to tetracycline, chloramphenicol, and norfloxacin, *Antimicrob. Agents Chemother. 38*, 1732–1741.

21. Li, X. Z., Ma, D., Livermore, D. M., and Nikaido, H. (1994) Role of efflux pump(s) in intrinsic resistance of *Pseudomonas aeruginosa*: active efflux as a contributing factor to beta-lactam resistance, *Antimicrob. Agents Chemother. 38*, 1742–1752.

22. Li, X. Z., Nikaido, H., and Poole, K. (1995) Role of *mexA–mexB–oprM* in antibiotic efflux in *Pseudomonas aeruginosa, Antimicrob. Agents Chemother. 39*, 1948–1953.

23. Fralick, J. A. (1996) Evidence that TolC is required for functioning of the Mar/AcrAB efflux pump of *Escherichia coli, J. Bacteriol. 178*, 5803–5805.

24. Ma, D., Cook, D. N., Alberti, M., Pon, N. G., Nikaido, H., and Hearst, J. E. (1995) Genes *acrA* and *acrB* encode a stress-induced efflux system of *Escherichia coli, Mol. Microbiol. 16*, 45–55.

25. Smith, I. C., Auger, M., and Jarrell, H. C. (1991) Molecular details of anesthetic–lipid interaction, *Ann. N.Y. Acad. Sci. 625*, 668–684.

26. Lomovskaya, O., and Lewis, K. (1992) Emr, an *Escherichia coli* locus for multidrug resistance, *Proc. Natl. Acad. Sci. USA 89*, 8938–8942.

27. Furukawa, H., Tsay, J. T., Jackowski, S., Takamura, Y., and Rock, C. O. (1993) Thiolactomycin resistance in *Escherichia coli* is associated with the multidrug resistance efflux pump encoded by *emrAB, J. Bacteriol. 175*, 3723–3729.

28. Kobayashi, N., Nishino, K., and Yamaguchi, A. (2001) Novel macrolide-specific ABC-type efflux transporter in *Escherichia coli, J. Bacteriol. 183*, 5639–5644.

29. Yum, S., Xu, Y., Shunfu Piao, S., Sim, S., Kim, H.-M., Jo, W.-S., Kim, K.-J., Kweon, H.-S., Jeong, M.-H., Jeon, H., et al. (2009) Crystal structure of the periplasmic component of a tripartite macrolide-specific efflux pump, *J. Mol. Biol. 387*, 1286–1297.

30. Tikhonova, E. B., Devroy, V. K., Lau, S. Y., and Zgurskaya, H. I. (2007) Reconstitution of the *Escherichia coli* macrolide transporter: the periplasmic membrane fusion protein MacA stimulates the ATPase activity of MacB, *Mol. Microbiol. 63*, 895–910.

31. Nikaido, H. (2009) Multidrug resistance in bacteria, *Annu. Rev. Biochem. 79*, 119–146.

32. Poole, K. (2005) Efflux-mediated antimicrobial resistance, *J. Antimicrob. Chemother. 56*, 20–51.

33. Murakami, S. (2008) Multidrug transporter, AcrB—the pumping mechanism, *Curr. Opin. Struct. Biol. 18*, 459–465.

34. Pos, K. M. (2009) Drug transport mechanism of the AcrB efflux pump, *Biochim. Biophys. Acta 1794*, 782–793.

35. Higgins, C. F. (2007) Multiple molecular mechanisms for multidrug resistance transporters, *Nature 446*, 749–757.

36. Nikaido, H., and Takatsuka, Y. (2009) Mechanisms of RND multidrug efflux pumps, *Biochim. Biophys. Acta 1794*, 769–781.

37. Nishino, K., Nikaido, E., and Yamaguchi, A. (2009) Regulation and physiological function of multidrug efflux pumps in *Escherichia coli* and *Salmonella, Biochim. Biophys. Acta 1794*, 834–843.

38. Zgurskaya, H. I., Yamada, Y., Tikhonova, E. B., Ge, Q., and Krishnamoorthy, G. (2009) Structural and functional diversity of bacterial membrane fusion proteins, *Biochim. Biophys. Acta 1794*, 794–807.

39. Nakajima, H., Kobayashi, K., Kobayashi, M., Asako, H., and Aono, R. (1995) Over-expression of the robA gene increases organic solvent tolerance and multiple antibiotic and heavy metal resistance in Escherichia coli, Appl. Environ. Microbiol. 61, 2302–2307.

40. White, D. G., Goldman, J. D., Demple, B., and Levy, S. B. (1997) Role of the acrAB locus in organic solvent tolerance mediated by expression of marA, soxS, or robA in Escherichia coli, J. Bacteriol. 179, 6122–6126.

41. Lomovskaya, O., Warren, M. S., Lee, A., Galazzo, J., Fronko, R., Lee, M., Blais, J., Cho, D., Chamberland, S., Renau, T., et al. (2001) Identification and characterization of inhibitors of multidrug resistance efflux pumps in Pseudomonas aeruginosa: novel agents for combination therapy, Antimicrob. Agents Chemother. 45, 105–116.

42. Zgurskaya, H. I., and Nikaido, H. (2000) Cross-linked complex between oligomeric periplasmic lipoprotein AcrA and the inner-membrane-associated multidrug efflux pump AcrB from Escherichia coli, J. Bacteriol. 182, 4264–4267.

43. Tikhonova, E. B., and Zgurskaya, H. I. (2004) AcrA, AcrB, and TolC of Escherichia coli form a stable intermembrane multidrug efflux complex, J. Biol. Chem. 279, 32116–32124.

44. Murakami, S., Nakashima, R., Yamashita, E., and Yamaguchi, A. (2002) Crystal structure of bacterial multidrug efflux transporter AcrB, Nature, 419, 2002, 587–593.

45. Sennhauser, G., Bukowska, M. A., Briand, C., and Grütter, M. G. (2009) Crystal structure of the multidrug exporter MexB from Pseudomonas aeruginosa, J. Mol. Biol. 389, 134–145.

46. Akama, H., Matsuura, T., Kashiwagi, S., Yoneyama, H., Narita, S., Tsukihara, T., Nakagawa, A., and Nakae, T. (2004) Crystal structure of the membrane fusion protein, MexA, of the multidrug transporter in Pseudomonas aeruginosa, J. Biol. Chem. 279, 25939–25942.

47. Higgins, M. K., Bokma, E., Koronakis, E., Hughes, C., and Koronakis, V. (2004) Structure of the periplasmic component of a bacterial drug efflux pump, Proc. Natl. Acad. Sci. USA 101, 9994–9999.

48. Mikolosko, J., Bobyk, K., Zgurskaya, H. I., and Ghosh, P. (2006) Conformational flexibility in the multidrug efflux system protein AcrA, Structure 14, 577–587.

49. Koronakis, V., Sharff, A., Koronakis, E., Luisi, B., and Hughes, C. (2000) Crystal structure of the bacterial membrane protein TolC central to multidrug efflux and protein export, Nature 405, 914–919.

50. Akama, H., Kanemaki, M., Yoshimura, M., Tsukihara, T., Kashiwagi, T., Yoneyama, H., Narita, S., Nakagawa, A., and Nakae, T. (2004) Crystal structure of the drug discharge outer membrane protein, OprM, of Pseudomonas aeruginosa: dual modes of membrane anchoring and occluded cavity end, J. Biol. Chem. 279, 52816–52819.

51. Tseng, T. T., Gratwick, K. S., Kollman, J., Park, D., Nies, D. H., Goffeau, A., and Saier, M. H., Jr., (1999) The RND permease superfamily: an ancient, ubiquitous and diverse family that includes human disease and developmental proteins, J. Mol. Microbiol. Biotechnol. 1, 107–125.

52. Zgurskaya, H. I., and Nikaido, H. (1999) AcrA is a highly asymmetric protein capable of spanning the periplasm, J. Mol. Biol. 285, 409–420.

53. Symmons, M. F., Bokma, E., Koronakis, E., Hughes, C., and Koronakis, V. (2009) The assembled structure of a complete tripartite bacterial multidrug efflux pump, *Proc. Natl. Acad. Sci. USA 106*, 7173–7176.

54. Ip, H., Stratton, K., Zgurskaya, H., and Liu, J. (2003) pH-induced conformational changes of AcrA, the membrane fusion protein of *Escherichia coli* multidrug efflux system, *J. Biol. Chem. 278*, 50474–50482.

55. Tamura, N., Murakami, S., Oyama, Y., Ishiguro, M., and Yamaguchi, A. (2005) Direct interaction of multidrug efflux transporter AcrB and outer membrane channel TolC detected via site-directed disulfide cross-linking, *Biochemistry 44*, 11115–11121.

56. Bavro, V. N., Pietras, Z., Furnham, N., Pérez-Cano, L., Fernández-Recio, J., Pei, X. Y., Misra, R., and Luisi, B. (2008) Assembly and channel opening in a bacterial drug efflux machine, *Mol. Cell 30*, 114–121.

57. Elkins, C. A., and Nikaido, H. (2003) Chimeric analysis of AcrA function reveals the importance of its C-terminal domain in its interaction with the AcrB multidrug efflux pump, *J. Bacteriol. 185*, 5349–5356.

58. Touzé, T., Eswaran, J., Bokma, E., Koronakis, E., Hughes, C., and Koronakis, V. (2004) Interactions underlying assembly of the *Escherichia coli* AcrAB–TolC multidrug efflux system, *Mol. Microbiol. 53*, 697–706.

59. Krishnamoorthy, G., Tikhonova, E. B., and Zgurskaya, H. I. (2007) Fitting periplasmic membrane fusion proteins to inner membrane transporters: mutations that enable *Escherichia coli* AcrA to function with *Pseudomonas aeruginosa* MexB, *J. Bacteriol. 190*, 691–698.

60. Lobedanz, S., Bokma, E., Symmons, M. F., Koronakis, E., Hughes, C., and Koronakis, V. (2007) A periplasmic coiled-coil interface underlying TolC recruitment and the assembly of bacterial drug efflux pumps, *Proc. Natl. Acad. Sci. USA 104*, 4612–4617.

61. Eswaran, J., Koronakis, E., Higgins, M. K., Hughes, C., and Koronakis, V. (2004) Three's company: component structures bring a closer view of tripartite drug efflux pumps, *Curr. Opin. Struct. Biol. 14*, 741–747.

62. Reffay, M., Gambin, Y., Benabdelhak, H., Phan, G., Taulier, N., Ducruix, A., Hodges, R. S., and Urbach, W. (2009) Tracking membrane protein association in model membranes, *PLoS One 4*, e5035.

63. Narita, S., Eda, S., Yoshihara, E., and Nakae, T. (2003) Linkage of the efflux-pump expression level with substrate extrusion rate in the MexAB–OprM efflux pump of *Pseudomonas aeruginosa, Biochem. Biophys. Res. Commun. 308*, 922–926.

64. Aires, J. R., and Nikaido, H. (2005) Aminoglycosides are captured from both periplasm and cytoplasm by the AcrD multidrug efflux transporter of *Escherichia coli, J. Bacteriol. 187*, 1923–1929.

65. Mima, T., Joshi, S., Gomez-Escalada, M., and Schweizer, H. P. (2007) Identification and characterization of TriABC–OpmH, a triclosan efflux pump of *Pseudomonas aeruginosa* requiring two membrane fusion proteins, *J. Bacteriol. 189*, 7600–7609.

66. Gristwood, T., Fineran, P. C., Everson, L., and Salmond, G. P. (2008) PigZ, a TetR/AcrR family repressor, modulates secondary metabolism via the expression of a putative four-component resistance–nodulation–cell-division efflux pump, *Mol. Microbiol. 69*, 418–435.

67. Zgurskaya, H. I., and Nikaido, H. (1999) Bypassing the periplasm: reconstitution of the AcrAB multidrug efflux pump of *Escherichia coli*, *Proc. Natl. Acad. Sci. USA 96*, 7190–7195.

68. Su, C. C., and Yu, E. W. (2007) Ligand-transporter interaction in the AcrB multidrug efflux pump determined by fluorescence polarization assay, *FEBS Lett. 581*, 4972–4976.

69. Ocaktan, A., Yoneyama, H., and Nakae, T. (1997) Use of fluorescence probes to monitor function of the subunit proteins of the MexA–MexB–oprM drug extrusion machinery in *Pseudomonas aeruginosa*, *J. Biol. Chem. 272*, 21964–21969.

70. Murakami, S., Tamura, N., Saito, A., Hirata, T., and Yamaguchi, A. (2004) Extra-membrane central pore of multidrug exporter AcrB in *Escherichia coli* plays an important role in drug transport, *J. Biol. Chem. 279*, 3743–3748.

70a. Bohnert, J., Karamian, B., and Nikaido, H. (2010) Optimized Nile Red efflux assay of AcrAB-Tolc multidrug efflux system shows competition between substrates. *Antimicrob. Agents Chemother. 54*, 3770–3775.

71. Nagano, K., and Nikaido, H. (2009) Kinetic behavior of the major multidrug efflux pump AcrB of *Escherichia coli*, *Proc. Natl. Acad. Sci. USA 106*, 5854–5858.

72. Mazzariol, A., Cornaglia, G., and Nikaido, H. (2000) Contributions of the AmpC β-lactamase and the AcrAB multidrug efflux system in intrinsic resistance of *Escherichia coli* K-12 to β-lactams, *Antimicrob. Agents Chemother. 44*, 1387–1390.

73. Nikaido, H., and Normark, S. (1987) Sensitivity of *Escherichia coli* to various β-lactams is determined by the interplay of outer membrane permeability and degradation by peri-plasmic β-lactamases: a quantitative predictive treatment, *Mol. Microbiol. 1*, 29–36.

73a. Lim, S. P., and Nikaido, H. (2010) Kinetic parameters of efflux of penicillins by the multidrug efflux transporter AcrAB-Tolc of *Escherichia coli*. *Antimicrob. Agents Chemother. 54*, 1800–1806.

74. Chang, T. Y., Chang, C. C., Ohgami, N., and Yamauchi, Y. (2006) Cholesterol sensing, trafficking, and esterification, *Annu. Rev. Cell Dev. Biol. 22*, 129–157.

75. Infante, R. E., Abi-Mosleh, L., Radhakrishnan, A., Dale, J. D., Brown, M.S., and Goldstein, J. L. (2008) Purified NPC1 protein: I. Binding of cholesterol and oxysterols to a 1278-amino acid membrane protein, *J. Biol. Chem. 283*, 1052–1063.

76. Infante, R. E., Radhakrishnan, A., Abi-Mosleh, L., Kinch, L. N., Wang, M. L., Grishin, N. V., Goldstein, J. L., and Brown, M. S. (2008) Purified NPC1 protein: II. Localization of sterol binding to a 240-amino acid soluble luminal loop, *J. Biol. Chem. 283*, 1064–1075.

77. Elkins, C. A., and Nikaido, H. (2002) Substrate specificity of the RND-type multidrug efflux pumps AcrB and AcrD of *Escherichia coli* is determined predominantly by two large periplasmic loops, *J. Bacteriol. 184*, 6490–6498.

78. Eda, S., Maseda, H., and Nakae, T. (2003) An elegant means of self-protection in gram-negative bacteria by recognizing and extruding xenobiotics from the periplasmic space, *J. Biol. Chem. 278*, 2085–2088.

79. Tikhonova, E. B., Wang, Q., and Zgruskaya, H. I. (2002) Chimeric analysis of the multicomponent multidrug efflux transporters from gram-negative bacteria, *J. Bacteriol. 184*, 6499–6507.

80. Wehmeier, C., Schuster, S., Fähnrich, E., Kern, W. V., and Bohnert, J. A. (2009) Site-directed mutagenesis reveals amino acid residues in the *Escherichia coli* RND efflux pump AcrB that confer macrolide resistance, *Antimicrob. Agents Chemother. 53*, 329–330.

81. Mao, W., Warren, M. S., Black, D. S., Satou, T., Murata, T., Nishino, T., Gotoh, N., and Lomovskaya, O. (2002) On the mechanism of substrate specificity by resistance nodulation division (RND)-type multidrug resistance pumps: the large periplasmic loops of MexD from *Pseudomonas aeruginosa* are involved in substrate recognition, *Mol. Microbiol. 46*, 889–901.

82. Bohnert, J. A., Schuster, S., Fähnrich, E., Trittler, R., and Kern, W. V. (2007) Altered spectrum of multidrug resistance associated with a single point mutation in the *Escherichia coli* RND-type MDR efflux pump YhiV (MdtF), *J. Antimicrob. Chemother. 59*, 1216–1222.

83. Middlemiss, J. K., and Poole, K. (2004) Differential impact of MexB mutations on substrate selectivity of the MexAB-OprM multidrug efflux pump of *Pseudomonas aeruginosa*, *J. Bacteriol. 186*, 1258–1269.

84. Goldberg, M., Pribyl, T., Juhnke, S., and Nies, D. H. (1999) Energetics and topology of CzcA, a cation/proton antiporter of the resistance–nodulation–cell division protein family, *J. Biol. Chem. 274*, 26065–26070.

85. Guan, L., and Nakae, T. (2001) Identification of essential charged residues in transmembrane segments of the multidrug transporter MexB of *Pseudomonas aeruginosa*, *J. Bacteriol. 183*, 1734–1739.

86. Takatsuka, Y., and Nikaido, H. (2006) Threonine-978 in the transmembrane segment of the multidrug efflux pump AcrB of *Escherichia coli* is crucial for drug transport as a probable component of the proton relay network, *J. Bacteriol. 188*, 7284–7289.

87. Yu, E. W., Aires, J. R., and Nikaido, H. (2003) AcrB multidrug efflux pump of *Escherichia coli*: composite substrate-binding cavity of exceptional flexibility generates its extremely wide substrate specificity, *J. Bacteriol. 185*, 5657–5664.

88. Yu, E. W., Aires, J. R., McDermott, G., and Nikaido, H. (2005) A periplasmic drug-binding site of the AcrB multidrug efflux pump: a crystallographic and site-directed mutagenesis study, *J. Bacteriol. 187*, 6804–6815.

89. Hearn, E. M., Gray, M. R., and Foght, J. M. (2006) Mutations in the central cavity and periplasmic domain affect efflux activity of the resistance–nodulation–division pump EmhB from *Pseudomonas fluorescens* cLP6a, *J. Bacteriol. 188*, 115–123.

90. Franke, S., Grass, G., Rensing, C., and Nies, D. H. (2003) Molecular analysis of the copper-transporting efflux system CusCFBA of *Escherichia coli*, *J. Bacteriol. 185*, 3804–3812.

91. Dastidar, V., Mao, W., Lomovskaya, O., and Zgurskaya, H. I. (2007) Drug-induced conformational changes in multidrug efflux transporter AcrB from *Haemophilus influenzae*, *J. Bacteriol. 189*, 5550–5558.

92. Ruetz, S., and Gros, P. (1994) Phosphatidylcholine translocase: a physiological role for the mdr2 gene, *Cell 77*, 1071–1081.

93. Mao, W., Warren, M. S., Lee, A., Mistry, A., and Lomovskaya, O. (2001) Mex-XY-OprM efflux pump is required for antagonism of aminoglycosides by divalent cations in *Pseudomonas aeruginosa*, *Antimicrob. Agents Chemother.* 45, 2001–2007.

94. Schmidt, T. G., and Skerra, A. (2007) The Strep-tag system for one-step purification and high-affinity detection or capturing of proteins, *Nat. Protocols 2*, 1528–1535.

95. Hanada, K., Yamato, I., and Anraku, Y. (1988) Purification and reconstitution of *Escherichia coli* proline carrier using a site specifically cleavable fusion protein, *J. Biol. Chem. 263*, 7181–7185.

96. Bagai, I., Liu, W., Rensing, C., Blackburn, N. J., and McEvoy, M. M. (2007) Substrate-linked conformational change in the periplasmic component of a Cu(I)/Ag(I) efflux system, *J. Biol. Chem. 282*, 35695–35702.

97. Bagai, I., Liu, W., Rensing, C., Blackburn, N. J., and McEvoy, M. M. (2008) Direct metal transfer between periplasmic proteins identifies a bacterial copper chaperone, *Biochemistry 47*, 11408–11414.

98. Yu, E. W., McDermott, G., Zgurskaya, H. I., Nikaido, H., and Koshland, D. E., Jr., (2003) Structural basis of multiple drug-binding capacity of the AcrB multidrug efflux pump, *Science 300*, 976–980.

99. Törnroth-Horsefield, S., Gourdon, P., Horsefield, R., Brive, L., Yamamoto, N., Mori, H., Snijder, A., and Neutze, R. (2007) Crystal structure of AcrB in complex with a single transmembrane subunit reveals another twist, *Structure 15*, 1663–1673.

100. Das, D., Xu, Q. S., Lee, J. Y., Ankoudinova, I., Huang, C., Lou, Y., Degiovanni, A., Kim, R., and Kim, S.-H. (2007) Crystal structure of the multidrug efflux transporter AcrB at 3.1 Å resolution reveals the N-terminal region with conserved amino acids, *J. Struct. Biol. 158*, 494–502.

101. Drew, D., Klepsch, M. M., Newstead, S., Flaig, R., de Gier, J.-W., Iwata, S., and Beis, K. (2008) The structure of the efflux pump AcrB in complex with bile acid, *Mol. Membr. Biol. 25*, 67–682.

102. Su, C. C., Li, M., Gu, R., Takatsuka, Y., McDermott, G., Nikaido, H., and Yu, E. W. (2006) Conformation of the AcrB multidrug efflux pump in mutants of the putative proton relay pathway, *J. Bacteriol. 188*, 7290–7296.

103. Murakami, S., Nakashima, R., Yamashita, E., Matsumoto, T., and Yamaguchi, A. (2006) Crystal structures of a multidrug transporter reveal a functionally rotating mechanism, *Nature 443*, 173–179.

104. Seeger, M. A., Schiefner, A., Eicher, T., Verrey, F., Diederichs, K., and Pos, K. M. (2006) Structural asymmetry of AcrB trimer suggests a peristaltic pump mechanism, *Science 313*, 1295–1298.

105. Sennhauser, G., Amstutz, P., Briand, C., Storchenegger, O., and Grütter, M. G. (2007) Drug export pathway of multidrug exporter AcrB revealed by DARPin inhibitors, *PLoS Biol. 5*, e7.

106. Seeger, M. A., von Ballmoos, C., Verrey, F., and Pos, K. M. (2009) Crucial role of Asp408 in the proton translocation pathway of multidrug transporter AcrB: evidence from site-directed mutagenesis and carbodiimide labeling, *Biochemistry 48*, 5801–5812.

107. Bohnert, J. A., Schuster, S., Seeger, M. A., Fähnrich, E., Pos, K. M., and Kern, W. V. (2008) Site-directed mutagenesis reveals putative substrate binding residues in the *Escherichia coli* RND efflux pump AcrB, *J. Bacteriol. 190*, 8225–8229.

108. Lomovskaya, O., and Totrov, M. (2005) Vacuuming the periplasm, *J. Bacteriol. 187*, 1879–1883.

108a. Husain, F., and Nikaido, H. (2010) Substrate path in the AcrB multidrug efflux pump of *Escherichia coli. Mol. Microbiol. 78*, 320–330.

108b. Takatsuka, Y., Chen, C., and Nikaido, H. (2010) Mechanism of recognition of compounds of diverse structures by the multidrug efflux pump AcrB of *Escherichia coli. Proc. Nat. Acad. Sci. U.S.A. 107*, 6559–6565.

109. Bohnert, J. A., and Kern, W. V. (2005) Selected arylpiperazines are capable of reversing multidrug resistance in *Escherichia coli* overexpressing RND efflux pumps, *Antimicrob. Agents Chemother. 49*, 849–852.

110. Takatsuka, Y., and Nikaido, H. (2007) Site-directed disulfide cross-linking shows that cleft flexibility in the periplasmic domain is needed for the multidrug efflux pump AcrB of *Escherichia coli, J. Bacteriol. 189*, 8677–8684.

111. Bardwell, J. C., McGovern, K., and Beckwith, J. (1991) Identification of a protein required for disulfide bond formation in vivo, *Cell 67*, 581–589.

112. Kamitani, S., Akiyama, Y., and Ito, K. (1992) Identification and characterization of an *Escherichia coli* gene required for the formation of correctly folded alkaline phosphatase, a periplasmic enzyme, *EMBO J. 11*, 57–62.

113. Kenyon, G. L., and Bruice, T. W. (1977) Novel sulfhydryl reagents, *Methods Enzymol. 47*, 407–430.

114. Seeger, M. A., von Ballmoos, C., Eicher, T., Brandstatter, L., Verrey, F., Diederichs, K., and Pos, K. M. (2008) Engineered disulfide bonds support the functional rotation mechanism of multidrug efflux pump AcrB, *Nat. Struct. Mol. Biol. 15*, 199–205.

115. Takatsuka, Y., and Nikaido, H. (2009) Covalently linked trimer of AcrB multidrug efflux pump provides support for the functional rotating mechanism, *J. Bacteriol. 191*, 1729–1737.

115a. Kim, H. S., Nagore, D., and Nikaido, H. (2010) Multidrug efflux pump MdtBC of *Escherichia coli* is active only as a B_2C heterotrimer, *J. Bacteriol. 192*, 1377–1386.

116. Baranova, N., and Nikaido, H. (2002) The BaeSR two-component regulatory system activates transcription of the *yegMNOB* (*mdtABCD*) transporter gene cluster in *Escherichia coli* and increases its resistance to novobiocin and deoxycholate, *J. Bacteriol. 184*, 4168–4176.

117. Nagakubo, S., Nishino, K., Hirata, T., and Yamaguchi, A. (2006) The putative response regulator BaeR stimulates multidrug resistance of *Escherichia coli* via a novel multidrug exporter system, MdtABC, *J. Bacteriol. 184*, 4161–4167.

118. Kumar, A., and Worobec, E. A. (2005) Cloning, sequencing, and characterization of the SdeAB multidrug efflux pump of *Serratia marcescens, Antimicrob. Agents Chemother. 49*, 1495–1501.

119. Crossman, L. C., Gould, V. C., Dow, J. M., Vernikos, G. S., Okazaki, A., Sebaihia. M., Saunders, D., Arrowsmith, C., Carver, T., Peters, N., et al. (2008) The complete

genome, comparative and functional analysis of *Stenotrophomonas maltophilia* reveals an organism heavily shielded by drug resistance determinants, *Genome Biol. 9*, R74.

120. Lakaye, B., Dubus, A., Lepage, S., Groslambert, S., and Frère, J.-M. (1999) When drug inactivation renders the target irrelevant to antibiotic resistance: a case story with β-lactams, *Mol. Microbiol. 31*, 89–101.

121. Tal, N., and Schuldiner, S. (2009) A coordinated network of transporters with overlapping specificities provides a robust survival strategy, *Proc. Natl. Acad. Sci. USA 106*, 9051–9056.

122. Adam, Y., Tayer, N., Rottem, D., Schreiber, G., and Schuldiner, S. (2007) The fast release of sticky protons: kinetics of substrate binding and proton release in a multidrug transporter, *Proc. Natl. Acad. Sci. USA 104*, 17989–17994.

123. Rotem, D., Steiner-Mordoch, S., and Schuldiner, S. (2006) Identification of tyrosine residues critical for the function of an ion-coupled multidrug transporters, *J. Biol. Chem. 281*, 18715–18722.

124. Schuldiner, S. (2009) EmrE, a model for studying evolution and mechanism of ion-coupled transporters, *Biochim. Biophys. Acta 1794*, 748–762.

125. Lovmar, M., Nilsson, K., Lukk, E., Vimberg, V., Tenson, T., and Ehrenberg, M. (2009) Erythromycin resistance by L4/L22 mutations and resistance masking by drug efflux pump deficiency, *EMBO J. 28*, 736–744.

126. Moore, S. D., and Sauer, R. T. (2008) Revisiting the mechanism of macrolide-antibiotic resistance mediated by ribosomal protein L22, *Proc. Natl. Acad. Sci. USA 105*, 18261–18266.

127. Cagliero, C., Mouline, C., Cloeckaert, A., and Payot, S. (2006) Synergy between efflux pump CmeABC and modifications in the ribosomal proteins L4 and L22 in conferring macrolide resistance in *Campylobacter jejuni* and *Campylobacter coli*, *Antimicrob. Agents Chemother. 50*, 3893–3896.

128. Elf, J., Nilsson, K., Tenson, T., and Ehrenberg, M. (2006) Bistable bacterial growth rate in response to antibiotics with low membrane permeability, *Phys. Rev. Lett. 97*, 258104.

129. Fange, D., Nilsson, K., Tenson, T., and Ehrenberg, M. (2009) Drug efflux pump deficiency and drug target resistance masking in growing bacteria, *Proc. Natl. Acad. Sci. USA 106*, 8215–8220.

130. Jeannot, K., Elsen, S., Köhler, T., Attree, I., van Delden, C., and Plésiat, P. (2008) Resistance and virulence of *Pseudomonas aeruginosa* clinical strains overproducing the MexCD–OprJ efflux pump, *Antimicrob. Agents Chemother. 52*, 2455–2462.

131. Zhao, Q., Li, X. Z., Mistry, A., Srikumar, R., Zhang, L., Lomovskaya, O., and Poole, K. (1998) Influence of the TonB energy-coupling protein on efflux-mediated multidrug resistance in *Pseudomonas aeruginosa*, *Antimicrob. Agents Chemother. 42*, 2225–2231.

132. Zhao, Q., and Poole, K. (2002) Differential effects of mutations in *tonB1* on intrinsic multidrug resistance and iron acquisition in *Pseudomonas aeruginosa*, *J. Bacteriol. 184*, 2045–2049.

133. Pettersen, E. F., Goddard, T. D., Huang, C. C., Couch, G. S., Greenblatt, D. M., Meng, E. C., and Ferrin, T. E. (2004) UCSF Chimera: a visualization system for exploratory research and analysis, *J. Comput. Chem. 25*, 1605–1612.

EFFLUX PUMPS OF GRAM-NEGATIVE BACTERIA: GENETIC RESPONSES TO STRESS AND THE MODULATION OF THEIR ACTIVITY BY pH, INHIBITORS, AND PHENOTHIAZINES

By LEONARD AMARAL, *Unit of Mycobacteriology, Instituto de Higiene e Medicina Tropical, Universidade Nova de Lisboa, Lisbon, Portugal,* SEAMUS FANNING, *Centre for Food Safety, UCD Veterinary Sciences Centre, University College Dublin, Dublin 4, Ireland,* and JEAN-MARIE PAGÈS, *UMR-MD1 Facultés de Médecine et de Pharmacie, Université de la Méditerranée, Marseille, France*

CONTENTS

Advances in Enzymology and Related Areas of Molecular Biology, Volume 77
Edited by Eric J. Toone   Copyright © 2011 John Wiley & Sons, Inc.

I.   Introduction to an RND Efflux Pump and Its Mode of Action

Multidrug resistance (MDR) in clinical Gram-negative isolates may arise following the accumulation of mutations (1), a reduction in membrane permeability due to an alteration in lipopolysaccharide (LPS) structure (2), modification of porins (down-regulation or functional alteration) (3), the activation of a master mutator gene (4, 5), and/or from overexpression of efflux pumps that extrude antimicrobial compounds of two or more unrelated classes before they reach their intended targets (6, 7). During the past decade, evidence has accumulated suggesting that the main mechanism underpinning the MDR phenotype of clinical Gram-negative isolates is due to the over-expression of efflux pumps, primarily of the resistance–nodulation–division (RND) genetic family of transporters (8, 9).

RND efflux pumps are tripartite units consisting of a tranporter that recognizes chemically unrelated molecules present on occasion in the periplasm (10), cytoplasm (11), or the outer leaflet of the plasma membrane (12). The transporter component of the efflux pump transports these molecules to TolC, which provides a conduit for the agent to the surface of the cell (13). Flanking the transporter unit are fusion proteins that anchor the main transporter to the plasma membrane (12, 13). These fusion proteins are currently believed to assist the movement of the chemical agents through the transporter by peristaltic action (12).

The source of energy required to power RND efflux pumps is obtained from the proton-motive force (PMF), which is derived from pH gradient arising primarily from catabolism in the cell (14–16). The PMF, $-\Delta p$ $(PMF) = -\Delta\mu_{H^+}/F = Z\Delta pH - \Delta\Psi$, is usually assumed to be an electro-chemical gradient that exists between the cytoplasm and the outer surface of the cell envelope (17, 18). This electrochemical gradient is generated from the activity of the terminal electron transport chain and translocates protons to the surface of the cell, where they are distributed in an orderly fashion (19). Gram-negative bacteria such as *Salmonella* have a pitted outer cell membrane believed to provide wells wherein protons are concentrated (20–22). High concentrations of surface-bound protons may bind to components of the LPS layer, thereby reducing pH in the immediate vicinity of the surface of the cell by two or more pH units lower than that of the bulk medium (19). This distribution of surface protons afforded a revision of the chemiosmotic theory which in its absence would have predicted that PMF would soon collapse, due to the escape of protons to the "vast ocean" of the bulk medium (19, 23).

RND efflux pumps are known to recognize and bind a large variety of unrelated chemical substrates, such as antimicrobial compounds, dyes, bile salts, and others (24, 25). Although the manner by which this chemical recognition is achieved remains to be elucidated, the purified RND trans-porter AcrB encoded from *Escherichia coli* will bind a variety of ligands in vitro (26). The K_d value for substrate binding is pH dependent, such that at low pH, K_d is high, and at high pH it is low. This observation is important if an inhibitor of a RND efflux pump is to be effective at physiological pH (i.e., K_d must be sufficiently low that its binding to the transporter is prolonged or even irreversible).

The precise mechanisms involved in the activation of RND efflux pumps and the subsequent transport steps involved in expelling a chemical molecule via TolC remain to be determined (12). Periplasmic proton concentration is related to the metabolic activity of the cell. Although bacteria can maintain a PMF in media of extreme pH (27, 28), when presented with an RND efflux pump substrate such as ethidium bromide, the pH modulates the activity of the efflux pump (29, 30). For example, at pH 5, *E. coli* readily extrudes the fluorochrome substrate ethidium bromide without the need for metabolic energy, whereas at pH 8, extrusion is dependent on metabolic energy (Figure 1A). Moreover, agents considered to be inhibitors of the efflux pump have little effect at pH 5, with major inhibitory effects being observed at pH 8. An example of the modulating activity of pH on efflux is demonstrated in Figure 1B. Similar results have been reported for other Gram-negative bacteria, such as *Salmonella* (31) and *Enterobacter aerogenes* (32). These results suggest that at low pH, dissociation of protons from the surface of the cell is discouraged by the high concentration of protons in the bulk medium (20). Consequently, this concentration of surface-bound protons assures the antiport required to power the RND efflux pump, thereby obviating, at least temporarily, any need for metabolic energy and its direct

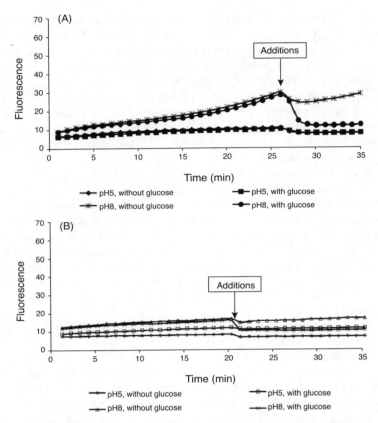

Figure 1.   Effect of pH and glucose on the accumulation and efflux of ethidium bromide (EB) by antibiotically susceptible (A) and multidrug-resistant (B) *E. coli. E. coli* grown in Mueller–Hinton broth until it reaches an optical density of 0.6 at 600 nm, washed, OD adjusted to 0.6, aliquot of 50 μL transferred to 100-μL microtubes containing glucose-free saline of pH 5 and 8 and EB to yield a final concentration of 1.0 μg/mL, the tubes transferred to the Corbett 3000 thermocycler, programmed for a temperature of 37°C, number of 1-min cycles, and total cycles. After 20 min the instrument is paused and 50 μL of pH-matched saline containing or lacking glucose to yield 0.4% is added and the instrument restarted. The amount of fluorescence in the first phase of the assay in the absence of metabolic energy affords the accumulation of EB at pH 5 and 8. This accumulation by antibiotic-susceptible wild-type *E. coli* is not affected by glucose-containing saline of pH 5. In contrast, accumulation of EB and its efflux by multidrug-resistant *E. coli*, whose RND AcrAB efflux pump is 14-fold overexpressed over that of the wild type, is not affected by glucose or by pH.

contribution to establishment and/or maintenance of the PMF. At pH 8, the dissociation of protons from the surface of the cell proceeds, and the number of protons is thereby limited. The source of protons on this occasion is provided through the activity of an ATP synthase, which at pH 8 proceeds in the direction of hydrolysis (33–38). It is assumed that the source of ATP for hydrolysis derives from metabolism inasmuch as at pH 8, extrusion of EB is dependent on metabolic energy (29–32).

II. Genetic Control and Response of Efflux Pumps to Environmental Stress

Today, a major challenge in the study of infectious disease is the identification of specific elements: the genetic and biochemical factors involved in the triggering of regulation cascades governing the expression of efflux pump genes, or, alternatively, in the selection of strains that overproduce efflux systems, promoting the extrusion of an antibiotic before it reaches its intended target (39, 40). For a bacterial cell, antimicrobial agents represent an environmental stress, and this external stimulus induces a complex response, involving a genetic regulatory cascade, leading to altered gene expression and assembly of specific proteins, and biochemical activation of metabolic and energy pathways to support the physiological adaptation to the antibiotic stress imposed within an ecological niche during bacterial colonization or infection (human, animal) (41–45).

To reduce the intracellular accumulation of antimicrobial compounds, a key bacterial response is to stringently control membrane transporters involved in the extrusion of an antibiotic and in the diffusion of drugs through the cell envelope, the *in-and-out flux* (3). This control of membrane transporters (e.g., efflux pumps and porins) may be carried out at different cellular levels:

- Activation by general or global transcriptional activators that coordinate the expression of several downstream genes
- Negative regulation by local repressors that directly block the expression of specific genes (e.g., efflux pump components or porins)
- Triggering off–on of one or more steps involved in interlinking regulatory cascades that may lead to a general alteration of membrane physiology
- Emergence and selection of mutations in key genes

Polyspecific bacterial efflux transporters confer a general resistance phenotype that can promote the acquisition of additional mechanisms contributing to antibiotic resistance, such as mutations in antibiotic gene targets or the secretion of enzymes that degrade antibiotics (e.g., β-lactamases), and can also reinforce the effects of these acquired mechanisms (3, 6). It has been reported that AcrAB–TolC overexpression is an important prerequisite for the selection of fluoroquinolone (FQ)-resistant mutants that exhibit mutated targets (DNA gyrase) in various Gram-negative bacteria involved in severe human diseases (3, 6, 9). Furthermore, a recent study reported an increase in the number of hospital *E. aerogenes* isolates that express an efflux mechanism involved in antibiotic resistance during the 1995 to 2003 period (8). These data show clearly the dynamic and evolutionary aspects of resistance mechanisms associated with membrane permeability and the important role of effectors in promoting the selection or emergence of such resistant clinical strains.

In 1998, the first complex bacterial response associated with expression of the mechanical barrier (3) that includes the down-regulation of porins and the activation of efflux pumps in clinical strains under antibiotic stress was reported in *E. aerogenes* (46). In 2000, during the treatment of various patients with imipenem, several isolates exhibiting various susceptibility levels to different antibiotic classes were cultured (47). The mechanisms associated indicated that imipenem-resistant isolates exhibited an alteration in the bacterial membrane, reducing the synthesis of a porin and up-regulating an efflux pump which responded to the efflux pump inhibitor (EPI) phenylalanine-arginine-β-naphthylamide (PAβN) (48). A similar outcome was noted when imipenem treatment was mimicked in vitro using increasing concentration of the carbapenem to select in vitro resistant variants (49, 50). Interestingly, these in vitro–selected strains had a sequential adaptative phenotype involving two steps: the first corresponding to the overexpression of efflux pumps, and the second corresponding to a decrease in porin expression. When the expression of the *marA* was considered, it showed two steps that corresponded to the two observed phenotypes (50). Modulating the number of membrane transporters in these isolates and their antibiotic resistant variants is reversible in the absence of antibiotic selective pressure. Therefore these strains rapidly recover their normal susceptibility patterns including the production of normal porin, suggesting that the regulation of this phenomenon may be associated with fitness costs (49–51).

III.   Efflux Pumps: Selection, Advantages, and Fitness Cost

The selection of resistant mutants using increasing concentrations of various antimicrobial compounds results in heterogeneous bacterial responses associated with a range of phenotypes. For example, from a similar *Enterobacter* ATCC strain, imipenem can select for a phenotype exhibiting porin deficiency along with the overexpression of an efflux pump; in contrast, chloramphenicol favors the selection of a mutant that overexpresses an efflux pump but retains normal porin synthesis (50, 52). This panel of resistant strains may arise due to differences in the complex regulation of these selection events, which includes sensors, effectors, local regulators (AcrR, MarR), global regulators (MarA, RamA, SoxS), and possible redundant regulatory systems (3). Reversion of the mechanism observed associated with drug resistance (e.g., the restoration of porin synthesis) (42) was noted in clinical isolates cultured following the cessation of antibiotic treatment or its alteration, suggesting a prominent role for regulation over a mutational event (48, 52, 53). However, as reported recently for integrons and mobile genes, the balance of responsive regulation indicates that genes which are involved in resistance profile can be silenced at no biological cost, whereas other adaptive traits continue to be expressed (54). In the case of mutations altering permeability, a clinical strain has been described that produces a mutated porin, the induced cost of which is related to a significant diminution in nutrient uptake via this altered channel (42). Resistant enterobacterial isolates with altered porin function are restricted of nutrients, consistent with a reduction in the rate of bacterial growth (42, 49). In contrast, efflux pump activity has been reported to be an important element during bacterial colonization (6, 9, 54).

IV.   Sensor-Dosage Hypothesis

Several hypotheses have been put forward to explain the factors that control alterations in membrane permeability. Among them, the role of trigger metabolites was recently described (55), which requires involvement of the regulation cascade, leading to the overexpression of efflux pumps (Figure 2). Interestingly, when the intracellular concentration of trigger metabolites reaches a certain threshold concentration, as reported previously for the regulation of quorum sensing (56), it is proposed that they activate the transcriptional regulators MarA, SoxS, and Rob (55, 57). In the case of Rob, chemical inducers that may mimic trigger metabolites

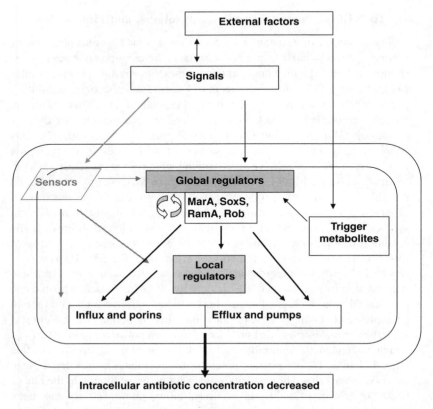

Figure 2.   Scheme of interactive factors resulting in extrusion of antibiotic.

such as dipyridyl and decanoate (55) are able to promote (1) desequestra-
tion of aggregated Rob from restricted loci and (2) dispersal of active forms
within the cytoplasm (58). In these cases, the on–off switch of efflux
activity is controlled by the concentration of effectors, which in turn
regulates the expression of activator cascades (59). The effect of salicylate,
which induces expression of the *mar* regulon, is an example of the
chemically induced MDR cascade driving the up-regulation of efflux
components while simultaneously decreasing porin expression (59–61).
Furthermore, a link has been reported between quorum sensing and the
expression of an efflux pump (56, 62). Moreover, a concentration-dependent

activity for some antibiotics on various transcription levels has recently been reported (54, 56).

By analogy, β-lactam antibiotic treatment (including carbapenem) may favor the accumulation of similar trigger metabolites in the cytoplasm after abortive peptidoglycan synthesis. These metabolites may subsequently be involved in the activation of MarA, RamA, and possibly other global regulators. Interestingly, this concentration-dependent process may explain the bimodal effects observed during imipenem treatment (47, 48).

The role of an outer membrane sensor has recently been suggested following a study involving *asm* mutants. Prieto et al. (63) reported that the general mechanism by which *asmA* mutations enhance bile resistance in different genetic backgrounds involved transcriptional activation of the *marRAB* regulon. Moreover, OmpX, a small outer membrane protein controlling the level of porin synthesis and modulating β-lactam susceptibility (63–66), is overexpressed following certain drug treatments. Overexpression of OmpX is induced by norfloxacin, dipyridyl, nalidixic acid, or novobiocin, and this production has a knock-on effect on membrane permeability. Expression of OmpX is also modulated by MarA and H-NS (66). Bacteria have developed various signaling systems which enable them to sense environmental stresses. These adaptive responses are usually mediated by two-component regulatory systems involving a histidine kinase sensor located in the inner membrane and a cognate response regulator in the cytoplasm or small RNAs controlling the post-transcriptional expression of various OMPs (67–69). Similar systems control the expression of genes for nutrient acquisition, virulence, antibiotic resistance, and several other pathways in different bacteria (65). However, despite extensive investigation, at this time the dialogue between bacterial pathogens and antibiotics remains unclear, especially with respect to the role played by membrane proteins in the signaling process (3, 70–72).

V.   Mar, Ram, Sox, and Rob: Redundancy and Overlapping Control

The regulation of efflux pump expression is multilayered (3, 6, 61, 73–77), with several regulation pathways overlapping and sharing similar effectors (local or global). For example, the *mar* and *sox* regulons exhibit interaction and mutual control (55, 57, 76, 78). In addition, specific genes are described as key coordinators, such as *ramA*, present in *Salmonella*, *Klebsiella pneumoniae*, and *Enterobacter* spp. but absent in *E. coli* (79–82).

Since efflux systems comprise membrane-bound proteins downstream post-translational regulation, including maturation, folding, and membrane assembly, is also required to promote functional assembly of the export machinery (Figure 3) (83–87). In addition, when a primary pump acts together with a secondary pump, such as EmrE and MdfA and AcrAB and TolC (88), some kinetic and dynamic elements govern the final combined activity of this sophisticated complex. Currently, it is quite difficult to present in a reasonable-sized document a global integrated view that would include all regulators (local and global), all chemical inducers, all stress events, and all possible sensors involved in the cascade. For this reason, we decided to focus sections on specific regulators and genes that can possibly act and interact in the complex process of organizing the bacterial defense against antibiotic attack.

A.   mar AND sox OPERONS

The *marRAB* operon exhibits a genetic organization preserved among the *Enterobacteriaceae* (3, 77, 89). This regulon is repressed constitutively by MarR, and the expression (or derepression) of *marRAB* is the consequence of either (1) mutations in the MarR binding (operator) sites; (2) modification of MarR at the protein level, preventing its binding to *mar*-responsive operators; or (3) the direct action of inductors (3).   MarA is a key regulator in *E. coli* involved in adaptation to the environment and protection against external drug-related stress, by inducing the direct or indirect action of more than 60 genes (3). A *marA* box has been detected in several genes, and MarA is reported to mediate the transcription of several other regulators, such as *rob* and *soxS* (3, 77). The expression of *marA* triggers a cascade that leads to an MDR phenotype by simultaneous reduction of influx and an increase in efflux of antibiotics. Furthermore, MarA induces a decrease in OmpF and OmpC porin families, the major outer membrane porins through which hydrophilic antibiotics gain entry in Enterobacteriaceae. The down-regulation of these porins is mediated through the transcription of a small nontranslated antisense RNA, *micF*, located upstream of the OmpC gene (3). The antisense *micF* hybridizes with *ompF* mRNA and thus prevents translation. In addition, MarA activates the expression of OmpX involved in the control of porin expression (3). The second effect of the MarA cascade is to up-regulate the expression of *acrAB*. In addition, other efflux pumps, including AcrD and AcrF, also respond to the MarA (3). In this way, the bacterial cell can overexpress efflux pumps and rapidly reduce the intracellular concentration of an antibiotic (6, 9, 39).

Two major regulatory systems in *E. coli*, OxyR and SoxRS, are involved in the response to oxidative stress and contribute to increased bacterial survival against oxidative attack by the host immune system (90). SoxR serves as a sensor for superoxide and nitric oxide ions and activates *soxS* transcription following oxidation (91). Up-regulation of *soxS* subsequently activates the expression of target genes that repair damaged DNA, maintain redox balance, and defend against toxic radicals. An extensive overlap exists between oxidative stress and antibiotic resistance (74, 92–94). Interestingly, a recent report demonstrated that the killing mechanism involving bactericidal antibiotics occurs following oxidative damage, a common mechanism involved in bacterial death (95, 96). SoxS can also activate MarA expression, and together they activate many of the same genes (3, 77). Mutations in the *soxR* activator have been identified in clinical isolates of *E. coli* and *S. enterica* from patients undergoing quinolone treatment, and these have been shown to confer a MDR phenotype (3, 77). Both of these regulators, SoxS and MarA, recognize common elements contained within the corresponding promoters, termed *soxbox* and *marbox*, respectively (94).

More than 80 genes targeted by SoxS and MarA have been reported, with diverse functions involving drug resistance, membrane transport, iron homeostasis, LPS modifications, reducing oxidants, and others. The overall effect of the *mar* and *sox* operons is to reduce the intracellular concentrations of antibiotics that enter the bacterial cell through porins and are substrates of active efflux via AcrAB–TolC (Figure 3).

B. RamA

RamA is a small regulatory protein belonging to the AraC/XylS transcriptional activator family, first described in *Enterobacter cloacae* (92). This regulator has also been identified in *Klebsiella pneumoniae*, *Salmonella enterica* serovars Typhimurium and Enteritidis and in *E. aerogenes* (79–81). Surprisingly, no *ramA* locus was identified in the *E. coli* genome. Genomic comparisons indicate that the gene organization surrounding *ramA* in the former bacteria and the corresponding region in *E. coli* are quite similar, with an open reading frame known as *Yi81-2* in *E. coli* in place of *ramA* (85). This regulator is involved in the modification of outer mem-brane permeability and in the active extrusion of intracellular antibiotics (81). RamA has also been demonstrated to bind to the *mar* operator, suggesting an overlap between these two regulatory circuits (81, 97). The

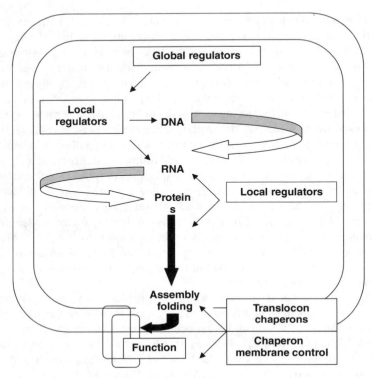

Figure 3.   Scheme for interaction of regulators, resulting in the expression of genes, syntheses of the components of efflux pump, and pump assembly.

putative *marbox* located within the *ram* promoter is conserved (77, 81), suggesting that MarA can also regulate the transcription of *ramA*. Constitutive expression of RamA results in an MDR phenotype even in the absence of a functioning *mar* locus. Conversely, RamA is also a transcriptional activator of the Mar regulon (80, 81).

RamA plays a role in the oxidative stress response in partnership with *soxRS*, but seems to be functionally different than *marA* and *soxS* in the development of MDR in *Salmonella* spp. (74, 98–100). The AcrAB induction pathway in *Salmonella* differs from that in *E. coli*. Bile induces *acrAB* in both *Salmonella* and *E. coli*. In *E. coli*, the transcriptional factor Rob plays a major role in inducing *acrAB* expression in response to bile (101). These data suggest that RamA is a prominent regulator of *acrAB* in

Salmonella and may modulate the contribution of other *acrAB* effectors, such as MarA, Rob, and SoxS (85). An additional regulatory mechanism has recently been proposed via direct interaction between RamA and bile salts, converting a preinactivated form to an activated RamA state, which in turn triggers the pump expression (85). This is similar to a mechanism described previously for the binding of bile to the Rob protein in *E. coli* (102). These various mechanisms suggest that RamA could play the role of a pivotal *sensor* in some bacterial species (3), similar to the role known for two component regulatory system (TCS) (78) and for which one element can detect an environmental stress and forward the information to an effector. This may be analogous to Mar A playing the role of the effector controlling the expression of several genes involved in drug transport, such as porins or pumps. Nevertheless, a question remains regarding the absence of the *ram* operon in *E. coli*. This feature suggests a special (adaptive) role for RamA in *Salmonella*, *Klebsiella*, and *Enterobacter*, as a booster of the regulation cascade triggering an enhanced response to counter an environmental stress. A repressor gene, *ramR*, has recently been identified in *Salmonella* and *E. aerogenes*, and belongs to the TetR family of transcriptional repressors (A. Molitor et al., unpublished results). Several mutations have been described in this gene, some of which occur in a helix–turn–helix region of the protein and may affect a repressor's binding abilities.

It is interesting to note that a number of regulators and regulatory cascades are involved in the control of drug transporters, and that these have been conserved in enterobacterial pathogens (with the sole expection of RamA and *E. coli* at this time) (7). This observation suggests that the balance between fitness cost and evolutionary advantage shows a benefit for the cell. Fine regulatory control is essential, as these transporters are involved in other crucial physiological processes. This is certainly evident in the case of efflux pump components and for porins, as discussed recently in several excellent reviews (9, 42, 72, 73). Further, this may support the role of redundancy and structural conservation between regulators that control solute flux through the membrane in environments wherein these bacteria must adapt rapidly to adverse environmental stress. Rather than evolving mutations in these genes, which can later alter the corresponding protein function as reported previously for porins (4), a change in the balance of regulation represents a less significant cost to physiological fitness (40, 41, 51, 52). Moreover, once the stress imposed has been removed, overproduction of efflux components can be down-regulated further, alleviating any fitness costs. It is not unreasonable to view these positive adaptive features favoring redundancy when one considers the competition with other bacteria sharing the same ecological niches.

VI. Inducing In Vitro MDR: Genes Involved and Simulation in the Infected Patient

When exposed to increasing concentrations of an antibiotic (74, 83, 103, 104) or to a noxious nonantimicrobial agent (105), bacteria become increasingly resistant to the antibiotic or agent. Accompanying this induced resistance is resistance to two or more nonstructurally related antibiotics (defined as *multidrug resistance*; if resistance is limited to the same antibiotic family, this is *cross-resistance*). In the case of tetracycline resistance in an MDR *E. coli*, expression of genes that regulate and code for transporters indicates that the local regulators *marA*, *marB*, stress genes *rob* and *soxS*, the transporter gene that codes for AcrB, the main efflux pump of this organism, as well as transporters that code for eight other efflux pumps are all increased manyfold over that of the unexposed control (104). Moreover, the regulator of permeability *ompX* is increasingly overexpressed following prolonged exposure to tetracycline, whereas the expression of outer membrane proteins (OMPs), normally part of the tri-barrel porins C and F, is unchanged, although the amount of these porins retrieved from the induced organism are almost practically absent (83). Because the genes that code for four main proteases are all overexpressed, the reduced amount of porins is apparently due to degradation of the OMPs prior to their assembly as porin structures (83). The response to serial exposure to the given antibiotic is one type of adaptation and not one that involves mutations inasmuch as when the antibiotic-induced MDR strain is transferred to a drug-free medium, the susceptibility of the strain to the antibiotics that constituted the MDR phenotype assumes that of the initial, nonexposed control. The MDR phenotype can also be reverted to that of the control by exposure to the inhibitor of Gram-negative efflux pumps PAβN, a feature which again suggests that the induced MDR phenotype and overexpressed efflux pumps are not the result of a mutation(s). Similar serial exposure of *Salmonella* to increasing concentrations of a quinolone induces an MDR phenotype that is accompanied by a sixfold-increased expression of *acrB* (98). However, the MDR phenotypes of quinolone-induced strains are not due to the same mechanisms, even though all of the strains show increased expression of *acrB*; one of the MDR strains accumulated successive mutations in *ramR*, and the phenotype could not be reversed to that of the nonexposed parent by transfer to a drug-free medium. The latter results suggest that for *Salmonella*, there are alternative pathways by which antibiotic-induced MDR phenotypes can arise (98).

Laboratory demonstrations of induced MDR phenotypes are believed by the authors to represent what takes place in the patient when treated with one or more antibiotics for a prolonged period of time (see Section I). The demonstration of alternative pathways for induced MDR phenotypes—one by increased expression of genes that regulate and code for proteins that control the permeability of the cell to antibiotics and the other accompanied by mutations in a regulator gene such as *soxS* or *ramR*—are representative of MDR clinical strains isolated from patients who have been treated with antibiotics for a prolonged period. Although studies that evaluate MDR Gram-negative clinical isolates show that the use of agents that inhibit efflux pumps reduce high-level resistance to these antibiotics, reduced resistance to a level comparable to that of a wild-type reference strain is rarely achieved for most strains. The reason is that these strains have accumulated mutations or that resistance is plasmid mediated (106), or both. Nevertheless, as mentioned in Section I, retrospective evaluation of many Gram-negative clinical strains over a period of time has correlated a MDR phenotype with an overexpressed efflux pump as well as genes that regulate the expression of such pumps (8, 92).

Clinical Gram-negative isolates often show resistance to given antibiotics at levels that exceed 100-fold that of a wild-type strain. Because the dose of an antibiotic administered to the patient is held constant over time, sometimes in excess of 30 days, the extremely high-level resistance noted for these strains must have arisen via a mechanism that was not reproduced by serial exposure to increasing concentrations of antibiotics. Consequently, a study was undertaken whereby the *E. coli* that had been induced to an MDR phenotype by serial exposure to increasing concentrations of tetracycline (83, 103) was serially transferred a total of 63 times to medium containing a constant concentration of tetracycline (10 mg/L). By the end of the tenth serial transfer, the levels of expressed *marA*, *soxS*, *ompX*, and *acrB* had returned to those of the wild type. However, the resistance of the strain to tetracycline had increased to 64 mg/L, and by the end of 60 serial passages in 10 mg/L tetracycline, resistance to tetracycline exceeded 256 mg/L. Whereas prior to passage 10 resistance to tetracycline could be reversed to that of the wild type either by exposure to the EPI PAβN or by transfer to a drug-free medium, reduction or reversal of resistance was not possible by these routes (5).

Evaluation by phenotypic array as the strain was serially transferred to a medium containing a constant concentration of tetracycline, revealed altered growth when challenged with β-lactams, fluoroquinolones, and macrolides. These studies suggest that when the organisms is faced with an unchanging noxious environment, rather than maintaining an energy-consuming cost such as that associated with an overexpressed genotype that decreases permeability to antibiotics as well as to nutrients, it switches on a mutator gene systems as described by Chopra et al. (4). In other words, as is true for all matter in the universe, the organism obeys the second law of thermodynamics: conservation of energy.

VII.   Inhibitors of Efflux Pumps

A.   CARBONYL CYANIDE m-CHLOROPHENYLHYDRAZONE

The demonstration of an active RND-type efflux pump of a Gram-negative bacterium is usually conducted with a PMF uncoupler such as carbonyl cyanide m-chlorophenylhydrazone (CCCP) at pH 7 in the absence of a metabolic energy source (e.g., glucose). Given the demonstration that at pH 8 metabolic energy optimizes efflux, the activity of varying concentrations of this agent at pH 5 and 8 on the efflux of ethidium bromide (EB) after the fluorochrome has accumulated in the absence of glucose has been studied and the results obtained described in Figure 4 for *E. coli* AG100$_{TET}$, which overexpresses its AcrAB efflux pump (8, 103). At pH 5 and 8, CCCP immediately prevents efflux and increases the rate and extent of accumulation of EB in a concentration-dependent manner. However, comparison of the slopes of accumulation at pH of 5 and pH 8 indicates that at pH 8 the effect of CCCP is considerably greater. Because these assays were conducted at the same time and under the same conditions with pH as the only variable, the lesser effects of CCCP at pH 5 may be due to the large contribution of protons at this pH, wherein their concentration exceeds the proton binding capacity of CCCP (30). At pH 8 and a much lower concentration of available protons, the concentration of CCCP essentially blocks all the proton antiport and thereby completely inhibits the efflux of EB. The same effects were produced by CCCP and its modulation of EB efflux by the intrinsic efflux pump system of the *E. coli* AG100 wild-type strain (data not shown). The observation that CCCP promotes the retention of EB at pH 5 demonstrates that the reduced accumulation of EB at this pH is due to efflux, as opposed to a decrease in the permeability to EB (30).

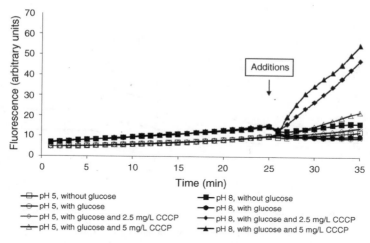

Figure 4.   Effect of CCCP on the accumulation of EB at pH 5 and 8.

B.   PHENYLALANINE-ARGININE-β-NAPHTHYLAMIDE

Pheylalanine-arginine-β-naphtylamide (PAβN) has been used to demonstrate the presence of an active efflux pump that renders some Gram-
negative bacteria susceptible to antibiotics. This demonstration involves the
reduction in the minimum inhibitory concentration (MIC) of a given
antibiotic over a 16-h culture time. However, our previous studies suggested that PAβN is not an inhibitor of an efflux pump but, rather, a
competitor of other efflux pump substrates for extrusion (107), a suggestion
also made by others (108). The preferential extrusion of PAβN postulated is
believed to result in the increased concentration of the antibiotic, which
eventually reaches a level that inhibits replication of the bacterium.
Moreover, if PAβN is an inhibitor of an RND efflux pump, it should
inhibit the efflux of EB at pH 5 inasmuch as the efflux of EB at pH 5 is
independent of metabolic energy and dependent upon the PMF. The effect
of PAβN on the accumulation and efflux of EB by the AG100$_{TET}$ strain at
pH 5 is described by Figure 5. As evident from this figure, the addition of
PAβN has no effect on the efflux of EB at either pH. However, because
PAβN competes with EB, as the concentration of PAβN is increased, more
and more EB would be expected to accumulate. This anticipated relationship was exploited for the derivation of a K_m for PAβN relative to EB at
pH 5 inasmuch as at this pH metabolic energy is not needed and PAβN has

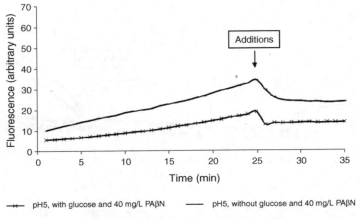

Figure 5.   Effect of PAβN on the efflux of EB.

no effect on the efflux of EB. Moreover, the dissociation constant of EB from the AcrB transporter is lowest at pH 5 (109), a condition that is necessary for continuous efflux of EB. As described by the composite Figure 6, as the concentration of PAβN is increased from 1 mg/L to 40 mg/L, the amount of EB accumulation is increased proportionately. Employing the Michaelis–Menten formula, the K_m for PAβN representing competitive inhibition between PAβN and EB was calculated to be 4.21 µg/µg of EB (30).

C. VERAPAMIL

Verapamil inhibits ABC transporters of *Staphylococcus aureus* (110) and mycobacteria (111) in a manner that is different from that of PAβN. However, there is little information regarding the effects of verapamil on efflux activity of gram-negative bacteria such as *E. coli*. Considering the possibility that the study of agents for inhibitory activity against efflux pumps is always conducted at neutral or nearly neutral pH, and because at pH 8 the efflux of EB by the *E. coli* strains employed in this study is dependent on metabolic energy, suggesting the involvement of an ABC-type transporter, we evaluated the effects of varying concentrations of verapamil on the efflux of EB (30). Figure 7 describes these data and consistent with data published previously, at pH 8 the efflux of EB is dependent on the provision of metabolic energy. The addition of verapamil in the absence of metabolic energy promotes a concentration-dependent

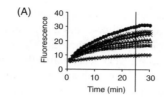

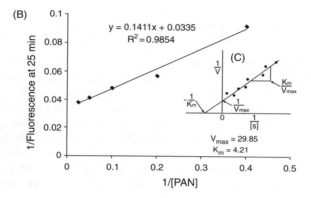

Figure 6. K_m value for PAβN relative to EB.

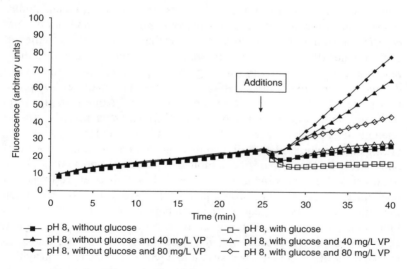

Figure 7. Effects of verapamil on the accumulation of EB at pH 8.

inhibition of efflux that causes proportional increases in accumulation of EB. The inhibitory effect on efflux exerted by verapamil is decreased significantly when metabolic energy is present. These data suggest that at pH 8, the efflux of EB, is at the very least, partially mediated by an ABC-type transporter. Similar results were obtained with the *E. coli* AG100$_{TET}$ strain, which overexpresses the AcrAB efflux pump (30).

D. PHENOTHIAZINES AND THEIR EFFECT ON THE EXPRESSION OF GENES THAT REGULATE THE AcrAB EFFLUX PUMP IN *Salmonella*

Approximately 95% of all medicinal compounds are derivatives of phenothiazines. Therefore, it is not surprising that this group of compounds are active against bacteria, wherein they mediate effects such as direct inhibition of replication (112), reduced antimicrobial resistance via increased drug efflux (103), enhanced killing of intracellular bacteria (113, 114), inhibition of bacterial motility (115), elimination of plasmids from *E. coli* (116), curing of mice infected with *Mycobacterium tuberculosis* and with Gram-negative bacteria (117, 118), and inhibition of efflux pumps of Gram-positive and gram-negative bacteria (107, 119). Although most of these activities have been known for decades, some even dating back to the time of Paul Ehrlich (120), the mechanism(s) by which these wide-ranging effects occur remain relatively obscure. However, recent studies suggest that exposure of *Staphylococcus aureus* (121) and *Salmonella enterica* serovar Enteritidis to a phenothiazine inhibits the activities of certain genes (99). With respect to *Salmonella*, exposure to chlorproma-zine increases the activity of a global regulatory gene *ramA* and decreases *acrB* expression, which codes for the main efflux pump transporter AcrB in a time-dependent manner (99). Interestingly, from the latter study one would expect that *Salmonella* is sensitive to the phenothiazine, as is the case for *S. aureus* (119) and *M. tuberculosis* (122). Furthermore, other earlier studies showed that at first, exposure of *Salmonella* to chlorpromazine inhibited replication, and that after approximately 8 h, the organism became resistant to over 100 mg/L of phenothiazine (123, 124). Consequently, the study that demonstrated inhibition of the *acrB* gene (99) is supported by a study that shows inhibition of replication during early exposure to the same agent, chlorpromazine (123). However, the question as to why the organism is at first sensitive and later is resistant to a phenothiazine is an intriguing one and has been answered in part (111).

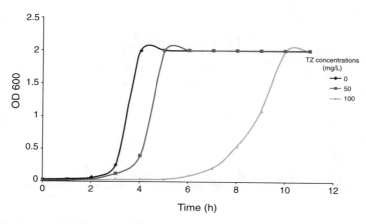

Figure 8.   Effect of thioridazine on the growth of a *Salmonella* strain. An MIC inoculum of clinical strain *Salmonella* Enteritidis 104, genetically characterized (144), was cultured in Mueller–Hinton broth containing increasing concentrations of TZ and the growth of the organism monitored at 600 nm, pH 7.4. Note the transient inhibition of growth during the first 5 h by 100 mg/L of TZ.

The effect of varying concentrations of thioridizine (TZ) against *S.* Enteritidis 104 is described in Figure 8, and the qualitative responses noted for a number of *Salmonella* so far studied are similar. As evident from this figure, whereas a concentration of 50 mg/L of TZ produces an approximate lag phase of 2 h, the higher concentration of 100 mg/L TZ inhibits growth completely for 5 h, after which time the organism begins to replicate, reaching the stationary phase of growth by the end of 10 h. The MIC of TZ for this strain is 200 mg/L and is similar to that of other *S.* Enteritidis strains employed in this study as well as those employed by others (99). It is important to note that unless a similar growth curve is performed for other Gram-negative bacteria that are deemed resistant to TZ, the transient effects of TZ and indeed any other phenothiazine, such as chlorpromazine (123), could be missed. The transient effect of TZ, at first one of inhibition, followed by a period when the inhibitory effects are dissipated, suggests that perhaps during the inhibitory phase, the activity of an efflux pump system is affected, as claimed by others (99). The expression of genes that code for regulators of the AcrAB efflux pump PmrA/B, PhoP/Q and both TCSs when *Salmonella* are cultured in the presence of 100 mg/L TZ at intervals during which the growth of the organism is at first inhibited and after it has fully grown is described by Figure 9.

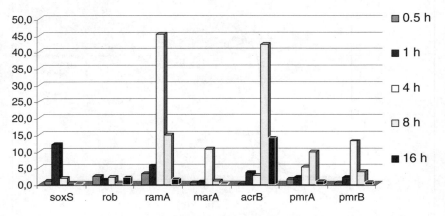

Figure 9.   Expression of genes that regulate and code for the AcrB transporter of *Salmonella*. Quantification of relative expression of regulator genes, stress genes, and the *acrB*, *pmrA*, and *pmrB* genes of *S*. Enteritidis 104 after 0, 0.5, 1, 4, 8, and 16 h of exposure to TZ 100 mg/L. Changes in gene expression are relative to the untreated cells and normalized against the housekeeping gene 16S rRNA.

After 30 min, there is a small overexpression of the stress response gene *rob* (2.3-fold) and the global regulator *ramA* (3.2-fold). However, at the end of 60 min, the expression of *rob* has returned to the levels of the untreated control, whereas *ramA* was increased 5.7-fold. Two other genes showed an increase in expression at this time point also, the stress response gene *soxS* (12.1-fold) and *acrB* (3.5-fold). After 4 h, *ramA* showed the highest level of overexpression (45.3-fold), followed by regulator *marA* (10.6-fold), the transporter *acrB* (2.6-fold), *pmrA* (5.3-fold), and *pmrB* (13.0-fold). After 8 h of exposure to TZ, *acrB* was the most active gene (42.2-fold), followed by the *ramA*, whose activity decreased from 45.3- to 14.9-fold, and the lipid A synthesis regulator *pmrA* increased (9.8-fold), whereas *pmrB* decreased (3.7). These results suggest that the stress response gene *soxS* plays a more important role than *rob* in the short-term response to exposure to TZ. Moreover, the activation of *ramA* precedes that of *marA*, with both genes returning to normal expression levels after the bacteria become resistant to TZ (16 h). Relative to the *pmrAB* two-component system, the sensor kinase *pmrB* is activated at the time the bacteria begins to become less inhibited by TZ (4 and 8 h) and the response regulator *pmrA* follows *pmrB* in activation, and remains more active than *pmrB*, by the end of 8 h. Activation of *pmrAB* is possibly mediated by the sensor kinase PmrB as a response to an activating trigger such as low pH (pH 5.5) (70). However, our results demonstrate that the *pmrAB* system can be activated at pH 7.4, and thus the response is not dependent on low pH but seems instead to be related to a regulatory cascade triggered by the presence of a stress agent (in this case, TZ). After 16 h of exposure to TZ, with the exception of *acrB*, which remained highly active (13.9-fold), all the genes tested returned to untreated levels. Interestingly, the activity of *phoQ* and *phoP* was not altered by exposure to TZ.

As noted above, an early response to the presence of TZ is the activation of *ramA*, suggesting that this global regulator of *Salmonella* may be essential if the organism is to respond to the inhibitory properties of the phenothiazine. To test the need for *ramA*, *S.* Enteritidis strains 104 and 5408 had their *ramA* genes deleted via the introduction of a kanamycin-resistant encoding gene cassette. The *ramA*-deleted mutants were then cultured in MH containing increasing concentrations of TZ, and the response of these mutants is depicted in Figure 10. As noted in Figure 10A, the *ramA*-deleted strains responded in a manner similar to that of all of the other *ramA*-intact *Salmonella* strains. Therefore, although the TZ-promoted increased activity of *ramA* is impressive, it appears that the deletion of this global regulator has little effect on the resistance developed by *Salmonella* to phenothiazine.

The effect of TZ on the growth of the same *Salmonella* strains whose *soxS* gene has been deleted is similar to that demonstrated for the *ramA*-deleted mutant exposed to TZ: that after 6 to 8 h of exposure to a subinhibitory concentration of TZ, the organism becomes resistant to the agent (Figure 11). All *soxS*-deleted strains showed similar patterns of growth in the presence of TZ. Therefore, we may conclude that neither *ramA* or *soxS* is required for the adaptation of *Salmonella* to TZ.

The question of whether the genes involved in the TZ transient inhibition of *Salmonella* strains are also involved when the strains are deleted for *ramA* or *soxS* genes was investigated by assessing the expression of *ramA*, *marA*, *soxS*, *rob*, *acrB*, *pmrA*, and *pmrB* by real-time RT-PCR. As shown by Figure 12, with the exception of *marA*, *soxS*, and *acrB*, the expression of *rob*, *pmrA*, and *pmrB* in the *ramA*-deleted mutant is essentially at levels similar to that of the unexposed control at the time the strain begins to grow (after 7 h of culture). By the end of the culture period, when the strain is fully grown, only *acrB* is overexpressed relative to the unexposed control. Interestingly, *ramA* and *marA* of the *soxS*-deleted strain are greatly expressed at the point when the strain begins to grow (after 7 h of culture

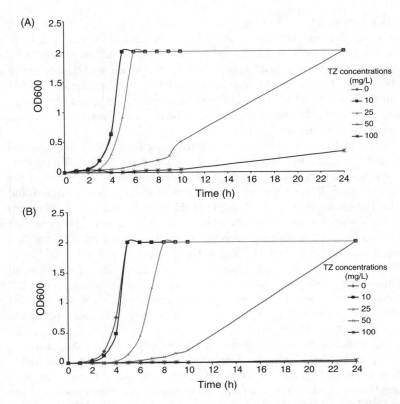

Figure 10.   Growth curve of *Salmonella* serovar Enteritidis 104 (A) and 5408 (B) ram A deleted in a medium containg increasing concentrations of TZ.

in the presence of TZ). By the end of the culture, the strain has reached full growth, and at this time, only *acrB* is significantly overexpressed. These results suggest that in the absence of *ramA*, the strain can still respond to TZ and overexpress *acrB*, and that it is *soxS* that induces the overexpression of *acrB*. Moreover, *marA* is not required, inasmuch as its expression remains at the level of the unexposed control. The absence of *soxS* does not affect the activation of *ramA*, *marA*, and *acrB*. These results show that a phenothiazine by an unknown mechanism affects the expression of genes involved in the process by which the *Salmonella* strain assumes eventual resistance to the agent.

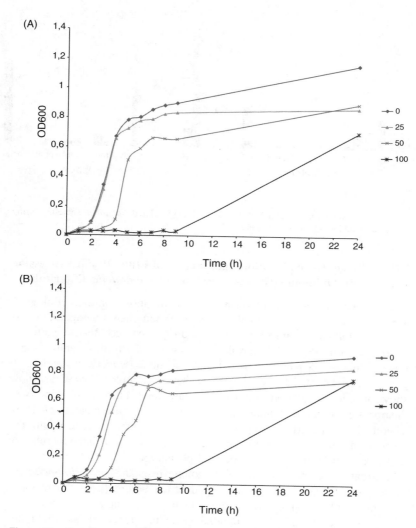

Figure 11. Growth curve of *Salmonella* serovar Enteritidis 104 (A) and 5408 (B) *sox*S deleted in a medium containg increasing concentrations of TZ.

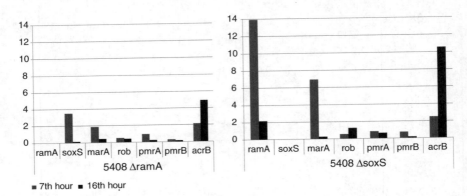

Figure 12. Expression of genes of *ramA*- and *soxS*-deleted *Salmonella* strains during exposure to subinhibitory concentrations of TZ.

VIII. Target Gene Mutations and Increased Efflux Pump Expression in Quinolone-Resistant Isolates of *Salmonella* Enteritidis

Salmonella enterica serovar Enteritidis is the most common etiological agent of foodborne salmonellosis globally. When chemotherapeutic intervention is indicated, ciprofloxacin is the drug of choice. Resistance to the latter is uncommon (MIC ciprofloxacin $\geq 4\,\mu g/mL$) in this genus. Nonetheless, increasing nalidixic acid resistance in *Salmonella* isolates from both human and food animal sources has been noted. These are now showing reduced susceptibility to fluoroquinolones (125, 126). Reports of therapeutic failure in humans are becoming more frequent (127–129). Therefore, *S.* Enteritidis represents a good model to investigate the mechanisms of resistance to these drugs and the associated regulatory network contributing to the associated phenotype.

Currently, the mechanisms of quinolone resistance in *Salmonella* include target gene mutations, increased efflux activity, and the recently recognized plasmid-mediated protection of target topoisomerases (130, 131). Involvement of other potential resistance mechanisms, including changes in the cell envelope, loss of membrane porins, or alterations in lipopolysaccharide (LPS) structure, remains to be elucidated (132–134). Nalidixic acid resistance and reduced susceptibility to ciprofloxacin is associated with *gyrA* mutations at codons S83 or D87 (131, 134, 135). Mutations at both of these residues have been reported in resistant isolates, often in association with mutations in other topoisomerase genes (135–137).

Similarly, overexpression of AcrAB–TolC contributes directly to fluoroquino lone and multidrug resistance (MDR) in *Salmonella* (136–139).

IX. Regulation of Efflux Pump Activity in *Salmonella*

Our current knowledge on the regulation of expression of AcrAB in *Salmonella* comes from work carried out in *E. coli* (9) (see Section I). At a local level, AcrR negatively modulates the expression of *acrAB*. Mutations in *acrR* lead to the overexpression of *acrB* (140). In contrast, at a global level, MarA, SoxS, and Rob, all of which belong to the AraC/XylS family of transcriptional regulators (141), can modulate the expression of this operon along with *tolC*. These global regulators positively regulate *micF*, an antisense RNA that inhibits synthesis of the OmpF (Figure 13), an outer membrane porin (9).

The *mar* locus consists of two transcription units, *marC* and *marRAB*, which are divergently transcribed from *marO* (142, 143). MarA plays a dual regulatory role, regulating its own transcription along with that of the *mar* regulon. MarR acts by repressing *marRAB* transcription. The functions

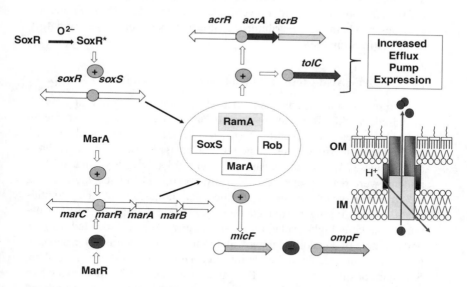

Figure 13.   Schematic of part of the network of global regulatory control. (See insert for color representation.)

of MarB and MarC are unknown. Importantly, *marRAB* transcription can also be activated by the MarA homologs, SoxS and Rob (141–143).

SoxS is the effector of the *soxRS* global superoxide response regulon. SoxR is a constitutively expressed homodimeric transcriptional regulator with redox-active iron–sulfur clusters (2Fe–2S). Oxidation of these clusters activates SoxR (Figure 13), which in turn activates transcription of *soxS* (144, 145). Increased expression of these global regulators may be associated with mutations in the regulatory genes of the associated operons (91, 146, 147) or the selective binding of inducers or *trigger molecules* (see Figure 2) (148, 149). As in *E. coli*, increased expression of *marA* and *soxS* has been associated with fluoroquinolone and MDR in *Salmonella* (135, 150). However, the contribution of these global regulators to overexpression of *acrAB* in fluoroquinolone and MDR *Salmonella* phenotypes remains to be fully determined. Furthermore, little is known about *rob* and its role in antibiotic resistance in *Salmonella*.

RamA, a homolog of MarA, was identified in *Salmonella* and a number of other organisms. This regulator was implicated in MDR in *Salmonella* and these other bacteria (102, 151–153). Overexpression of *ramA* has been associated with increased expression of *acrB* in *Salmonella* and other Enterobacteriaceae (102, 152, 153). Figure 13 provides a schematic describing the interplay between these regulatory elements in *Salmonella*.

X.   Nalidixic Acid–Resistant Isolates and Selection for FQ Resistance

Two recent studies, confirmed that a D87 mutation in *gyrA*, along with increased AcrAB–TolC activity, contributed to quinolone resistance in *Salmonella* serovars Typhimurium and Enteritidis isolates displaying high-level nalidixic acid resistance and decreased susceptibility to fluoroquinolones (74, 135). In the case of *S.* Enteritidis, ciprofloxacin and MDR mutants were readily selected in vitro from two field isolates, highlighting the ease with which resistance to fluoroquinolones and other antibiotics could emerge during prolonged fluoroquinolone therapy in infected patients. The latter high-level ciprofloxacin resistance was associated with the development of an additional *gyrA* mutation in *S.* Enteritidis104-cip isolate and hitherto undocumented mutations in *gyrB* and *parE* in a second isolate, 5408-cip, of the same serovar. Both overexpressed *acrB* and displayed an MDR phenotype along with a reduction in OmpF expression and altered LPS (74).

Double mutations in *gyrA* have been widely reported in ciprofloxacin-resistant *Salmonella* isolates, whereas mutations in *gyrB* and *parE* in *Salmonella* are rarely detected (131). Following complementation of the latter mutants with wild-type genes, O'Regan and colleagues demonstrated a role for the double *gyrA* mutations (D87Y and S83F) in a quinolone-resistant *S.* Enteritidis isolate and for a *gyrA* (D87Y) and a novel *gyrB* (E466D) mutation in another ciprofloxacin-resistant isolate of the same serovar. Interestingly, in a second Enteritidis isolate, when the *parE* mutation was complemented with a wild-type gene, no altered phenotype could be detected. Nonetheless, the possibility that *parE* mutations may contribute indirectly to high-level fluoroquinolone resistance in isolates already containing target gene mutations cannot be ruled out (154).

XI.   Choosing a Regulatory Pathway

Increased expression of *acrB* can arise in several ways one, of which includes the differential expression of global regulators (74, 98). The choice of a regulatory pathway appears to be a unique feature that may be strain dependent and may reflect previous genetic imprinting. *Salmonella* Enteritidis isolates 104-cip (cultured from a clinical source) and 5408-cip (cultured from an animal) demonstrated this, wherein increased expression of both *soxS* and *marA* was detected in the former, while 5408-cip overexpressed *ramA*. Direct evidence of their roles in the activation of *acrB* was confirmed by deletion of each regulatory element in the corresponding genetic background. The contribution of each regulator was different, since compared to the *marA* mutant, the *soxS* attenuated mutant showed a greater down-regulation of *acrB* and displayed lower MIC values to ciprofloxacin and other antibiotics tested. These data suggest that *soxS* plays a greater role than *marA* in MDR in 104-cip (74). Expression of *marA* decreased following deletion of *soxS* and vice versa, highlighting the cross-regulation that exists between these transcriptional factors (150, 155).

Recently, *ramA* has been reported to contribute to fluoroquinolone and MDR in *S.* Typhimurium through activation of *acrB* (98, 99, 102, 139) and also in *S.* Enteritidis (74). In *S.* Enteritidis, *ramA* activates the MDR cascade independent of *marA*. In *E. coli*, overexpression of Rob confers a MDR phenotype via the up-regulation of *acrB* (156). Reduced expression of *rob* in *S.* Enteritidis 104-cip is probably due to down-regulation mediated by *soxS* and *marA* (155, 157). The nature and extent of the crosstalk between *ramA* and these other global regulators in *Salmonella* and other

RamA-containing organisms is currently unknown. Based on data obtained for *S.* Enteritidis, RamA appears to down-regulate both *soxS* and *rob* (74).

In *E. coli*–increased expression of *soxS* and *marA* has been linked to mutations in the *soxR* gene, rendering SoxR active and independent of oxidative stress or mutations in *marR* that alleviate its repression of *marA* (91, 147, 158) (see Section I). Two reports documenting the contribution of a mutation in *soxR* to increased *soxS* expression and MDR in *Salmonella* have been published (74, 146). The nature of this effect is unclear at present but may involve a feedforward activation mechanism via SoxS. Sequence analysis also revealed a hitherto unreported mutation within *ramR* (G25A). Mutations in *ramR* have been reported to play a role in up-regulation of *ramA* and AcrAB and, consequently, the efflux-mediated MDR phenotype in *S.* Typhimurium (102, 139).

A reduction in the expression of OmpF (see Figure 13) is consistent with the role of *marRAB* and *soxRS* (74, 159, 160). Few studies investigated OmpF expression in fluoroquinolone-resistant *Salmonella*; therefore, its contribution to resistance remains to be elucidated (132, 133, 161). One study to date has documented alterations in the LPS profile in fluoroquinolone-resistant *Salmonella* and an increase in the proportion of long O-chain LPSs was observed which could lead to reduced levels of drug accessing the porins (132).

XII. Fitness Costs Associated with the Overexpression of AcrAB in *Salmonella*

S. Enteritidis is the predominant serotype associated with egg-borne salmonellosis in humans. It is invasive in poultry and therefore has the potential to contaminate eggs via transovarian transmission (162). The mechanisms involved are not fully understood. The necessary steps could include proliferation of the bacterium in the intestinal contents, followed by growth in or migration through the mucus layer, and finally, attachment to and invasion of the epithelium.

The MDR bacterium may lack the fitness to complete this process effectively, due to the burden imposed by their drug-resistant phenotype. Fitness costs associated with high-level ciprofloxacin resistance in *Salmonella* isolates have been reported only in the serovars Typhimurium (132) and Enteritidis (163). These effects include reduced growth rates, altered morphology, decreased motility and invasiveness in Caco-2 cells, and increased sensitivity to environmental stresses such as pH and osmotic stimuli, all of which could affect intestinal colonization negatively, thereby limiting the spread of resistant clones. In vitro derived high-level ciprofloxacin-resistant *S.* Typhimurium mutants having reduced growth rates and altered bacterial cell morphology were unable to colonize the gut of chickens (164).

XIII. Clues from Transcriptomics Assays

DNA microarray expression analysis can be used to assess whether the fitness burden of high-level ciprofloxacin resistance was associated with global effects on gene expression. In two *S.* Enteriditis mutants studied, consistent down-regulation of invasion genes within the *Salmonella* pathogenicity island-1 (SPI-1) encoding a type III secretion system (TTSS) and flagellar biosynthesis genes, correlated with the observed phenotypes of decreased motility and Caco-2 cell invasion. In *S.* Typhimurium attenuated for components of the AcrAB–TolC efflux pump, there was a similar reduction in the expression of the same genes (165) based on transcriptomic analysis alone. Bacterial adherence, followed by invasion, is a highly coordinated sequence of events, involving fimbriae, flagella, and the TTSS. Adherence is an essential step in the successful colonization by *Salmonella* and is mediated by bacterial fimbrial adhesins (166). Motility afforded by flagella and assisted by chemotactic responses plays an important role during the early stages of invasion, facilitating the interaction between the bacterium and the host cell (167). Invasion is then mediated by a TTSS encoded on SPI-1 (168, 169). In *S.* Enteriditis there appeared to be a greater reduction in motility and invasiveness of some isolates, a feature that correlated with the decreased expression of a large number of genes involved in flagellar biosynthesis and chemotaxis. LPS can also play a role in the adherence and/or invasion of epithelial cells (170, 171). The possibility that the altered LPS profile may also have influenced its adherence and invasiveness cannot be ruled out.

XIV. DNA Supercoiling

Mechanisms widely associated with high-level fluoroquinolone resistance (as described earlier) have been associated with fitness costs (172–175). Precisely how these resistance traits affect bacterial fitness remains to be elucidated. Decreased growth of high-level fluoroquinolone-resistant isolates possessing multiple topoisomerase mutations has been associated with decreased DNA supercoiling (173, 174). DNA in bacterial cells is maintained in a negatively supercoiled state (176) by the opposing actions of DNA gyrase, which introduces negative supercoils, and DNA topoisomerase I (topA), which prevents supercoiling from reaching unacceptably high levels (177). Decreased supercoiling indicates the presence of a less efficient DNA gyrase (178), which could in turn slow replication and thereby affect growth. Numerous studies have provided evidence that the expression of many genes, including those associated with bacterial virulence such as invasion and flagellar genes, is influenced by changes in DNA supercoiling (179). In S. Enteritidis, topoisomerase mutations may alter the supercoiling of DNA, which then may affect bacterial virulence negatively. To test whether the mutations characterized in high-level-resistant S. Enteritidis strains affects DNA supercoiling, purified pUC18 from two isogenic strains was electrophoresed in agarose gels in the presence of 2.5% chloroquine (Figure 14). Analysis of the topoisomers harvested from the bacterial cells at the mid-logarithmic growth phase shows clear differences in the patterns of topoisomers recovered. These early data suggest that toposiomerase mutations have altered DNA supercoiling.

XV. Efflux of Metabolites Other Than Antimicrobial Compounds

Reduced growth could result from the inopportune efflux of nutrients and other metabolic intermediates by overexpressed MDR efflux pumps (180). In addition to antimicrobial and other toxic compounds, efflux pumps can expel quorum-sensing (QS) signals (181, 182). Overexpression of MDR efflux pumps in P. aeruginosa correlates with decreased production of QS-regulated virulence determinants (183). In E. coli, genes encoding the EHEC TTSS, the expression and assembly of flagella, motility, chemotaxis and attaching, and effacing are activated by QS (184, 185). In the latter case, QS is mediated by the poorly characterized autoinducer-3 (AI-3). In S. Typhimurium, a role for quorum sensing in the activation of SPI-I gene expression (186) and in motility and the colonization of pigs has been reported (187). In S. Enteritidis 104-cip, sdiA, a homolog of luxR was down-regulated, as determined by analysis of the transcriptome (S. Fanning, unpublished data). This feature may contribute to reduced virulence.

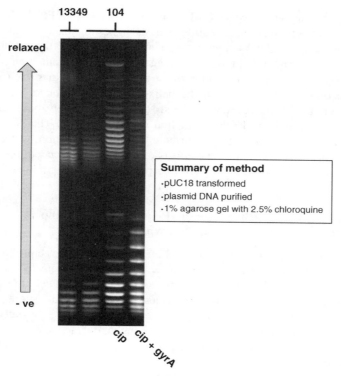

Figure 14. DNA supercoiling in *Salmonella* Enteritidis grown to mid-logarithmic phase. Isogenic 104, 104-cip, and 104-cip (*gyrA*) containing pUC18. A 1% agarose gel containing 2.5 mg/L chloroquine, showing the topoisomers separated after electrophoresis. Note the differences in the topoisomers in 104-cip (which contains mutations in *gyrA* along with an overexpressed AcrB) compared to the other lanes.

XVI. Restoration of Fitness

A key factor in determining the rate of emergence and spread of antibiotic resistance is the biological fitness costs it imposes (188). Fitness costs can often be reduced by compensatory mutations, resulting in the stabilization of resistant bacteria in the population (173). When this was investigated in two *S.* Enteritidis ciprofloxacin-resistant mutants (S. Fanning, unpublished data), no restoration of growth rates or decrease in the level of resistance was observed in one strain, while a second strain reverted to an intermediate-level ciprofloxacin-resistant phenotype after 20 passages, along with reversal of all fitness costs except motility. In the latter case, the expression of *acrB* was decreased, suggesting that overexpression of this pump makes a significant contribution to the fitness burden. Decreased expression of *acrB* was also linked to a reduction in the expression of *soxS*. In the latter strain a new mutation in *soxR* (E16G) was acquired, consistent with the helix–turn–helix region of the SoxR protein in *E. coli*, as reported previously (189). Interestingly, both reverted mutants acquired an identical mutation in *parC* (D79N). The possibility that this *parC* mutation, may contribute to fitness restoration by compensating for the potential negative effects of resistance-associated topoisomerase mutations on global supercoiling remains to be determined.

XVII. Celluler Energy Requirements to Support an Overexpressed AcrB

When the transcriptome of *S.* Enteritidis high-level ciprofloxacin-resistant mutants was determined, several signals were detected in genes associated with energy production and conservation in the cell. In particular, many of the genes encoding enzymes of the citric acid cycle were up-regulated, in an effort to maintain the energy requirements associated with the the the proton motive force (PMF) supporting an overactive AcrB. One of these genes, *acnA* (encoding aconotate hydratase), is activated directly by SoxS (data not shown).

XVIII. Conclusions

Control of the regulatory circuits governing the expression of membrane transport systems in bacterial cells is both complex and demonstrated a degree of redundancy. Efforts to unravel this are currently under way and are leading to the recognition of potentially new drug target sites. More recently, the role of a new global regulator RamA in *Salmonella* has been recognized. Together with SoxS, these appear to be important first responders to the effects of trigger molecules, which leads to pleiotrophic effects within a cell. The role of phenothiazines in the modulation of efflux activity is now recognized. Early evidence suggests a role for RamA and SoxS but that a new, hitherto unrecognized regulatory molecule is also possible. *Salmonella* appears to demonstrate a resistance to this drug following long-term culture exposure. In the serovar Enteritidis, a detailed understanding of the regulatory framework supporting the control of efflux activity in different genetic backgrounds is beginning to emerge. Important roles for both RamA and SoxS, motility function, and QS have been described and confirmed using data from transcriptomic and cell culture experiments. The role of TTSS linked to overexpression of AcrB has been described, and there appears to be an energy cost, which can, at least in part, be attributed to the fitness costs associated with some of these phenotypes. The DNA supercoiling function appears to be an important indicator of fitness, as strains with altered topoisomerase genes were less well adapted for growth and virulence.

An important aspect related to the location and organization of transporters involved in bacterial drug is susceptibility. For porins and efflux pump components, the membrane location of these transporters requires specific steps, beginning with export export from cytoplasm to final assembly in bacterial outer membrane (porin) or in the tripartite complex inserted in a Gram-negative cell wall (efflux pump). Any change in membrane composition (e.g., the protein ratio in the outer membrane, the lipopolysaccharide structure) may alter the assembly of these components and down-regulate their level in the bacterial membrane. In addition, overproduction of efflux channels may also have some indirect effect on the internal concentration of certain metabolites involved in specific physiological events. The versatility and flexibility of biological functions associated with the efflux transporters and their subsequent polyselectivity may also concur with the complexity of the regulation of membrane transporters. The sophisticated steps that are involved in the addressing of membrane proteins reinforce the control level and participate in the complex regulation and organization of influx and efflux transporters. At this moment, various data are available regarding the protein components of transporters, but the involvement of lipopolysaccharide in the diffusion of solutes through the membrane has not yet been deciphered completely.

Acknowledgements

We would like to thank Ana Martins for her excellent technical assistance in the construction of the manuscript and its figures. We wish to acknowledge the support provided to the authors by COST Action BM0701 (ATENS), which made the interaction among the authors possible.

References

1. Perdigão, J., Macedo, R., João, I., Fernandes, E., Brum, L., and Portugal, I. (2008) Multidrug-resistant tuberculosis in Lisbon, Portugal: a molecular epidemiological perspective, *Microb. Drug Resist. 14*, 133–143.

2. Wiese, A., Brandenburg, K., Ulmer, A. J., Seydel, U.,and Müller-Loennies, S. (1999) The dual role of lipopolysaccharide as effector and target molecule, *Biol. Chem. 380*, 767–784.

3. Davin-Régli, A., Bolla, J. M., James, C. E., Lavigne, J. P., Chevalier, J., Garnotel, E., Molitor, A.,and Pagès, J. M. (2008) Membrane permeability and regulation of drug "influx and efflux" in enterobacterial pathogens, *Curr. Drug Targets 9*, 750–759.

4. Chopra, I., O'Neill, A. J.,and Miller, K. (2003) The role of mutators in the emergence of antibiotic-resistant bacteria, *Drug Resist. Update 6*, 137–145.

5. Martins A., Iversen C., Rodrigues L., Spengler G., Ramos J., Kern WV., Couto I., Viveiros M., Fanning S., Pages JM., Amaral L.(2010). An AcrAB-mediated multidrug-resistant phenotype is maintained following restoration of wild-type activities by efflux pump genes and their regulators, *Int. J. Antimicrob. Agents 34*, 602–604.

6. Li, X. Z., and Nikaido, H. (2004) Efflux-mediated drug resistance in bacteria, *Drugs 64*, 159–204.

7. Piddock, L. J. (2006) Multidrug-resistance efflux pumps—not just for resistance, *Nat. Rev. Microbiol. 4*, 629–636.

8. Chevalier, J., Mulfinger, C., Garnotel, E., Nicolas, P., Davin-Régli, A., and Pagès, J. M. (2008) Identification and evolution of drug efflux pump in clinical *Enterobacter aerogenes* strains isolated in 1995 and 2003, *PLoS One 3*(9); e3203.

9. Piddock, L. J. (2006) Clinically relevant chromosomally encoded multidrug resistance efflux pumps in bacteria, *Clin. Microbiol. Rev. 19*, 382–402.

10. Nagano, K., and Nikaido, H. (2009) Kinetic behavior of the major multidrug efflux pump AcrB of *Escherichia coli*, *Proc. Natl. Acad. Sci. USA 106*, 5854–5858.

11. Aires, J. R., and Nikaido, H. (2005) Aminoglycosides are captured from both periplasm and cytoplasm by the AcrD multidrug efflux transporter of *Escherichia coli, J. Bacteriol. 187*, 1923–1929.

12. Seeger, M. A., Diederichs, K., Eicher, T., Brandstatter, L., Schiefner, A., Verrey, F., and Pos, K. M. (2008) The AcrB efflux pump: conformational cycling and peristalsis lead to multidrug resistance, *Curr. Drug Targets 9*, 729–749.

13. Pietras, Z., Bavro, V. N., Furnham, N., Pellegrini-Calace, M., Milner-White, E. J., and Luisi, B.-F. (2008) Structure and mechanism of drug efflux machinery in gram negative bacteria, *Curr. Drug Targets 9*, 719–728.

14. Jormakka, M., Byrne, B., and Iwata, S. (2003) Proton motive force generation by a redox loop mechanism, *FEBS Lett. 545*, 25–30.

15. Michell, P. (1961) Coupling of phosphorylation to electron and hydrogen transfer by a chemi-osmotic type of mechanism, *Nature 191*, 144–148.

16. Williams, R. J. P. (1961) Possible functions of chains of catalysts, *J. Theor. Biol. 1*, 1–13.

17. Simon, J., van Spanning, R. J., and Richardson, D. J. (2008) The organisation of proton motive and non-proton motive redox loops in prokaryotic respiratory systems, *Biochim. Biophys. Acta 1777*, 1480–1490.

18. Nelson, N. (1992) The vacuolar H(+)-ATPase: one of the most fundamental ion pumps in nature, *J. Exp. Biol. 172*, 19–27.

19. Mulkidjanian, A. Y., Heberle, J., and Cherepanov, D. A. (2006) Protons at interfaces: implications for biological energy conversion, *Biochim. Biophys. Acta 1757*, 913–930.

20. Mulkidjanian, A. Y. (2006) Proton in the well and through the desolvation barrier, *Biochim. Biophys. Acta 1757*, 415–427.

21. Lengeler, J. W., Drews, G., and Schlegel, H. G. (1999) *Biology of the Prokaryotes*, Blackwell, New York.

22. Willams, R. J. (2001) The structures of organelles and reticula localized bioenergetics and metabolism, *Chembiochemistry 2*, 637–641.

23. Williams, R. J. P. (1975) Proton-driven phosphorylation reactions in mitochondrial and chloroplast membranes, *FEBS Lett. 53*, 445–502.

24. Zechini, B., and Versace, I. (2009) Inhibitors of multidrug resistant efflux systems in bacteria, *Recent Pat. Antiinfect. Drug Discov. 4*, 37–50.

25. Vila, J., and Martínez, J. L. (2008) Clinical impact of the over-expression of efflux pump in nonfermentative gram-negative bacilli, development of efflux pump inhibitors, *Curr. Drug Targets 9*, 797–807.

26. Su, C. C., and Yu, E. W. (2007) Ligand–transporter interaction in the AcrB multidrug efflux pump determined by fluorescence polarization assay, *FEBS Lett. 581*, 4972–4976.

27. Krulwich, T. A., Ito, M., Gilmour, R., Sturr, M. G., Guffanti, A. A., and Hicks, D. B. (1996) Energetic problems of extremely alkaliphilic aerobes, *Biochim. Biophys. Acta 1275*, 21–26.

28. Guffanti, A. A., Mann, M., Sherman, T. L., and Krulwich, T. A. (1984) Patterns of electrochemical proton gradient formation by membrane vesicles from an obligately acidophilic bacterium, *J. Bacteriol. 159*, 448–452.

29. Spengler G., Martins A., Schelz Z., Rodrigues L., Aagaard L., Martins M., Costa SS., Couto I., Viveiros M., Fanning S., Kristiansen JE., Molnar J., Amaral L. (2009). Characterization of intrinsic efflux activity of Enterococcus faecalis ATCC29212 by a semi-automated ethidium bromide method, *In Vivo 23*, 81–87.

30. Martins, A., Spengler, G., Rodrigues, L., Viveiros, M., Ramos, J., Martins, M., Couto, I., Fanning, S., Pagès, J. M., Bolla, J. M., et al. (2009) pH modulation of efflux pump activity of multi-drug resistant *E. coli*: protection during its passage and eventual colonization of the colon, *PLoS One 4*(8), e6656.

31. Spengler, G., Ramos, J., Martins, A., Rodrigues, L., Martins, M., McCusker, M., Viveiros, M., Couto, I., Fanning, S.and Amaral, L. (2009) Physiology of efflux pump activity of antibiotic susceptible and induced ciprofloxacin resistant *Salmonella*, Unpublished observations.

32. Martins A., Spengler G., Martins M., Rodrigues L., Viveiros M., Davin-Regli A., Chevalier J., Couto I., Pagès JM., Amaral L. (2010). Physiological characterisation of the efflux pump system of antibiotic-susceptible and multidrug-resistant Enterobacter aerogenes, *Int. J. Antimicrob. Agents 36*, 313–318.

33. Börsch, M., and Gräber, P. (2005) Subunit movement in individual H^+-ATP synthases during ATP synthesis and hydrolysis revealed by fluorescence resonance energy transfer, *Biochem. Soc. Trans. 33*, 878–882.

34. Dimroth, P., and Cook, G. M. (2004) Bacterial Na^+ - or H^+ -coupled ATP synthases operating at low electrochemical potential, *Adv. Microb. Physiol. 49*, 175–218.

35. Cross, R. L., and Müller, V. (2004) The evolution of A-, F-, and V-type ATP synthases and ATPases: reversals in function and changes in the H^+/ATP coupling ratio, *FEBS Lett. 576*, 1–4.

36. Futai, M., Noumi, T., and Maeda, M. (1989) ATP synthase (H^+-ATPase): results by combined biochemical and molecular biological approaches, *Annu. Rev. Biochem. 58*, 111–136.

37. Boyer, P. D. (1987) The unusual enzymology of ATP synthase, *Biochemistry 26*, 8503–8507.

38. Godinot, C., and Di Pietro, A. (1986) Structure and function of the ATPase-ATP synthase complex of mitochondria as compared to chloroplasts and bacteria, *Biochimie 68*, 367–374.

39. Alekshun, M. N., and Levy, S. B. (2007) Molecular mechanisms of antibacterial multidrug resistance, *Cell 128*, 1037–1050.

40. Johnsen, P. L., Townsend, J. P., Bøhn, T., Simonsen, G. S., Sundsfjord, A., and Nielsen, K. M. (2009) Factors affecting the reversal of antimicrobial-resistance, *Lancet Infect. Dis. 9*, 357–364.

41. Gefen, O., and Balaban, N. Q. (2009) The importance of being persistent: heterogeneity of bacterial populations under antibiotic stress, *FEMS Microbiol. Rev. 33*, 704–717.

42. Pagès, J. M., James, C. E., and Winterhalter, M. (2008) The porin and the permeating antibiotic: a selective diffusion barrier in gram-negative bacteria, *Nat. Rev. Microbiol. 6*, 893–903.

43. Girgis, H. S., Hottes, A. K., and Tavazoie, S. (2009) Genetic architecture of intrinsic antibiotic susceptibility, *PLoS One 4*(5), e5629.

44. Kelley, W. L. (2006) Lex marks the spot: the virulent side of SOS and a closer look at the LexA regulon, *Mol. Microbiol. 62*, 1228–1238.

45. Rowley, G., Spector, M., Kormanec, J., and Roberts, M. (2006) Pushing the envelope: extracytoplasmic stress responses in bacterial pathogens, *Nat. Rev. Microbiol. 4*, 383–394.

46. Malléa, M., Chevalier, J., Bornet, C., Eyraud, A., Davin-Régli, A., Bollet, C., and Pagès, J. M. (1998) Porin alteration and active efflux: two in vivo drug resistance strategies used by *Enterobacter aerogenes*, *Microbiology 144*, 3003–3009.

47. Bornet, C., Davin-Régli, A., Bosi, C., Pagès, J. M., and Bollet, C. (2000) Imipenem resistance of *Enterobacter aerogenes* mediated by outer membrane permeability, *J. Clin. Microbiol. 38*, 1048–1052.

48. Bornet, C., Chollet, R., Malléa, M., Chevalier, J., Davin-Régli, A., Pagès, J. M., and Bollet, C. (2003) Imipenem and expression of multidrug efflux pump in *Enterobacter aerogenes*, *Biochem. Biophys. Res. Commun. 301*, 985–990.

49. Pagès, J. M. (2004) Role of bacterial porins in antibiotic susceptibility of gram-negative bacteria, in *Bacterial and Eukaryotic Porins*, Benz, R., Ed., Wiley-VCH, Hoboken, NJ, pp. 41–59.

50. Ghisalberti, D., Masi, M., Pagès, J. M., and Chevalier, J. (2005) Chloramphenicol and expression of multidrug efflux pump in *Enterobacter aerogenes, Biochem. Biophys. Res. Commun. 328*, 1113–1118.

51. Andersson, D. I. (2006) The biological cost of mutational antibiotic resistance: Any practical conclusions? *Curr. Opin. Microbiol. 9*, 461–465.

52. Ferenci, T., and Spira, B. (2007) Variation in stress responses within a bacterial species and the indirect costs of stress resistance, *Ann NY Acad. Sci. 1113*, 105–113.

53. Guerin, E., Cambray, G., Sanchez-Alberola, N., Campoy, S., Erill, I., Da, R. S., Gonzalez-Zorn, B., Barbé, J., Ploy, M. C., and Mazel, D. (2009) The SOS response controls integron recombination, *Science 324*, 1034.

54. Martínez, J. L., Sánchez, M. B., Martínez-Solano, L., Hernández, A., Garmendia, L., Fajardo, A., and Alvarez-Ortega, C. (2009) Functional role of bacterial multidrug efflux pumps in microbial natural ecosystems, *FEMS Microbiol. Rev. 33*, 430–449.

55. Rosner, J. L., and Martin, R. G. (2009) An excretory function for the *Escherichia coli* outer membrane pore *TolC*: upregulation of *marA* and *soxS* transcription and Rob activity due to metabolites accumulated in *tolC* mutants, *J. Bacteriol., 19*, 5283–5292.

56. Jayaraman, A., and Wood, T. K. (2008) Bacterial quorum sensing: signals, circuits, and implications for biofilms and disease, *Annu. Rev. Biomed. Eng. 10*, 145–167.

57. Pomposiello, P. J., and Demple, B. (2002) Global adjustment of microbial physiology during free radical stress, *Adv. Microb. Physiol. 46*, 319–341.

58. Griffith, K. L., Fitzpatrick, M. M., Keen, E. F., 3rd, and Wolf, R. E., Jr., (2009) Two functions of the C-terminal domain of *Escherichia coli* Rob: mediating "sequestration-dispersal" as a novel off–on switch for regulating Rob's activity as a transcription activator and preventing degradation of Rob by Lon protease, *J. Mol. Biol. 388*, 415–430.

59. Martin, R. G., Bartlett, E. S., Rosner, J. L, and Wall, M. E. (2008) Activation of the *Escherichia coli marA/soxS/rob* regulon in response to transcriptional activator concentration, *J. Mol. Biol. 380*, 278–284.

60. Martin, R. G., Jair, K. W., Wolf, R. E., Jr., and Rosner, J. L. (1996) Autoactivation of the *marRAB* multiple antibiotic resistance operon by the MarA transcriptional activator in *Escherichia coli, J. Bacteriol. 178*, 2216–2223.

61. Pomposiello, P. J., Bennik, M. H., and Demple, B. (2001) Genome-wide transcriptional profiling of the *Escherichia coli* responses to superoxide stress and sodium salicylate, *J. Bacteriol. 183*, 3890–3902.

62. Fajardo, A., and Martínez, J. L. (2008) Antibiotics as signals that trigger specific bacterial responses, *Curr. Opin. Microbiol. 11*, 161–167.

63. Prieto, A. I., Hernandez, S. B., Cota, I., Pucciarelli, G. M., Orlov, Y., Ramos-Morales, F., García-del Portillo, and Casadesu, J. (2009) Roles of the outer membrane protein AsmA of *Salmonella enterica* in the control of *marRAB* expression and invasion of epithelial cells, *J. Bacteriol. 191*, 3615–3622.

64. Gayet, S., Chollet, R., Molle, G., Pagès, J. M., and Chevalier, J. (2003) Modification of outer membrane protein profile and evidence suggesting an active drug pump in *Enterobacter aerogenes* clinical strains, *Antimicrob. Agents Chemother. 47*, 1555–1559.

65. Hu, W. S., Li, P. C., and Cheng, C. Y. (2005) Correlation between ceftriaxone resistance of *Salmonella enterica* serovar *Typhimurium* and expression of outer membrane proteins OmpW and Ail/OmpX-like protein, which are regulated by BaeR of a two-component system, *Antimicrob. Agents Chemother. 49*, 3955–3958.

66. Dupont, M., James, C. E., Chevalier, J., and Pagès, J. M. (2007) An early response to environmental stress involves regulation of OmpX and OmpF, two enterobacterial outer membrane pore-forming proteins, *Antimicrob. Agents Chemother. 51*, 3190–3198.

67. Vogel, J., and Papenfort, K. (2006) Small non-coding RNAs and the bacterial outer membrane, *Curr. Opin. Microbiol. 9*, 605–611.

68. Valentin-Hansen, P., Johansen, J., and Rasmussen, A. A. (2007) Small RNAs controlling outer membrane porins, *Curr. Opin. Microbiol. 10*, 152–155.

69. Gooderham, W. J., and Hancock, R. E. (2009) Regulation of virulence and antibiotic resistance by two-component regulatory systems in *Pseudomonas aeruginosa*, *FEMS Microbiol. Rev. 33*, 279–294.

70. Gunn, J. S. (2008) The *Salmonella* PmrAB regulon: lipopolysaccharide modifications, antimicrobial peptide resistance and more, *Trends Microbiol. 16*, 284–290.

71. Bishop, R. E. (2008) Structural biology of membrane-intrinsic beta-barrel enzymes: sentinels of the bacterial outer membrane, *Biochim. Biophys. Acta 1778*, 1881–1896.

72. Delcour, A. H. (2009) Outer membrane permeability and antibiotic resistance, *Biochim. Biophys. Acta 1794*, 808–816.

73. Nishino, K., Nikaido, E., and Yamaguchi, A. (2009) Regulation and physiological function of multidrug efflux pumps in *Escherichia coli* and *Salmonella*, *Biochim. Biophys. Acta 1794*, 834–843.

74. O'Regan, E., Quinn, T., Pagès, J. M., McCusker, M., Piddock, L., and Fanning, S. (2009) Multiple regulatory pathways associated with high-level ciprofloxacin and multidrug resistance in *Salmonella enterica* serovar *enteritidis*: involvement of RamA and other global regulators, *Antimicrob. Agents Chemother. 53*, 1080–1087.

75. Nikaido, E., Yamaguchi, A., and Nishino, K. (2008) AcrAB multidrug efflux pump regulation in *Salmonella enterica* serovar *typhimurium* by RamA in response to environmental signals, *J. Biol. Chem. 283*, 24245–24253.

76. Blanchard, J. L., Wholey, W. Y., Conlon, E. M., and Pomposiello, P. J. (2007) Rapid changes in gene expression dynamics in response to superoxide reveal SoxRS-dependent and independent transcriptional networks, *PLoS One 2*(11), e1186.

77. Davin-Régli, A., and Pagès, J. M. (2006) Regulation of efflux pumps in Enterobacteriaceae: genetic and chemical effectors, in *Antimicrobial Resistance in Bacteria*, Amabiles-Cuevos, C. F., Ed., Horizon Biosciences, Horizon Scientific Press, Norwich, UK, pp. 55–75.

78. Zhang, A., Rosner, J. L, and Martin, R. G. (2008) Transcriptional activation by MarA, SoxS and Rob of two *tolC* promoters using one binding site: a complex promoter configuration for *tolC* in *Escherichia coli*, *Mol. Microbiol. 69*, 1450–1455.

79. George, A. M., Hall, R. M, and Stokes, H. W. (1995) Multidrug resistance in *Klebsiella pneumoniae*: a novel gene, ramA, confers a multidrug resistance phenotype in *Escherichia coli*, *Microbiology 141*, 1909–1920.

80. van der Straaten, T., Janssen, R., Mevius, D. J., and van Dissel, J. T. (2004) *Salmonella* gene *rma* (*ramA*) and multiple-drug-resistant *Salmonella enterica* serovar *typhimurium*, *Antimicrob. Agents Chemother. 48*, 2292–2294.

81. Chollet, R., Chevalier, J., Bollet, C., Pagès, J. M., and Davin-Régli, A. (2004) RamA is an alternate activator of the multidrug resistance cascade in *Enterobacter aerogenes*, *Antimicrob. Agents Chemother. 48*, 2518–2523.

82. Komatsu, T., Ohta, M., Kido, N., Arakawa, Y., Ito, H., and Kato, N. (1991) Increased resistance to multiple drugs by introduction of the *Enterobacter cloacae romA* gene into OmpF porin-deficient mutants of *Escherichia coli* K-12, *Antimicrob. Agents Chemother. 35*, 2155–2158.

83. Viveiros, M., Dupont, M., Rodrigues, L., Couto, I., Davin-Régli, A., Martins, M., Pagés, J. M., and Amaral, L. (2007) Antibiotic stress, genetic response and altered permeability of *E. coli*, *PLoS One 2*, e365.

84. Knowles, T. J., Scott-Tucker, A., Overduin, M., and Henderson, I. R. (2009) Membrane protein architects: the role of the BAM complex in outer membrane protein assembly, *Nat. Rev. Microbiol. 7*, 206–214.

85. Misra, R., and Bavro, V. N. (2009) Assembly and transport mechanism of tripartite drug efflux systems, *Biochim. Biophys. Acta 1794*, 817–825.

86. Bos, M. P., Robert, V., and Tommassen, J. (2007) Biogenesis of the gram-negative bacterial outer membrane, *Annu. Rev. Microbiol. 61*, 191–214.

87. Tal, N., and Schuldiner, S. (2009) A coordinated network of transporters with overlapping specificities provides a robust survival strategy, *Proc. Natl. Acad. Sci. USA 106*, 9051–9056.

88. Grkovic, S., Brown, M. H., and Skurray, R. A. (2002) Regulation of bacterial drug export systems, *Microbiol. Mol. Biol. Rev. 66*, 671–701.

89. Pomposiello, P. J., and Demple, B. (2001) Redox-operated genetic switches: the SoxR and OxyR transcription factors, *Trends Biotechnol. 19*, 109–114.

90. Koo, M. S., Lee, J. H., Rah, S. Y., Yeo, W. S., Lee, J. W., Lee, K. L., Koh, Y. S., Kang, S. O., and Roe, J. H. (2003) A reducing system of the superoxide sensor SoxR in *Escherichia coli*, *EMBO J. 22*, 2614–2622.

91. Koutsolioutsou, A., Peña-Lopis, S., and Demple, B. (2005) Constitutive *soxR* mutations contribute to multiple-antibiotic resistance in clinical *Escherichia coli* isolates, *Antimicrob. Agents Chemother. 49*, 2746–2752.

92. Bratu, S., Landman, D., George, A., Salvani, J., and Quale, J. (2009) Correlation of the expression of *acrB* and the regulatory genes *marA*, *soxS* and *ramA* with antimicrobial resistance in clinical isolates of *Klebsiella pneumoniae* endemic to New York City, *J. Antimicrob. Chemother., 64*, 278–283.

93. Lee, J. H., Lee, K. L., Yeo, W. S., Park, S. J., and Roe, J. H. (2009) SoxRS-mediated lipopolysaccharide modification enhances resistance against multiple drugs in *Escherichia coli*, *J. Bacteriol. 191*, 4441–4450.

94. Kohanski, M. A., Dwyer, D. J., Hayete, B., Lawrence, C. A., and Collins, J. J. (2007) A common mechanism of cellular death induced by bactericidal antibiotics, *Cell 130*, 797–810.

95. Kohanski, M. A., Dwyer, D. J., Wierzbowski, J., Cottarel, G., and Collins, J. J. (2008) Mistranslation of membrane proteins and two-component system activation trigger antibiotic-mediated cell death, *Cell 135*, 679–690.

96. Yassien, M. A., Ewis, H. E., Lu, C. D., and Abdelal, A. T. (2002) Molecular cloning and characterization of the *Salmonella enterica* serovar *paratyphi* B *rma* gene, which confers multiple drug resistance in *Escherichia coli*, *Antimicrob. Agents Chemother. 46*, 360–366.

97. Zheng, J., Cui, S., and Meng, J. (2009) Effect of transcriptional activators RamA and SoxS on expression of multidrug efflux pumps AcrAB and AcrEF in fluoroquinolone-resistant *Salmonella typhimurium*, *J. Antimicrob. Chemother. 63*, 95–102.

98. Ricci, V., and Piddock, L. J. (2009) Ciprofloxacin selects for multidrug resistance in *Salmonella enterica* serovar *typhimurium* mediated by at least two different pathways, *J. Antimicrob. Chemother. 63*, 909–916.

99. Bailey, A. M., Paulsen, I. T., and Piddock, L. J. (2008) RamA confers multidrug resistance in *Salmonella enterica* via increased expression of *acrB*, which is inhibited by chlorpromazine, *Antimicrob. Agents Chemother. 52*, 3604–3611.

100. Rosenberg, E. Y., Bertenthal, D., Nilles, M. L., Bertrand, K. P., and Nikaido, H. (2003) Bile salts and fatty acids induce the expression of *Escherichia coli* AcrAB multidrug efflux pump through their interaction with Rob regulatory protein, *Mol. Microbiol. 48*, 1609–1619.

101. Mitrophanov, A. Y., and Groisman, E. A. (2008) Signal integration in bacterial two-component regulatory systems, *Genes Dev. 22*, 2601–2611.

102. Abouzeed, Y. M., Baucheron, S., and Cloeckaert, A. (2008) *ramR* mutations involved in efflux-mediated multidrug resistance in *Salmonella enterica* serovar *typhimurium*, *Antimicrob. Agents Chemother. 52*, 2428–2434.

103. Viveiros, M., Jesus, A., Brito, M., Leandro, C., Martins, M., Ordway, D., Molnár, A. M., Molnár, J., and Amaral, L. (2005) Inducement and reversal of tetracycline resistance in *Escherichia coli* K-12 and the expression of proton gradient dependent multidrug efflux pump genes, *Antimicrob. Agents Chemother. 49*, 3578–3582.

104. Viveiros, M., Portugal, I., Bettencourt, R., Victor, T. C., Jordaan, A. M., Leandro, C., Ordway, D., and Amaral, L. (2002) Isoniazid-induced transient high-level resistance in *Mycobacterium tuberculosis*, *Antimicrob. Agents Chemother. 46*, 2804–2810.

105. Couto, I., Costa, S. S., Viveiros, M., Martins, M., and Amaral, L. (2008) Efflux-mediated response of *Staphylococcus aureus* exposed to ethidium bromide, *J. Antimicrob. Chemother. 62*, 504–513.

106. Nikaido, H. (2009) Multidrug resistance in bacteria, *Annu. Rev. Biochem. 78*, 119–146.

107. Viveiros, M., Martins, A., Paixão, L., Rodrigues, L., Martins, M., Couto, I., Fähnrich, E., Kern, W. V., and Amaral, L. (2008) Demonstration of intrinsic efflux activity of *Escherichia coli* K-12 AG100 by an automated ethidium bromide method, *Int. J. Antimicrob. Agents 31*, 458–462.

108. Lomovskaya, O., Zgurskaya, H. I., Totrov, M., and Watkins, W. J. (2007) Waltzing transporters and "the dance macabre" between humans and bacteria, *Nat. Rev. Drug Discov. 6*, 56–65.

109. Su, C. C., Nikaido, H., and Yu, E. W. (2007) Ligand-transporter interaction in the AcrB multidrug efflux pump determined by fluorescence polarization assay, *FEBS Lett. 581*, 4972–4976.

110. Thota, N., Koul, S., Reddy, M. V., Sangwan, P. L., Khan, I. A., Kumar, A., Raja, A. F., Andotra, S. S., and Qazi, G. N. (2008) Citral derived amides as potent bacterial NorA efflux pump inhibitors, *Bioorg. Med. Chem. 16*, 6535–6543.

111. Rodrigues, L., Wagner, D., Viveiros, M., Sampaio, D., Couto, I., Martina, V., Winfried, V. K., and Amaral, L. (2008) Thioridazine and chlorpromazine inhibition of ethidium bromide efflux in *Mycobacterium avium* and *Mycobacterium smegmatis, J. Antimicrob. Chemother. 61*, 1076–1082.

112. Amaral, L., Kristiansen, J. E., and Lorian, V. (1992) Synergistic effect of chlorpromazine on the activity of some antibiotics, *J. Antimicrob. Chemother. 30*, 556–558.

113. Martins, M., Bleiss, W., Marko, A., Ordway, D., Viveiros, M., Leandro, C., Pacheco, T., Molnár, J., Kristiansen, J. E., and Amaral, L. (2004) Clinical concentrations of thioridazine enhance the killing of intracellular methicillin-resistant *Staphylococcus aureus*: an *in vivo, ex vivo* and electron microscopy study, *In Vivo 18*, 787–794.

114. Ordway, D., Viveiros, M., Leandro, C., and Amaral, L. (2003) Clinical concentrations of thioridazine kill intracellular multi-drug resistant *Mycobacterium tuberculosis, Antimicrob. Agents Chemother. 47*, 917–922.

115. Molnár, J., Ren, J., Kristiansen, J. E., and Nakamura, M. J. (1992) Effects of some tricyclic psychopharmacons and structurally related compounds on motility of *Proteus vulgaris, Antonie Van Leeuwenhoek 62*, 319–320.

116. Spengler, G., Miczák, A., Hajdú, E., Kawase, M., Amaral, L., and Molnár, J. (2003) Enhancement of plasmid curing by 9-aminoacridine and two phenothiazines in the presence of proton pump inhibitor 1-(2-benzoxazolyl)-3,3-trifluoro-2-propanone, *Int. J. Antimicrob. Agents 22*, 223–228.

117. Martins, M., Viveiros, M., and Amaral, L. (2007) The curative activity of thioridazine on mice infected with *Mycobacterium tuberculosis, In Vivo 21*, 771–776.

118. Dastidar, S. G., Debnath, S., Mazumdar, K., Ganguly, K., and Chakrabarty, A. N. (2004) Trifluopromazine: a microbicide non-antibiotic compound, *Acta Microbiol. Immunol. Hung. 51*, 75–83.

119. Sabatini, S., Kaatz, G. W., Rossolini, G. M., Brandini, D., and Fravolini, A. (2008) From phenothiazine to 3-phenyl-1,4-benzothiazine derivatives as inhibitors of the *Staphylococcus aureus* NorA multidrug efflux pump, *J. Med. Chem. 51*, 4321–4330.

120. Kristiansen, J. E., and Amaral, L. (1997) The potential management of resistant infection with non-antibiotics, *J. Antimicrob. Chemother. 40*, 319–327.

121. Klitgaard, J. K., Skov, M. N., Kallipolitis, B. H., and Kolmos, H. J. (2008) Reversal of methicillin resistance in *Staphylococcus aureus* by thioridazine, *J. Antimicrob. Chemother. 62*, 1215–1221.

122. Amaral, L., Kristiansen, J. E., Abebe, L. S., and Millet, W. (1996) Inhibition of the respiration of multi-drug resistant clinical isolates of *Mycobacterium tuberculosis* by

thioridazine: potential use for the initial therapy of freshly diagnosed tuberculosis, *J. Antimicrob. Chemother. 38*, 1049–1053.

123. Amaral, L., Kristiansen, J. E., Thomsen, V. F., and Markowich, B. (2000) The effect of chlorpromazine on the outer cell wall constituents of *Salmonella typhimurium* ensuring resistance to the drug, *Int. J. Antimicrob. Agents 14*, 225–229.

124. Viveiros, M., Martins, M., Martins, A., Rodrigues, L., Couto, I., Fanning, S., and Amaral, L. (2008) Methods for the identification of efflux mediated mdr gram-negatives; assessment of efflux activity; evaluation of genes that regulate permeability of gram-negative pathogens, and evaluation of agents for activity against mdr efflux pumps, *Curr. Drug Targets 9*, 760–778.

125. Carrique-Mas, J. J., Papadopoulou, C., Evans, S. J., Wales, A., Teale, C. J., and Davies, R. H. (2008) Trends in phage types and antimicrobial resistance of *Salmonella enterica* serovar *enteritidis* isolated from animals in Great Britain from 1990 to 2005, *Vet. Rec. 162*, 541–546.

126. Meakins, S., Fisher, I. S., Berghold, C., Gerner-Smidt, P., Tschape, H., Cormican, M., Luzzi, I., Schneider, F., Wannett, W., Coia, J., et al. (2008) Antimicrobial drug resistance in human nontyphoidal *Salmonella* isolates in Europe 2000–2004: a report from the Enter-net International Surveillance Network, *Microb. Drug Resist. 14*, 31–35.

127. McCarron, B., and Love, W. C. (1997) Acalculous nontyphoidal *Salmonella cholecystitis* requiring surgical intervention despite ciprofloxacin therapy: report of three cases, *Clin. Infect. Dis. 24*, 707–709.

128. Molbak, K., Baggesen, D. L., Aarestrup, F. M., Ebbesen, J. M., Engberg, J., Frydendahl, K., Gerner-Smidt, P., Petersen, A. M., and Wegener, H. C. (1999) An outbreak of multidrug-resistant, quinolone-resistant *Salmonella enterica* serotype *typhimurium* DT104, *N. Engl. J. Med. 341*, 1420–1425.

129. Vasallo, F. J., Martin-Rabadan, P., Alcala, L., Garcia-Lechuz, J. M., Rodriguez-Creixems, M., and Bouza, E. (1998) Failure of ciprofloxacin therapy for invasive nontyphoidal salmonellosis, *Clin. Infect. Dis. 26*, 535–536.

130. Giraud, E., Baucheron, S., and Cloeckaert, A. (2006) Resistance to fluoroquinolones in *Salmonella*: emerging mechanisms and resistance prevention strategies, *Microbes. Infect. 8*, 1937–1944.

131. Hopkins, K. L., Davies, R. H., and Threlfall, E. J. (2005) Mechanisms of quinolone resistance in *Escherichia coli* and *Salmonella*: recent developments, *Int. J. Antimicrob. Agents 25*, 358–373.

132. Giraud, E., Cloeckaert, A., Kerboeuf, D., and Chaslus-Dancla, E. (2000) Evidence for active efflux as the primary mechanism of resistance to ciprofloxacin in *Salmonella enterica* serovar *typhimurium*, *Antimicrob. Agents Chemother. 44*, 1223–1228.

133. Miro, E., Verges, C., Garcia, I., Mirelis, B., Navarro, F., Coll, P., Prats, G., and Martínez-Martínez, L. (2004) (Resistance to quinolones and beta-lactams in *Salmonella enterica* due to mutations in topoisomerase-encoding genes, altered cell permeability and expression of an active efflux system), *Enferm. Infecc. Microbiol. Clin. 22*, 204–211.

134. Piddock, L. J., Ricci, V., McLaren, I., and Griggs, D. J. (1998) Role of mutation in the *gyrA* and *parC* genes of nalidixic-acid-resistant *Salmonella* serotypes isolated from animals in the United Kingdom, *J. Antimicrob. Chemother. 41*, 635–641.

135. Chen, S., Cui, S., McDermott, P. F., Zhao, S., White, D. G., Paulsen, I., and Meng, J. (2007) Contribution of target gene mutations and efflux to decreased susceptibility of *Salmonella enterica* serovar *typhimurium* to fluoroquinolones and other antimicrobials, *Antimicrob. Agents Chemother. 51*, 535–542.

136. Baucheron, S., Chaslus-Dancla, E., and Cloeckaert, A. (2004) Role of *tolC* and *parC* mutation in high-level fluoroquinolone resistance in *Salmonella enterica* serotype *typhimurium* DT204, *J. Antimicrob. Chemother. 53*, 657–659.

137. Baucheron, S., Imberechts, H., Chaslus-Dancla, E., and Cloeckaert, A. (2002) The AcrB multidrug transporter plays a major role in high-level fluoroquinolone resistance in *Salmonella enterica* serovar *typhimurium* phage type DT204, *Microb. Drug Resist. 8*, 281–289.

138. Baucheron, S., Tyler, S., Boyd, D., Mulvey, M. R., Chaslus-Dancla, E., and Cloeckaert, A. (2004) AcrAB–TolC directs efflux-mediated multidrug resistance in *Salmonella enterica* serovar *typhimurium* DT104, *Antimicrob. Agents Chemother. 48*, 3729–3735.

139. Zheng, J., Cui, S., and Meng, J. (2009) Effect of transcriptional activators RamA and SoxS on expression of multidrug efflux pumps AcrAB and AcrEF in fluoroquinolone-resistant *Salmonella typhimurium*, *J. Antimicrob. Chemother. 63*, 95–102.

140. Olliver, A., Valle, M., Chaslus-Dancla, E., and Cloeckaert, A. (2004) Role of an acrR mutation in multidrug resistance of in vitro–selected fluoroquinolone-resistant mutants of *Salmonella enterica* serovar *typhimurium*, *FEMS Microbiol. Lett. 238*, 267–272.

141. Alekshun, M. N., and Levy, S. B. (1997) Regulation of chromosomally mediated multiple antibiotic resistance: the *mar* regulon, *Antimicrob. Agents Chemother. 41*, 2067–2075.

142. Cohen, S. P., Hachler, H., and Levy, S. B. (1993) Genetic and functional analysis of the multiple antibiotic resistance (*mar*) locus in *Escherichia coli*, *J. Bacteriol. 175*, 1484–1492.

143. Sulavik, M. C., Dazer, M., and Miller, P. F. (1997) The *Salmonella typhimurium* mar locus: molecular and genetic analyses and assessment of its role in virulence, *J. Bacteriol. 179*, 1857–1866.

144. Li, Z., and Demple, B. (1994) SoxS, an activator of superoxide stress genes in *Escherichia coli*: purification and interaction with DNA, *J. Biol. Chem. 269*, 18371–18377.

145. Wu, J., Dunham, W. R., and Weiss, B. (1995) Overproduction and physical characterization of SoxR, a (2Fe-2S) protein that governs an oxidative response regulon in *Escherichia coli*, *J. Biol. Chem. 270*, 10323–10327.

146. Koutsolioutsou, A., Martins, E. A., White, D. G., Levy, S. B., and Demple, B. (2001) A *soxRS*-constitutive mutation contributing to antibiotic resistance in a clinical isolate of *Salmonella enterica* (serovar *typhimurium*), *Antimicrob. Agents Chemother. 45*, 38–43.

147. Oethinger, M., Podglajen, I., Kern, W. V., and Levy, S. B. (1998) Overexpression of the *marA* or *soxS* regulatory gene in clinical topoisomerase mutants of *Escherichia coli*, *Antimicrob. Agents Chemother. 42*, 2089–2094.

148. Prouty, A. M., Brodsky, I. E., Falkow, S., and Gunn, J. S. (2004) Bile-salt-mediated induction of antimicrobial and bile resistance in *Salmonella typhimurium*, *Microbiology 150*, 775–783.

149. Randall, L. P., and Woodward, M. J. (2001) Multiple antibiotic resistance (*mar*) locus in *Salmonella enterica* serovar *typhimurium* DT104, *Appl. Environ. Microbiol. 67*, 1190–1197.

150. Eaves, D. J., Ricci, V., and Piddock, L. J. (2004) Expression of *acrB*, *acrF*, *acrD*, *marA*, and *soxS* in *Salmonella enterica* serovar *typhimurium*: role in multiple antibiotic resistance, *Antimicrob. Agents Chemother. 48*, 1145–1150.

151. Hernandez-Urzua, E., Zamorano-Sanchez, D. S., Ponce-Coria, J., Morett, E., Grogan, S., Poole, R. K., and Membrillo-Hernandez, J. (2007) Multiple regulators of the Flavohae-moglobin (*hmp*) gene of *Salmonella enterica* serovar *typhimurium* include RamA, a transcriptional regulator conferring the multidrug resistance phenotype, *Arch. Microbiol. 187*, 67–77.

152. Keeney, D., Ruzin, A., and Bradford, P. A. (2007) RamA, a transcriptional regulator, and AcrAB, an RND-type efflux pump, are associated with decreased susceptibility to tigecycline in *Enterobacter cloacae*, *Microb. Drug Resist. 13*, 1–6.

153. Ruzin, A., Visalli, M. A., Keeney, D., and Bradford, P. A. (2005) Influence of transcriptional activator RamA on expression of multidrug efflux pump AcrAB and tigecycline suscepti-bility in *Klebsiella pneumoniae*, *Antimicrob. Agents Chemother. 49*, 1017–1022.

154. Ling, J. M., Chan, E. W., Lam, A. W., and Cheng, A. F. (2003) Mutations in topoisomerase genes of fluoroquinolone-resistant salmonellae in Hong Kong, *Antimi-crob. Agents Chemother. 47*, 3567–3573.

155. Schneiders, T., and Levy, S. B. (2006) MarA-mediated transcriptional repression of the *rob* promoter, *J. Biol. Chem. 281*, 10049–10055.

156. Tanaka, T., Horii, T., Shibayama, K., Sato, K., Ohsuka, S., Arakawa, Y., Yamaki, K., Takagi, K., and Ohta, M. (1997) RobA-induced multiple antibiotic resistance largely depends on the activation of the AcrAB efflux, *Microbiol. Immunol. 41*, 697–702.

157. Michan, C., Manchado, M., and Pueyo, C. (2002) SoxRS down-regulation of *rob* transcription, *J. Bacteriol. 184*, 4733–4738.

158. Webber, M. A., and Piddock, L. J. (2001) Absence of mutations in *marRAB* or *soxRS* in *acrB*-overexpressing fluoroquinolone-resistant clinical and veterinary isolates of *Escherichia coli*, *Antimicrob. Agents Chemother. 45*, 1550–1552.

159. Pomposiello, P. J., and Demple, B. (2000) Identification of SoxS-regulated genes in *Salmonella enterica* serovar *typhimurium*, *J. Bacteriol. 182*, 23–29.

160. Randall, L. P., and Woodward, M. J. (2002) The multiple antibiotic resistance (*mar*) locus and its significance, *Res. Vet. Sci. 72*, 87–93.

161. Piddock, L. J., Griggs, D. J., Hall, M. C., and Jin, Y. F. (1993) Ciprofloxacin resistance in clinical isolates of *Salmonella typhimurium* obtained from two patients, *Antimicrob. Agents Chemother. 37*, 662–666.

162. Thiagarajan, D., Saeed, A. M., and Asem, E. K. (1994) Mechanism of transovarian transmission of *Salmonella enteritidis* in laying hens, *Poult. Sci. 73*, 89–98.

163. O'Regan, E., Quinn, T., Frye, J.G., Pagès, J-M., Porwollik, S., Fedorka-Cray, P., McClelland, M. and Fanning, S. (2010) Fitness costs and stability of a high-level ciprofloxacin resistant phenotype in *Salmonella enterica* serovar Enteritidis. *Antimicrob. Agents Chemother. 54*, 367–374.

164. Giraud, E., Cloeckaert, A., Baucheron, S., Mouline, C., and Chaslus-Dancla, E. (2003) Fitness cost of fluoroquinolone resistance in *Salmonella enterica* serovar *typhimurium*, *J. Med. Microbiol. 52*, 697–703.

165. Webber, M. A., Bailey, A. M., Blair, J. M., Morgan, E., Stevens, M. P., Hinton, J. C., Ivens, A., Wain, J., and Piddock, L. J. (2009) The global consequence of disruption of the AcrAB–TolC efflux pump in *Salmonella enterica* includes reduced expression of SPI-1 and other attributes required to infect the host, *J. Bacteriol. 191*, 4276–4285.

166. Baxter, M. A., and Jones, B. D. (2005) The fimYZ genes regulate *Salmonella enterica* serovar *typhimurium* invasion in addition to type 1 fimbrial expression and bacterial motility, *Infect. Immun. 73*, 1377–1385.

167. La Ragione, R. M., Cooley, W. A., Velge, P., Jepson, M. A., and Woodward, M. J. (2003) Membrane ruffling and invasion of human and avian cell lines is reduced for aflagellate mutants of *Salmonella enterica* serotype *enteritidis*, *Int. J. Med. Microbiol. 293*, 261–272.

168. Galan, J. E. (1996) Molecular genetic bases of *Salmonella* entry into host cells, *Mol. Microbiol. 20*, 263–271.

169. Mills, D. M., Bajaj, V., and Lee, C. A. (1995) A 40 kb chromosomal fragment encoding *Salmonella typhimurium* invasion genes is absent from the corresponding region of the *Escherichia coli* K-12 chromosome, *Mol. Microbiol. 15*, 749–759.

170. Jones, B. D., Lee, C. A., and Falkow, S. (1992) Invasion by *Salmonella typhimurium* is affected by the direction of flagellar rotation, *Infect. Immun. 60*, 2475–2480.

171. Murray, G. L., Attridge, S. R., and Morona, R. (2003) Regulation of *Salmonella typhimurium* lipopolysaccharide O antigen chain length is required for virulence; identification of FepE as a second Wzz, *Mol. Microbiol. 47*, 1395–1406.

172. Alonso, A., Morales, G., Escalante, R., Campanario, E., Sastre, L., and Martínez, J. L. (2004) Overexpression of the multidrug efflux pump SmeDEF impairs *Stenotrophomonas maltophilia* physiology, *J. Antimicrob. Chemother. 53*, 432–434.

173. Bagel, S., Hullen, V., Wiedemann, B., and Heisig, P. (1999) Impact of gyrA and parC mutations on quinolone resistance, doubling time, and supercoiling degree of *Escherichia coli*, *Antimicrob. Agents Chemother. 43*, 868–875.

174. Kugelberg, E., Lofmark, S., Wretlind, B., and Andersson, D. I. (2005) Reduction of the fitness burden of quinolone resistance in *Pseudomonas aeruginosa*, *J. Antimicrob. Chemother. 55*, 22–30.

175. Sanchez, P., Linares, J. F., Ruiz-Diez, B., Campanario, E., Navas, A., Baquero, F., and Martínez, J. L. (2002) Fitness of in vitro selected *Pseudomonas aeruginosa* nalB and nfxB multidrug resistant mutants, *J. Antimicrob. Chemother. 50*, 657–664.

176. Dorman, C. J. (2006) DNA supercoiling and bacterial gene expression, *Sci. Prog. 89*, 151–166.

177. Drlica, K. (1992) Control of bacterial DNA supercoiling, *Mol. Microbiol. 6*, 425–433.

178. Drlica, K., and Zhao, X. (1997) DNA gyrase, topoisomerase IV, and the 4-quinolones, *Microbiol. Mol. Biol. Rev. 61*, 377–392.

179. Dorman, C. J., and Corcoran, C. P. (2009) Bacterial DNA topology and infectious disease, *Nucleic Acids Res. 37*, 672–678.

180. Nikaido, H. (1998) Multiple antibiotic resistance and efflux, *Curr. Opin. Microbiol. 1*, 516–523.

181. Chan, Y. Y., Bian, H. S., Tan, T. M., Mattmann, M. E., Geske, G. D., Igarashi, J., Hatano, T., Suga, H., Blackwell, H. E., and Chua, K. L. (2007) Control of quorum sensing by a *Burkholderia pseudomallei* multidrug efflux pump, *J. Bacteriol. 189*, 4320–4324.

182. Pearson, J. P., Van Delden, C., and Iglewski, B. H. (1999) Active efflux and diffusion are involved in transport of *Pseudomonas aeruginosa* cell-to-cell signals, *J. Bacteriol. 181*, 1203–1210.

183. Kohler, T., Van Delden, C., Curty, L. K., Hamzehpour, M. M., and Pechere, J. C. (2001) Overexpression of the MexEF–OprN multidrug efflux system affects cell-to-cell signaling in *Pseudomonas aeruginosa*, *J. Bacteriol. 183*, 5213–5222.

184. Moxley, R. A. (2004) *Escherichia coli* 0157:H7: an update on intestinal colonization and virulence mechanisms, *Anim. Health Res. Rev. 5*, 15–33.

185. Sperandio, V., Torres, A. G., and Kaper, J. B. (2002) Quorum sensing *Escherichia coli* regulators B and C (QseBC): a novel two-component regulatory system involved in the regulation of flagella and motility by quorum sensing in *E. coli*, *Mol. Microbiol. 43*, 809–821.

186. Choi, J., Shin, D., and Ryu, S. (2007) Implication of quorum sensing in *Salmonella enterica* serovar *typhimurium* virulence: the *luxS* gene is necessary for expression of genes in pathogenicity island 1, *Infect. Immun. 75*, 4885–4890.

187. Bearson, B. L., and Bearson, S. M. (2008) The role of the QseC quorum-sensing sensor kinase in colonization and norepinephrine-enhanced motility of *Salmonella enterica* serovar *typhimurium*, *Microb. Pathog. 44*, 271–278.

188. Marcusson, L. L., Frimodt-Moller, N., and Hughes, D. (2009) Interplay in the selection of fluoroquinolone resistance and bacterial fitness, *PLoS Pathog. 5*, e1000541.

189. Chander, M., Raducha-Grace, L., and Demple, B. (2003) Transcription-defective *soxR* mutants of *Escherichia coli*: isolation and in vivo characterization, *J. Bacteriol. 185*, 2441–2450.

EFFLUX PUMPS OF THE RESISTANCE–NODULATION–DIVISION FAMILY: A PERSPECTIVE OF THEIR STRUCTURE, FUNCTION, AND REGULATION IN GRAM-NEGATIVE BACTERIA

By MATHEW D. ROUTH, *Molecular, Cellular and Developmental Biology Interdepartmental Graduate Program, Iowa State University, Ames, Iowa,* YARAMAH ZALUCKI, *Department of Microbiology and Immunology, Emory University School of Medicine, Atlanta, Georgia,* CHIH-CHIA SU, *Department of Chemistry, Iowa State University, Ames, Iowa,* FENG LONG, *Molecular, Cellular and Developmental Biology Interdepartmental Graduate Program, Iowa State University, Ames, Iowa,* QIJING ZHANG, *Department of Veterinary Microbiology and Preventive Medicine, College of Veterinary Medicine, Iowa State University, Ames, Iowa,* WILLIAM M. SHAFER, *Department of Microbiology and Immunology, Emory University School of Medicine, Atlanta, Georgia, and Laboratories of Microbial Pathogenesis, VA Medical Center, Decatur, Georgia,* and EDWARD W. YU, *Molecular, Cellular and Developmental Biology Interdepartmental Graduate Program, Department of Chemistry, Department of Physics and Astronomy, and Department of Biochemistry, Biophysics and Molecular Biology, Iowa State University, Ames, Iowa*

CONTENTS

Advances in Enzymology and Related Areas of Molecular Biology, Volume 77
Edited by Eric J. Toone   Copyright © 2011 John Wiley & Sons, Inc.

I. Introduction

With the initial discovery of penicillin and the ensuing mass production of antibiotics in the 1940s, infectious bacteria quickly adapted and developed resistance to the deleterious molecules. In fact, a report published in 1947 found that of 100 *Staphylococcus* infections tested, 38 were classified as highly resistant to penicillin (1). The initial resistance was associated primarily with individual enzymes inactivating specific antibiotics, such as β-lactamases on penicillin. As novel antibiotics were implemented to combat resistant pathogens, selective pressure led to fundamentally new methods of drug resistance. Currently, there are roughly three major mechanisms utilized by bacteria to evade the toxic effects of biocidal agents. These mechanisms include enzymes that modify the drug, alteration of the antibacterial target, and reduced drug uptake due to the presence of efflux pumps or a decrease in porin expression. Enzymatic modification involves two classes of enzymes, including those that degrade specific antibiotics (2) and enzymes that chemically modify the antibacterial compound (3), resulting in inhibition of drug function. The second mechanism employed by bacteria is alteration of the drug target. Nearly all relevant fluoroquinolone resistance has been attributed to target alteration, whereby specific mutations inhibit drug interaction to DNA gyrase and topoisomerase IV (4). Although highly effective, these mechanisms are limited by inhibiting only specific classes of antibiotics. A more critical issue is the broad spectrum of antibiotic resistance associated with multidrug efflux pumps. Ubiquitous in most living cells, multidrug efflux transporters have gained recognition as the major contributor to drug resistance observed in many pathogenic microorganisms (5, 6). These transporters are capable of capturing and exporting toxic compounds before they reach their cellular target, thereby decreasing the effectiveness of the drug (6–8).

The first drug transporter, TetA, identified in 1980 (9), specifically conferred tetracycline resistance on its host. Following this breakthrough, ensuing discoveries identified protein families capable of eliminating various structurally unrelated toxic compounds.

Based on sequence and functional similarities, there are currently five families of multidrug-resistant (MDR) efflux pumps, including primary transporters of the ATP-binding cassette (ABC) family (10), and secondary transporters in the resistance–nodulation–division (RND) (11), multidrug and toxic compound extrusion (MATE) (12, 13), major facilitator (MF) (14–16), and small multidrug resistance (SMR) families (17). Of these protein families, RND transporters are considered the primary contributor to multidrug resistance associated with gram-negative bacteria (11, 18, 19). Transporters in the RND family are energized through the proton-motive force (PMF), with translocation of protons occurring in the transmembrane (TM) domain. These pumps generally function as a tripartite efflux complex in conjunction with a membrane fusion protein and an outer membrane channel to export substrates completely out of the bacterial cell (11). Strict regulation of these proteins is maintained at the transcription level through repressors and activators that respond to a similar library of compounds to initiate protein expression.

Recently, tremendous strides have been made toward understanding the mechanisms that govern the RND transporter function. These works have focused on both the transporters and the regulatory networks that control their expression. In this chapter we focus on the tripartite RND transporters that are the primary cause of biocidal resistance identified in the three gram-negative bacterial species. These tripartite systems include the AcrAB–TolC and CusABC efflux complexes of *Escherichia coli*, the CmeABC pump of *Campylobacter jejuni*, and the MtrCDE transporter of *Neisseria gonorrhoeae*. Further, we discuss the regulation of these efflux pumps, especially for the structural aspects of the *E. coli* AcrR and *C. jejuni* CmeR transcriptional regulators. With a detailed understanding of the nature of these protein machineries, it will be possible to generate novel therapeutics capable of inhibiting the function of these efflux transporters, thus making futile antibiotics viable once again.

II. *Escherichia coli* AcrAB–TolC Efflux System

Bacteria are highly adaptive microorganisms with an impressive ability to acquire resistance to each antibiotic they encounter. Broad-spectrum antibiotic resistance, a critical problem associated with the treatment of bacterial infections, is attributed primarily to multidrug-resistant (MDR) efflux pumps. Through sequence analysis, it was identified that *E. coli* harbor 37 putative multidrug efflux transporters (20, 21). Thus far, about 20 of these transporters have been identified as contributors to multidrug resistance (21). MDR proteins of the RND transporter family have been suggested to play the most prominent role in drug resistance, with the multidrug efflux pump AcrB providing a resistant phenotype to the widest range of substrates (22, 23). This inner membrane transporter recognizes and drives the extrusion of various structurally dissimilar antibacterial compounds, including commonly used antibiotics, bile salts, dyes, and detergents (22, 23). It functions as part of a tripartite efflux complex in conjunction with the membrane fusion protein AcrA (24, 25) and outer membrane channel TolC (26, 27). Together, the AcrAB–TolC complex spans both the inner and outer membranes of gram-negative bacteria to effectively lower the intracellular concentration of bactericidal compounds. Drug extrusion is catalyzed through the proton-motive force (PMF), whereby the RND transporter AcrB constitutes both the site of drug recognition and energy transduction for the entire protein complex (8, 28–30).

The AcrB transporter has been implicated in many clinically significant cases of drug-resistant bacterial infections (31–34). For example, fluoro-quinolone resistance in *Salmonella enteri*ca serovar *typhimurium* DT104 has been associated with increased efflux through the AcrB pump (31). Further, in drug-resistant, clinical *E. coli* isolates, AcrB overexpression was identified as the main cause of resistance to ciprofloxacin (34). To combat these pathogens and design better treatments against drug-resistant dis-eases, it is important to delineate the mechanisms of multiple drug recognition and expulsion associated with these transporters. Recently, the crystal structures of individual components of both the *E. coli* AcrAB–TolC (27, 35–40) and *Pseudomonas aeruginosa* MexAB–OprM (41–45) tripartite efflux complexes have been made available to us. These structures have provided invaluable insights into the efflux mechanisms of these RND transport systems.

The structures of *E. coli* AcrB (35–39) and *P. aeruginosa* MexB (41) suggested that these two proteins span the entire width of the inner membrane and protrude approximately 70 Å into the periplasm. The crystal structures of the outer membrane channels, *E. coli* TolC (27) and *P. aeruginosa* OprM (42), have also been determined. TolC is highlighted by a 100-Å-long α-helical tunnel that extends from the outer membrane anchored β-sheet domain and protrudes into the α-helical periplasmic domain. The *P. aeruginosa* OprM channel possesses a similar elongated α-helical tunnel that projects into the periplasmic space. Recently, the structures of the two periplasmic membrane fusion proteins, *E. coli* AcrA (40) and *P. aeruginosa* MexA (43–45), associated with these RND transporters have been solved. The structures suggest that these two periplasmic proteins are folded into elongated secondary structures consisting of an ~47-Å-long α-hairpin domain that presumably interacts with the α-helical tunnels of their corresponding outer membrane channels. Furthermore, the N- and C-terminal ends of these membrane fusion proteins are thought to contact their respective inner membrane transporters, creating a functional complex that spans both membranes (46).

In 2009, the full-length MexA membrane fusion protein structure was made available (45). The structure revealed that this protein consists of four distinct regions, which include the α-hairpin, lipoyl, β-barrel, and membrane proximal domains. These structural features are in good agreement with our recently determined crystal structure of the CusB membrane fusion protein, in which the full-length CusB protein also consists of four distinct domains (47). Based on the full-length structure of MexA and cross-linking experiments between AcrA, AcrB, and TolC, the assembled structure of a complete tripartite efflux complex in the form $TolC_3$–$AcrA_3$–$AcrB_3$, of size ~600 kDa, has been generated (45). Figure 1 illustrates a proposed model of this assembled tripartite drug efflux pump. This model revealed that the α-hairpin domain of AcrA makes coiled-coil interactions with helices H3 and H4 of the TolC entrance coils. Presumably, the flexible nature of AcrA permits the TolC channel to shift from the closed state to open conformation, which allows substrates to exit the protein complex. In addition, the membrane proximal domain of AcrA, formed by both the N- and C-termini, directly contacts the PN2 and PC1 subdomains of AcrB in the periplasm. This structural model confirmed the earlier work that the C-terminal end of AcrA is important for AcrA and AcrB interaction (46). The model also indicated that AcrA does not contact the cleft region between the PC1 and PC2 subdomains of AcrB. Thus, the cleft is opened to the periplasm in the assembled structure, presumably allowing substrates to enter the pump. A possible pathway for ligand binding and extrusion in the AcrB pump has been suggested by the structure of the asymmetric trimer (37–39), in which the external cleft may serve as an entrance for AcrB substrates. Indeed, the structures of AcrB in complexes with various substrates have confirmed that these molecules are located inside the cleft (48, 49).

Figure 1.   Model of the assembled tripartite drug efflux pump of AcrAB–TolC in the form of TolC₃–AcrA₃–AcrB₃. This possible model is generated based on the crystal structures of individual components of the complex in addition to cross-linked data between AcrA, AcrB, and TolC. The TolC trimer [orange, red, and yellow subunits with gray equatorial domains and outer membrane (OM) regions] was docked onto AcrA (green)-docked AcrB trimer (blue/light blue subunits with gray inner membrane (IM) regions). [From (45), with permission from the National Academy of Sciences and V. Koronakis.] (See insert for color representation.)

The asymmetric model of AcrB (37–39) suggests that the three proto-mers of the AcrB trimer may go through a cyclic conformational change, from an *access* form through a *binding* mode and finally, to the *extrusion* conformation. This mechanism is needed to couple with the process of proton translocation through the proton relay system in the transmembrane domain (29). The periplasmic domain of the binding protomer, containing an expanded binding pocket surrounded by several aromatic and hydrophobic amino acids, has also been found to bind drugs (37). A detailed description of the AcrAB–TolC system is provided in Chapter 1.

A. CRYSTAL STRUCTURE OF THE AcrR REGULATOR

As a drug/proton antiporter with a wide substrate range, transcriptional regulation of AcrB is an important factor in maintaining cell viability. Unregulated overexpression of RND transporters can lead to deleterious side effects, including a loss of membrane integrity and potential export of important metabolites (50–52). Furthermore, expression of the multidrug efflux complex must be induced rapidly while the natural substrates are at subinhibitory concentrations.

The *acrAB* operon is globally regulated by the transcriptional activators MarA, SoxS, and Rob. These are well-characterized transcriptional activators that initiate expression of a group of 40 promoters (known as the *marA/soxS/rob* regulon), including *acrAB* (53, 54) and *tolC* (55), by interacting with the *marbox* operator sequence. Local regulation of *acrAB* is achieved through the transcriptional regulator AcrR, which is located 141 bp upstream of the *acrAB* operon and is transcribed divergently. AcrR is a prototypical member of the TetR family of transcriptional regulators and is responsible for fine-tuning the transcription of *acrAB* by repressing transcription initiation until the appropriate ligands are present (56).

Transcription of *acrR* gives rise to a 215-amino acid regulatory protein (56). Similar to other members of the TetR family, it is characterized by a homologous N-terminal three-helix DNA binding domain composed of the hallmark helix–turn–helix (HTH) motif (57). This regulator presumably functions as a dimer of dimers to repress transcription of *acrAB* by interacting with the 24-base pair inverted repeat (IR) operator sequence, 5′-TACATACATTTGTGAATGTATGTA-3′. The IR sequence is positioned between the *acrR* and *acrAB* genes and overlapping with the *acrAB* promoter (56, 58, 59). In addition, the C-terminal domain, a region highly diverse among the TetR family, functions primarily in ligand recognition. As the regulator of AcrAB, AcrR responds to the same array of ligands as AcrB. In fact, recent studies indicate that both AcrB and AcrR bind these ligands with strikingly similar affinities. It appears that AcrR binds ethidium (Et), proflavin (Pf), and rhodamine 6G (R6G) with dissociation constants of 4.2, 10.1, and 10.7 µM (59), while AcrB interacts with these ligands with K_d values of 8.7, 14.5, and 5.5 µM, respectively (60). Presumably, ligand interaction may initiate a conformational shift that results in the dissociation of AcrR from its cognate operator (57). Together, this suggests an effective mechanism to initiate AcrB expression and expulsion of substrates, while the toxic compounds are at subinhibitory levels. Recently, we have determined two distinct crystal structures of AcrR in space groups of $P222_1$ (61) and $P3_1$ (62). Figure 2 illustrates a superimposition of these two AcrR structures. The structures have provided valuable insights into the mechanisms of ligand recognition and transcriptional regulation.

The crystal structures of AcrR reveal a dimeric protein with secondary structures composed entirely of α-helices and an overall architecture similar to other members of the TetR family, including TetR (63, 64),

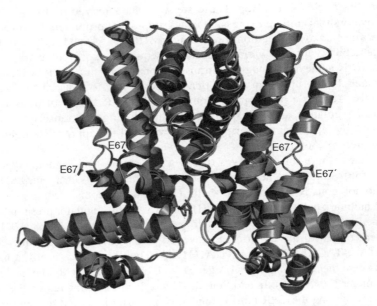

Figure 2.   Structural comparison of the $P3_1$ and $P222_1$ structures of AcrR. Superimposition of the dimeric AcrR structures was performed using the program ESCET (green, $P3_1$ structure; orange, $P222_1$ structure). The conformational differences highlighted in these two crystal structures provide a plausible model to describe transcriptional regulation by AcrR. Residue E67 in each subunit is shown as a stick model. (See insert for color representation.)

QacR (65, 66), CmeR (67), CprB (68), and EthR (69, 70). Each protomer of the AcrR dimer consists of helices ($\alpha 1$ to $\alpha 9$ and $\alpha 1'$ to $\alpha 9'$), with helices $\alpha 1$ to $\alpha 3$ making up the N-terminal DNA-binding domain and helices $\alpha 4$ to $\alpha 9$ forming the larger C-terminal ligand-binding region. Dimerization of AcrR occurs mostly through contacts made between helices $\alpha 8$ and $\alpha 9$, which creates a contact region of 2002 $\mathring{A}^2$ per monomer. The structures also demonstrate that a 350-$\mathring{A}^3$ internal hydrophobic pocket, surrounded by helices $\alpha 4$ through $\alpha 8$ of each monomer, forms a multifaceted ligand-binding pocket. Docking of known inducers, including Pf, Et, and cipro-floxacin (Cip), into this ligand-binding pocket suggests that W63, I170, and F114 are supposed to perform hydrophobic contacts with the bound drugs. Additionally, residues E67 and Q130 are predicted to make important electrostatic interactions to the inducing compounds (61). Further confir-mation of the binding site was identified with an E67A AcrR mutant that abolished the binding of Pf, Et, and rhodamine 6G (R6G) in the regula-tor (61). Evidence to support the multifaceted multidrug binding pocket was provided through fluorescence polarization experiments, which indi-cated that Pf and Et bind noncompetitively within the hydrophobic cavity (59).

Intriguing structural changes were observed when comparing the two conformations of AcrR (Figure 2), with the most significant changes occurring in the stretch of amino acids between helix $\alpha 1$ and the N-terminal half of helix $\alpha 4$ (residues 7–65) in the DNA-binding region [root-mean-square deviation (RMSD) of 2.8 Å]. Conformational changes in this region seem to be predominately rigid-body translation and rotation, resulting in a downward shift by 2.6 Å and a rotational movement of 10° of the N-terminal domain in the $P3_1$ structure with respective to that of $P222_1$. As a result of these movements, the N-terminal domains of the dimer move closer together by about 3 Å (42 Å in the $P222_1$ structure and 39 Å in the $P3_1$ conformation) (71). In both QacR (64) and TetR (66), similar changes have been observed in which the distances between N-terminal domains decrease by 11 Å and 3 Å, respectively, when converting from a drug-bound state to the DNA-bound conformation. Presumably, these changes allow the regulators to bind consecutive major grooves in B-form DNA. In addition to the overall movement of the N-terminal domain, the distance between two R45 residues in the dimeric AcrR shifts from 40 Å in the $P222_1$ structure to 35 Å in the $P3_1$ structure (62). As AcrR and QacR share a similar DNA-binding mode, in which two dimeric regulators bind one double-stranded DNA, a speculative model of DNA-bound AcrR was generated by aligning individual domains of AcrR to those of QacR-DNA (61). The model suggests that two AcrR dimers interact with the DNA through amino acids, R45, G46, and W50. A study of fluoroquinolone-resistant *E. coli* strains highlights the critical role of R45. Six of the 36 isolated drug-resistant strains of *E. coli* examined had increased levels of AcrA and AcrB due to a mutation at R45 (arg → cys) that presumably inhibited AcrR–DNA binding (34).

When examining the changes in the C-terminal domain between the $P222_1$ and $P3_1$ structures, the most striking distinction involves the shift of amino acid E67. In the $P3_1$ conformation, E67 is expelled from the ligand-binding pocket and faces the exterior of the protein, which contrasts the $P222_1$ form, whereby E67 is completely buried within the binding cavity. These intriguing shifts of E67 suggest that this amino acid may act as a molecular switch that drives conformational change during ligand binding (71). The movement of E67 out of the binding pocket in the $P3_1$ structure initiates considerable changes in the C-terminal domain of AcrR, including helix $\alpha4a$ shifting toward the N-terminal domain by 2.3 Å and a local unwinding of N-terminus of $\alpha6$, which shortens the helix by one turn. Moreover, an important hydrogen-bonded network between residues R105, Q14, and D18 identified in the $P222_1$ structure is disrupted in the $P3_1$ crystal form (62).

With the evidence indicated previously, it is likely that the DNA-bound state of AcrR resembles the $P3_1$ crystal structure, whereby E67 is positioned outside the ligand-binding site. Upon ligand binding, E67 presumably flips into the cavity and makes appropriate electrostatic interactions with the ligand creating a conformation closely related to the $P222_1$ structure. Thus, transmission of the signal from the C-terminal ligand-binding domain to the N-terminal DNA-binding region is thought to occur through the hydrogen-bonded network, R105, Q14, D18, between helices $\alpha1$ and $\alpha4$. The crystal structures of the DNA- and ligand-bound forms of AcrR are required to confirm the proposed model for transcriptional regulation of AcrR.

III.   *Escherichia coli* CusABC Efflux System

E. coli contains seven different RND efflux transporters. These transporters can be categorized into two distinct subfamilies, the hydrophobic and amphiphilic efflux RND (HAE-RND) and heavy-metal efflux RND (HME-RND) families (11, 72). Six of these transporters—AcrB (35–39,

73), AcrD (74), AcrF (75), MdtB (76, 77), MdtC (76, 77), and YhiV (21, 78)—are multidrug efflux pumps which belong to the HAE–RND protein family (11). In addition to these multidrug efflux pumps, *E. coli* consists of only one HME–RND transporter, CusA, which specifically recognizes and confers resistance to Ag(I) and Cu(I) ions (79, 80). These two metal ions are highly toxic to prokaryotes and have been used widely for centuries as effective antimicrobial agents to combat pathogens. The two subfamilies of these RND transporters share relatively low protein sequence homology. For example, alignment of protein sequences suggests that CusA and AcrB possess only 19% identity. Because of this low protein sequence homology, the structural model of AcrB may not be precise enough to describe the conformation of the CusA transporter.

As an RND transporter, CusA works in conjunction with a periplasmic component, belonging to the membrane fusion protein (MFP) family, and an outer membrane channel to form a functional protein complex. CusA is a large PMF-dependent inner membrane efflux pump that contains 1047 amino acid residues (79, 80). CusC, however, is a 457-amino acid polypeptide that forms an outer membrane channel (79, 80). The membrane fusion protein CusB consists of 379 amino acids and contacts both the inner membrane CusA and outer membrane CusC proteins (79, 80). Presumably, the three components of this HME–RND system form a tripartite efflux complex that resembles the AcrAB–TolC complex, whereby heavy-metal efflux in CusABC is driven by proton import and catalyzed through CusA.

Between the *cusC* and *cusB* genes, there is a small chromosomal gene that produces a periplasmic protein CusF (79, 80). This small periplasmic protein is also involved in Cu^+ and Ag^+ resistance. The crystal structure of CusF suggests that this protein forms a five-stranded β-barrel, and its conserved residues H36, W44, M47, and M49 form a Cu^+ or Ag^+ binding site (81–83). CusF probably functions as a chaperone that carries Cu^+ or Ag^+ to the CusABC heavy-metal efflux pump. In fact, it has recently been shown that CusF is able to transfer bound Cu^+/Ag^+ ions directly to the membrane fusion CusB (84). The entire CusAB(F)C system is controlled by a two-component sensory circuit, which includes the histidine kinase CusS and the response regulator CusR (85).

Currently, little structural information is available for any components of the HME–RND tripartite efflux complexes. Different from the HAE–RND family, members of the HME–RND family are highly substrate specific, with the ability to differentiate between monovalent and divalent ions. Thus, there is a strong rationale to understand the structural aspect of this efflux system. As an initial step to examine the mechanisms used by the CusABC efflux system to facilitate recognition and extrusion of Ag(I) and Cu(I) ions, we recently determined the crystal structure of the periplasmic membrane fusion protein CusB (47).

A. CRYSTAL STRUCTURE OF THE MEMBRANE FUSION PROTEIN CusB

The crystal structure of CusB (47) suggests that this protein is folded into an elongated molecule with each protomer being divided into four different domains (Figure 3). The first three domains of the protein are mostly β-strands. However, the fourth domain forms an entirely α-helical domain featuring a three-helix bundle secondary structure.

The first β-domain (domain 1) is formed by the N- and C-terminal ends of the polypeptide (residues 89–102 and 324–385). Presumably, this domain is located directly above the outer leaflet of the inner membrane and interacts with the CusA efflux pump. Overall, domain 1 is a β-barrel domain. It is composed of six β-strands, with the N-terminal end forming one of the β-strands, while the C-terminus of the protein constitutes the remaining five strands (Figure 3).

The second β-domain (domain 2) of CusB is formed by residues 105–115 and 243–320. This domain consists of six β-strands and one short α-helix. Again, the N-terminal residues form one of the β-strands that is incorporated into this domain. The C-terminal residues contribute a β-strand, an α-helix, and four antiparallel β-sheets.

Domain 3 is another globular β-domain adjacent to the second domain of CusB. This domain consists of residues 121–154 and 207–239, with a majority of these residues folding into eight β-strands.

Perhaps the most interesting motif appears to be in the fourth domain (domain 4) of CusB. This region forms an all-helical domain, which comprises residues 156–205. Surprisingly, this α-domain is folded into an antiparallel, three-helix bundle. This structural feature, not found in other known protein structures in the MFP family, highlights the uniqueness of the CusB protein. The helix bundle creates an ~27-Å-long helix–turn–helix–turn–helix secondary structure, making it at least 20 Å shorter than the two-helical hairpin domains of MexA (43–45) and AcrA (40). To date, CusB is the only periplasmic protein in the MFP family that possesses this three-helical domain instead of a two-helical hairpin motif. The overall structure of CusB is quite distinct from the known structures of other membrane fusion proteins.

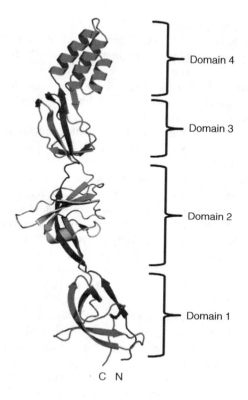

Figure 3.   Crystal structure of the CusB membrane fusion protein. The structure can be divided into four distinct domains. Domain 1 is formed by the N- and C-termini and is located above the inner membrane. The loops between domains 2 and 3 appear to form an effective hinge to allow the molecule to shift from an open conformation to a more compact structure. Domain 4 is folded into an antiparallel, three-helix bundle, which is thought to be located near the outer membrane. (See insert for color representation.)

B.   CusB–Cu(I) AND CusB–Ag(I) COMPLEXES

To identify the metal binding sites of CusB, we prepared the CusB–Cu(I) and CusB–Ag(I) crystal complexes by soaking these metal ions into the apo-CusB crystals. The overall structures of these complexes are very similar to that of apo–CusB. For example, the structures of CusB–Cu(I) and apo–CusB can easily be superimposed, giving an overall RMSD of 0.8 Å. It appears that we have found two Cu^+ (designated sites C1 and C2) and one Ag^+ (designated site A1) binding sites in this protein (47). To our knowledge, these are the first structures of any membrane fusion proteins that have been determined with their ligands. The structures suggest an unusual metal binding mode in which each metal-binding site consists of only one methionine residue to facilitate metal binding.

Cu^+ in site C1 is located in domain 1, which is formed by the N- and C-termini of the protein. Coordinating with the bound Cu^+ ion at this site are M324, F358, and R368. Site C1 is located near the bottom of the elongated CusB molecule. Presumably, this region may interact directly with the periplasmic domain of the CusA efflux pump. The binding of Cu^+ in site C2 is located close to the center of the three-helix bundle in domain 4. This α-helical domain may make a direct contact with the outer membrane channel CusC. Cu^+ in this location is bound by M190, W158, and Q162.

For the Ag^+ binding, site A1 is found right next to M324 of CusB. It appears that the location of this Ag^+-binding site is the same as that of site C1 for Cu^+ binding. Thus, the bound Ag^+ at site A1 is coordinated with M324, F358, and R368.

There is evidence that members of the MFP family play a functional role in the efflux of substrates. It has been found that the MFP EmrA is able to bind different transported drugs directly (86). Recently, the CusB protein has also been shown to interact with Ag(I) (87). The crystal structures of the CusB–Cu(I) and CusB–Ag(I) complexes provide direct evidence that this protein specifically interacts with and contacts Cu(I) and Ag(I). Thus, in addition to their role as adaptors to bridge the inner and outer membrane efflux components, these membrane fusion proteins may participate in recognizing and extruding their substrates.

C.   INTERACTION BETWEEN CusA AND CusB

To determine how CusB interacts with the CusA efflux pump and the relative orientation of CusB in the efflux complex, we cross-linked the purified CusA and CusB proteins in vitro using the lysine–lysine cross-linker disuccinimidyl suberate (DSS) (47). The resulting product was digested with trypsin and examined using LC-MS/MS. Analysis of the mass spectral data suggests that the lysine residue of the polypeptide β (IDPTQTQNLGVKTATVTR), originating from the N-terminal residues (84–101) of CusB, interacts directly with the lysine residue of peptide α (SGKHDLADLR), which belongs to the periplasmic domain (residues 148–157) of the CusA efflux pump. Although the CusA and AcrB efflux pumps share only 19% protein sequence identity, we generated a structural model of the CusA transporter based on the crystal structure of AcrB and alignment of protein sequences of these two transporters (Figure 4). The model indicates that polypeptide α (residues 148–157 of CusA) is located

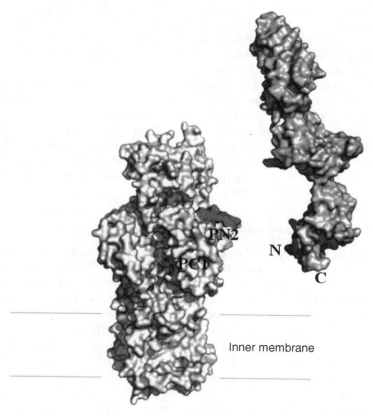

Figure 4.   Specific interaction between CusA and CusB. The model of CusA (gray) was created based on protein sequence alignment and the crystal structure of AcrB. Mass spectral data suggest that the periplasmic domain of CusA specifically interacts with the N-terminus of CusB (light brown). Polypeptides α, SGKHDLADLR (from CusA), and β, IDPTQTQNLGVKTATVTR (from CusB), are shown in red and blue, respectively. (See insert for color representation.)

directly above the vestibule region of CusA, facing the periplasm. This location should correspond to the PN2 region of AcrB. If this is the case, the C-terminus of CusB should interact with CusA at a position corresponding to the PC1 region in AcrB (Figure 4). According to the most recently determined MexA structure, it suggested that both the N- and C-terminal ends of MexA are located close to the MexB transporter (45). In addition, in vivo cross-linking studies also demonstrated that the N- and C-termini of AcrA interact directly with PN2 and PC1 of the periplasmic domain of AcrB, respectively (45). Together with the crystal structure of CusB and the mass spectrometric data, it has been suggested that domain 1, formed by the N- and C-terminal ends, of CusB should interact with the periplasmic domain of the CusA transporter (47).

As CusB is folded into a distinct secondary structure compared with the current crystal structures of other membrane fusion proteins, this may imply that its tripartite partners, the inner membrane transporter CusA and the outer membrane channel CusC, may also possess unique secondary structural features that distinguish them from the existing structures of their homologous proteins. Exactly how these individual heavy-metal efflux components assemble into a functional complex must await elucidation of the CusA and CusC structures.

IV. *Campylobacter jejuni* CmeABC Efflux System

Campylobacter jejuni is a major causative agent of human enterocolitis and is responsible for more than 400 million cases of diarrhea each year worldwide (88). *Campylobacter* infection may also trigger an autoimmune response, which is associated with the development of Guillain–Barré syndrome, an acute flaccid paralysis caused by degeneration of the peripheral nervous system (89). *C. jeuni* is widely distributed in the intestinal tracts of animals and is transmitted to humans via contaminated food, water, or raw milk. For antibiotic treatment of human campylobacteriosis, fluoroquinolones and macrolides are frequently prescribed (90). Unfortunately, *Campylobacter* has developed resistance to both classes of antimicrobials, especially to fluoroquinolones (91–93). Resistance of *Campylobacter* to antibiotics is mediated by multiple mechanisms (94), including (1) synthesis of antibiotic-inactivating enzymes, (2) alteration or protection of antibiotic targets, and (3) active extrusion of drugs from *Campylobacter* cells through drug efflux transporters. Different from the first two mechanisms that are usually involved in the resistance to a specific class of drugs, antibiotic efflux pumps in *Campylobacter* contribute to both intrinsic and acquired resistance to a broad range of antimicrobials and toxic compounds.

   C. jejuni harbors multiple drug efflux transporters of different families (94). Among them, CmeABC, an RND-type efflux pump, is the primary antibiotic efflux system and is the best functionally characterized transporter in *Campylobacter* (95, 96). CmeABC consists of three components, including an outer membrane protein (CmeC), an inner membrane drug transporter (CmeB), and a periplasmic fusion protein (CmeA). CmeABC contributes significantly to the intrinsic and acquired resistance of *Campylobacter* to structurally diverse antimicrobials, including fluoroquinolones and macrolides, by reducing the accumulation of drugs in *Campylobacter* cells (95–100). It has been found that CmeABC functions synergistically with target mutations in conferring and maintaining high-level resistance to fluoroquinolones and macrolides (97, 98, 100–102). This efflux pump also plays an important role in the emergence of fluoroquinolone-resistant *Campylobacter* under selection pressure (103). Inactivation of CmeABC reduced the frequency of emergence of fluoroquinolone-resistant mutants, while overexpression of CmeABC increased the frequency of emergence of the mutants (103). This contributing effect of CmeABC is due to the fact that many of the spontaneous *gyrA* mutants cannot survive selection by ciprofloxacin in the absence of CmeABC.

   In addition to conferring antibiotic resistance, CmeABC also has important physiological functions. It has been shown that CmeABC is functionally interactive with CmeDEF, another RND-type efflux pump, in maintaining optimal cell viability in *Campylobacter*, possibly by extruding endogenous toxic metabolites (104). Double mutations in CmeABC and CmeDEF appeared to be lethal to *C. jejuni* strain 11168 and significantly reduced the growth of strain 81–176 in conventional media. Another important function of CmeABC is bile resistance. As an enteric pathogen, *C. jejuni* must possess means to adapt in the animal intestinal tract, where bile acids are commonly present. Mutations of CmeB in *C. jejuni* resulted in a drastic increase in the susceptibility to various bile acids and a severe growth defect in bile-containing media or in chicken intestinal extracts (105). When inoculated into chickens, the CmeB mutant failed to colonize the inoculated birds. These findings provide compelling evidence that by mediating the resistance to bile acids, CmeABC is essential for *Campylobacter* adaptation to the intestinal environment. These findings also strongly suggest that bile resistance is a natural function of this RND-type efflux pump.

CmeABC is subject to regulation by a transcriptional factor named CmeR (106). The *cmeR* gene is located immediately upstream of the *cmeABC* operon and encodes a 210-amino acid protein that shares N-terminal sequence homology with members of the TetR family of transcriptional repressors (57, 107). Similar to other members in the TetR family, the N-terminal region of CmeR contains a DNA-binding α-helix–turn–α-helix (HTH) motif, while the C-terminal region is involved in the interaction with inducing ligands (67). *cmeR* is transcribed in the same direction as *cmeABC* and the intergenic region between *cmeR* and *cmeA* contains the promoter (P$_{cmeABC}$) for *cmeABC*. As a transcriptional repressor, CmeR binds directly to an inverted repeat in P$_{cmeABC}$ and inhibits the transcription of *cmeABC* (106). Deletion of *cmeR* or mutations in the inverted repeat of P$_{cmeABC}$ releases the repression and results in overexpression of CmeABC. Recently, it was shown by DNA microarray that CmeR functions as a pleiotropic regulator and modulates the expression of multiple genes in *C. jejuni*, occurring by direct or indirect methods (108). One of the newly identified CmeR-regulated genes is Cj0561c, which is predicted to be a periplasmic protein. The promoter of Cj0561c contains two CmeR-binding sites and is strongly repressed by CmeR. Although the exact function of Cj0561c was unknown, inactivation of this gene led to reduction of *Campylobacter* colonization in the intestinal tract of chicken (108), suggesting that Cj561c is important for *Campylobacter* physiology.

As a major mechanism in bile resistance, *cmeABC* is inducible by various bile conjugates (109). Thus, the expression level of *cmeABC* is influenced by bile. The induction of *cmeABC* is due to the inhibitory effect of bile on the binding of CmeR to P$_{cmeABC}$, which promotes the release of CmeR from the promoter DNA (109). Since CmeR represses the expression of *cmeABC*, dissociation of CmeR from P$_{cmeABC}$ results in elevated expression of *cmeABC*. Not surprisingly, bile salts also strongly induce the expression of Cj0561c, which is also regulated by CmeR (108). These in vitro findings were consistent with the results from in vivo studies, in which DNA microarray and real-time RT-PCR revealed that expression of *cmeABC* and Cj0561c was greatly up-regulated in animal intestinal tracts (110). This suggests that bile-mediated induction of CmeR-regulated genes also occurs in animal hosts. Based on these findings, it is conceivable that CmeR senses the presence of bile compounds in the environment and accordingly, modulates the expression levels of its target genes.

The information discussed above indicates that the CmeR regulatory network plays an important role in *Campylobacter* physiology and in its resistance to various antimicrobials. To understand the structural basis of CmeR regulation and facilitate the development of anti-*Campylobacter* therapeutics, we have initiated our work to determine the three-dimensional structure of CmeR. Our protein crystallization studies confirmed the two-domain structure of CmeR and showed that CmeR functions as a homodimer (67).

A. CRYSTAL STRUCTURE OF THE CmeR REGULATOR

The crystal structure of dimeric CmeR, a member of the TetR family of regulators, is shown in Figure 5. This structure revealed that each subunit of CmeR is composed of nine α-helices, in which the characteristic short-

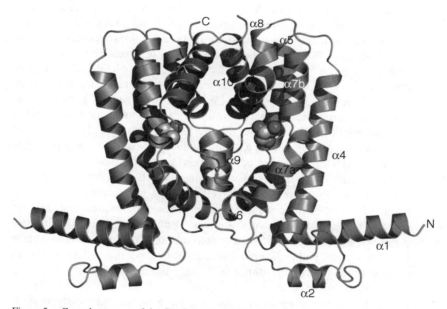

Figure 5. Crystal structure of the CmeR regulator. The dimeric structure of CmeR indicates that CmeR is an all-helical protein (α1-α10 and α1'-α10', respectively) which can easily be divided into two domains (the N-terminal DNA-binding and C-terminal ligand-binding domains). The bound glycerol molecule in each subunit of CmeR is represented as a hard-sphere model. (See insert for color representation.)

recognition α3 helix, presumably formed by residues 47–53, is replaced by an intriguing random coil (67). This is, perhaps, the most striking feature that distinguishes CmeR from the other TetR family members. To date, the CmeR regulator is the only case observed of a random coil replacing helix α3 in a TetR family member. Presumably, the TetR regulators possess a HTH DNA-binding motif formed by helices α2 and α3. Owing to its important role in recognizing target DNA, helix α3 is named the *recognition helix* (57). Since CmeR is a pleiotropic regulator of a large set of genes and is predicted to bind multiple operator sites, with many of those not being of the consensus IR sequence located in the promoter region of *cmeABC* (108). It could be postulated that the flexibility of the DNA-binding domain, illustrated by the random coil in place of the helix α3, permits CmeR to recognize multiple cognate DNA sites. One other unique feature of the CmeR structure is its large center-to-center distance between the two N-termini of the dimer. This center-to-center distance (according to the separation between Cα atoms of Y51 and Y51′) was measured to be 54 Å (67). The corresponding distances are 39 Å and 35 Å in the apo forms of QacR (65) and TetR (64). These center-to-center distances increase upon ligand binding. For the ligand-bound dimers of QacR (65), TetR (63), EthR (70), and YfiR (111), these distances become 41, 38, 52, and 54 Å, respectively. Thus, the relatively large center-to-center distance observed for CmeR reflected the fact that CmeR was liganded (67). Indeed, the crystal structure indicated that a fortuitous glycerol molecule was bound in each subunit of the CmeR dimer (Figure 5) (67). Although glycerol has not previously been identified as a natural inducer of *cmeABC*, it is plausible that it mimics the binding mode of other CmeR substrates.

The C-terminal domain of CmeR consists of helices α4 through α10, with helices α4, α5, α7, α8, and α10 forming an antiparallel five-helix bundle. In view of the crystal structure, helices α6, α8, α9, and α10 are involved in the formation of the dimer. Dimerization occurs mainly by couplings between pairs of helices (α6 and α9′, α8 and α10′, and their identical counter pairs). A surface area of 1950 Å^2 per monomer is buried in the contact region of the dimer (67). This interaction surface is mostly hydrophobic in character.

The C-terminal domain also forms a large tunnel-like cavity in each subunit of CmeR. This tunnel, surrounded by mostly hydrophobic residues of helices α4 to α9, opens horizontally from the front to the back of each protomer. The length of this tunnel is approximately 20 Å. Helices α7 and α8 from one subunit and α9′ from the other subunit of the regulator make the entrance of the tunnel. Helices $\alpha4$ to $\alpha6$, however, contribute to form the end of this hydrophobic tunnel. Each hydrophobic tunnel, occupying a volume of about $1000\,\text{Å}^3$, spans horizontally across the C-terminal domain and can be seen through from the front to the back of the dimer without obstruction. This unique feature, not found in other structures of the TetR family of regulators, highlights the flexibility of the CmeR regulator (67). As indicated above, the crystal structure of CmeR revealed the presence of a glycerol molecule inside this large ligand-binding tunnel (Figure 5). Glycerol binds identically in each subunit, as indicated by the crystallographic twofold symmetry of the CmeR dimer (67). This ligand-binding mode is different from that of QacR, in which one dimer of QacR binds one drug (65), but is similar to that of TetR, which interacts with tetracycline in the manner of a 1 : 1 monomer-to-drug molar ratio (63). The volume of the ligand-binding tunnel of CmeR is large enough to accommodate a few of the ligand molecules. Additional water molecules fill the portion of the large tunnel that is unoccupied by ligand. The structure suggests that CmeR might be able to bind more than one drug molecule at a time, or possibly to accommodate significantly larger ligands, which spans across the entire binding tunnel.

Indeed, a docking study showed that the hydrophobic tunnel of CmeR should be able to accommodate large, negatively charged bile acid molecules, such as taurocholate and cholate (67). The bound bile acids are predicted to anchor to several hydrophobic, polar, and positively charged residues, including H72, F99, F103, F137, S138, Y139, V163, C166, T167, K170, and H174. These anionic ligands were predicted to span almost the entire length of the ligand-binding tunnel of the regulator, respectively. The large tunnel, possibly consisting of multiple minipockets that may be employed to interact with different ligands, is rich in aromatic residues and contains three positively charged amino acids (two histidines and one lysine). It is very likely that these positively charged residues are crucial for CmeR to recognize negatively charged ligands. Site-directed mutagenesis is needed for an understanding of the detailed function of these charged residues.

V. Efflux Pumps of *Neisseria gonorrhoeae*: Repertoire and Contributions to Antimicrobial Resistance

The strict human pathogen *N. gonorrhoeae* expresses four drug efflux pumps (112), which belong to the resistance–nodulation–division (RND) family (MtrCDE), the major facilitator (MF) family (FarAB–MtrE), the ABC transporter family (MacAB), and the multidrug toxic compound extrusion (MATE) family (NorM). These pumps are also possessed by *N. meningitides* (112), but in this chapter we concentrate on their structure, function, and regulation in gonococci. In addition to these four efflux pumps, some clinical isolates have acquired the *mef* gene, which encodes a pump that recognizes macrolides (113).

There is evidence that gonococcal efflux pumps can contribute to levels of bacterial resistance to classical antibiotics since inactivation of efflux pump-encoding genes can enhance susceptibility to pump substrates (114–116). Moreover, mutations that increase efflux pump gene expression can also increase antimicrobial resistance of *N. gonorrhoeae*. From a clinical perspective, the important question is whether efflux pumps can influence the efficacy of antibiotic treatment. In this respect, work on the MtrCDE efflux pump in clinical isolates indicates that this is indeed the case. As an example, overexpression of the *mtrCDE* operon due to mutations in the *mtrR*-coding sequence, which encodes a repressor of *mtrCDE* expression (see below), or its promoter can provide gonococci with a twofold increase in resistance to penicillin (116). However, when strains have co-resident mutations in other chromosomal genes that influence the affinity of penicillin for penicillin-binding proteins (PBPs) or drug influx, resistance can become clinically significant ($\geq 2.0\,\mu g/mL$). The outbreak of penicillin-resistant gonorrhea that occurred in Durham, North Carolina in the 1980s (117), due to a strain (termed FA6140) that had *mtrR* mutations as well as other mutations that both decreased the binding of penicillin to PBP-1 and PBP-2 and the influx of penicillin (118), is an example of the impact that efflux can have on gonococcal resistance to antibiotics. Thus, while introduction of the *mtrR* mutations from penicillin-resistant strain FA6140 by transformation into highly penicillin-sensitive strain FA19 resulted in only a twofold increase in resistance, inactivation of the *mtrD* gene, which encodes the inner membrane RND transporter protein, in resistant strain FA6140 decreased resistance from 4 to $0.25\,\mu g/mL$. This decrease in resistance, due to the loss of efflux activity, was intriguing, as it represented a transition from clinical resistance to sensitivity and provides support for the notion that inhibitors of efflux pumps could reverse antibiotic resistance exhibited in pathogenic organisms. In addition to penicillin, gonococcal clinical isolates bearing *mtrR* mutations can express decreased susceptibility to macrolides and tetracycline (119). In fact, a cluster of azithromycin-resistant gonococci identified in a cohort of patients in Kansas City, Missouri were found to have a Correia insertion within the DNA sequence that intervenes the divergently transcribed *mtrR* and *mtrCDE* genes (120).

It has been suggested (112) that efflux pumps endow bacteria with the ability to resist natural or manufactured antimicrobial agents in their local environment and that such resistance is important for their survival in ecosystems. For strict human pathogens such as gonococci that do not naturally exist for long periods of time outside the human body, these antimicrobial agents would be compounds (e.g., antimicrobial peptides, long-chain fatty acids, bile salts, certain hormones) that are at the frontline of the innate host defense system. In this respect, the MtrCDE efflux pump appears to recognize certain antimicrobial peptides (121), progesterone (115) and bile salts (115), while the FarAB–MtrE pump recognizes long-chain fatty acids (122). In support of the hypothesis that efflux pumps can promote bacterial survival during infection, Jerse et al. (123) found that the MtrCDE efflux pump is required for survival of gonococci in the lower genital tract of experimentally infected female mice (115). More recently, this group reported (124, 125) that the degree of in vivo fitness expressed by gonococci is related to the presence of the MtrR repressor or MtrA activator, which modulate levels of *mtrCDE* expression (see below). This is a unique example of how a mechanism of antibiotic resistance can actually increase in vivo fitness and is probably due to the ability of the MtrC–MtrD–MtrE pump to recognize both classical antibiotics (e.g., penicillin) and host-derived antimicrobials.

A. STRUCTURE OF THE MtrCDE EFFLUX SYSTEM

The *mtr* (multiple transferable resistance) system was first identified by Maness and Sparling (126) when they isolated a spontaneous mutant that exhibited increased resistance to multiple structurally diverse antimicrobial hydrophobic compounds. It was originally thought that *mtr* modified outer membrane permeability (127). However, subsequent cloning and sequencing experiments (114, 115, 128, 129) showed that the mutation was located within a gene encoding a transcriptional repressor (MtrR) of a downstream but transcriptionally divergent operon (*mtrCDE*) encoding the tripartite MtrCDE efflux pump. Similar to other RND-type pumps of gram-negative bacteria, the three proteins are a cytoplasmic membrane transporter (MtrD), a membrane fusion protein (MtrC), and an outer membrane channel protein (MtrE). Directly or indirectly, other proteins also participate in efflux mediated by the pump. Veal and Shafer (130) identified an accessory protein (MtrF) which, for reasons that are not yet clear, is required for efflux activity when the host strain is expressing high levels of the pump during stressful conditions. Energy supplied by the TonB–ExbB–ExbD system is also needed for inducible antimicrobial resistance mediated by MtrC–MtrD–MtrE (131).

B. REGULATION OF THE *mtrCDE* SYSTEM

Expression of the *mtrCDE* operon is controlled by both *cis-* and *trans-*acting regulatory elements that negatively or positively control production of the pump proteins at the level of transcription. The DNA-binding proteins that are involved in regulation of this system and their biological importance are described in Figure 6. The degree of *mtrCDE* gene expression that is controlled by these elements corresponds to constitutive or inducible levels of gonococcal resistance to antimicrobials recognized by the pump. The important *trans*-acting factors that control *mtrCDE* expression directly or indirectly in gonococci are MtrR, MpeR, and MtrA (Figure 6). Recent studies indicate (132) that the two-component regulatory system termed MisR–MisS (133) also controls *mtrR* expression and, as a consequence, *mtrCDE*. In meningococci, regulation of *mtrCDE* appears to be controlled by a Correia element (CE) within the *mtrR* and *mtrCDE* intervening region that contains a binding site for integration host factor (134).

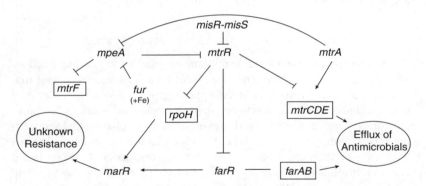

Figure 6. Proposed model of the regulatory system in *N. gonorrhoeae.* Shown is a proposed regulatory system in *N. gonorrhoeae* strain FA19 that affects levels of antimicrobial resistance via drug efflux pumps. The lines with bars indicate transcriptional repression; the lines with arrows indicate transcriptional activation.

This element was found to be similarly positioned in azithromycin-resistant gonococcal clinical isolates in Kansas City, Missouri (120), but the presence of the CE at this site is apparently unique to these strains.

The MtrR repressor, which is similar to the TetR family of repressors (128), binds as two homodimers to a DNA sequence that lies just upstream of the *mtrCDE* operon (135). The 15-bp MtrR-binding site is within the *mtrCDE* promoter and is characterized by two pseudo-direct 7-bp repeats. Mutations in the *mtrR* helix–turn–helix coding region that reduce or abrogate DNA binding, as well as missense mutations in the C-terminal-encoding region that probably impact homodimer formation, can enhance transcription of *mtrCDE* and elevate gonococcal resistance to antimicrobials (114, 136, 137). This regulation has important biological consequences. First, by dampening *mtrCDE* expression levels of gonococcal, resistance to hydrophobic antimicrobials is decreased. Second, and probably linked to the first point, production of an active MtrR decreases gonococcal fitness in vivo (but not in vitro), as assessed by the use of an experimental murine model of lower genital tract infection (124, 125). Expression of *mtrR* seems to be regulated by both direct and indirect processes involving DNA-binding proteins (Figure 6) and a *cis*-acting 13-bp inverted repeat within the *mtrR* promoter (see below). We reported on an AraC-like transcriptional regulator that represses *mtrF* (Figure 6) expression and termed the protein MpeR (138). In the same year, Dyer's group reported that *mpeR* (NGO0025) is maximally expressed under iron-limiting conditions (139), and recent studies showed that iron-replete conditions dampen *mtrR* expression by an MpeR-dependent mechanism during late-log (140), a phase of growth likely to be iron limiting. It may be that *mpeR* expression is negatively controlled by the ferric uptake regulator (Fur) in the presence of iron (141), which provides a connection between iron limitation and the regulation of genes encoding transcriptional modulators of efflux pump genes. *mtrR* expression is also negatively regulated by a two-component regulatory system termed MisR–MisS (142) that is functionally similar (133), but not identical, to the PhoP–PhoQ system in *Salmonella enterica* serovar *typhimurium*, which is known to modulate levels of bacterial susceptibility to cationic APs (143).

In addition to regulating *mtrCDE*, we found (144) through microarray analysis that MtrR can positively or negatively regulate the expression of nearly 70 genes, including genes known or presumed to be involved in the regulation of antimicrobial resistance (*marR* and *farR*), response to stress conditions (*rpoH*), polyamine uptake (*potF*), glutamine biosynthesis (*glnA and glnE*) and transport (*glnM* and *glnQ*), peroxide detoxification (*ccp*), and sodium-glutamate symporter activity (*gltS*). FarR is a transcriptional repressor of the *farAB* efflux pump system (145), which exports antimicrobial fatty acids (122), but can also activate *glnA* and *marR* (Figure 6). Thus, we propose that a regulatory circuit exists in gonococci as demonstrated in Figure 6, which controls not only levels of bacterial resistance to host antimicrobials and antibiotics, but also expression levels of important metabolic processes (e.g., glutamine biosynthesis).

Thus, apart from its ability to control levels of MtrCDE, MtrR may control other processes important for basic metabolism and pathogenesis. Given this hypothesis, it is fair to ask: Why did evolution not select against *mtrR* if its loss increases fitness during infection? We believe that the answer is, in part, because MtrR can directly or indirectly activate certain genes important for metabolism and other cellular processes. For example, MtrR appears to up-regulate expression of *glnE*, which encodes the enzymatic regulator of glutamine synthetase (GlnA). Thus, the initial advantage afforded by loss of MtrR (resistance to host antimicrobials) might be negated later during infection if biosynthetic processes are affected negatively due to loss of MtrR. This, in part, may explain why natural *mtrR* mutants represent <25% of all clinical isolates (112) and why *mtrR* expression is subject to transcriptional regulatory processes.

While *mtrR*-coding mutations can enhance transcription of *mtrCDE* and elevate HA resistance in gonococci, high-level resistance requires a *cis*-acting mutation within the overlapping *mtrR* and *mtrCDE* promoters (114, 133, 146). The *mtrR* promoter contains a 13-bp inverted repeat sequence between the −10 and −35 hexamers (114, 146). A single bp deletion (114, 146) or a dinucleotide insertion within this inverted repeat is sufficient to reduce transcription of *mtrR* dramatically. These *cis*-acting promoter mutations are observed in clinical isolates expressing high levels of HA resistance. We proposed (146) that because the *mtrR* and *mtrCDE* promoters overlap on opposite strands at their −35 hexamer region, the single bp deletion or the dinucleotide insertion enhance *mtrCDE* expression both by reducing *mtrR* expression and by making the *mtrCDE* promoter more available for interacting with RNA polymerase. This model provides a mechanism to explain why *mtrR* promoter mutants express higher levels of antimicrobial resistance and *mtrCDE* than those of strains with *mtrR* point mutations that cause radical amino acid changes, inactivating MtrR function (114, 125, 137). High-level antimicrobial resistance mediated by overexpression of the *mtrCDE* operon can also be afforded by a point mutation 120 nucleotides upstream of the *mtrC* translational start (125). This mutation, found in strain MS11, when transferred to strain FA19 could result in levels of antimicrobial resistance and in vivo fitness similar to that endowed by the single bp deletion in the *mtrR* promoter. Warner et al. (125) found that this point mutation could enhance the half-life of the *mtrCDE* transcript, and recent evidence (147) indicates that it generates a new promoter element that is used preferentially for *mtrCDE* transcription.

Transcriptional activation of efflux pump genes is common in bacteria and can lead to inducible resistance to antibiotics and other antimicrobials recognized by the specific pump (112). For example, environmental and antimicrobial stimuli have been shown to modulate expression of the *acrAB* efflux pump system in *E. coli* through the action of several DNA-binding proteins (MarR, Rob, and SoxS) (148). This inducible resistance process is typically transient and allows bacteria to respond quickly to antimicrobials in their local environment. Expression of the *mtrCDE* efflux pump locus can be up-regulated when gonococci are grown in the presence of a sublethal level of an antimicrobial agent recognized by the pump (149). This induction requires the presence of an AraC/XylS-like protein termed MtrA (149) and energy supplied by the TonB–ExbB–ExbD system (131). MtrA can activate the expression of *mtrCDE* even in the presence of an active MtrR, and we (150) found that it can bind to a DNA sequence upstream of *mtrCDE*. Moreover, MtrA can bind certain substrates (e.g., TX-100) that induce expression of the *mtrCDE* operon, and these interactions enhance its binding to target DNA sequences. Like MtrR, MtrA can regulate directly or indirectly (positively or negatively) several genes in gonococci, including *mpeR* (Figure 6). The capacity of MtrA to regulate gonococcal genes is probably of importance during infection because an *mtrA* null mutant of strain FA19 was less fit than its *mtrA*$^{+}$ parent strain in the lower genital tract of experimentally infected female mice.

VI. Concluding Remarks

It has been well established that overexpression of RND multidrug efflux pumps led to a resistant phenotype in pathogenic organisms. This problem is exacerbated by the ease with which many of these resistant genes can pass from one microorganism to another through plasmid transfer (151). The availability of the three-dimensional structures of these efflux transporters potentially allows us rationally to design agents that block the function of these pumps. However, there is still quite a mountain to climb in the development toward achieving this goal. To date, no efflux pump inhibitor has been licensed for use in the clinical treatment of bacterial infections. As RND pumps assemble as tripartite complexes, one method to inhibit the drug-resistant phenotype is to prevent the assembly of complex formation by blocking the different subunits from forming a functional complex. The possibility of this approach has been demonstrated by the crystal structure of AcrB in complex with the designed ankyrin repeat protein inhibitor (DARPin), in which the inhibitor is bound in such a way that AcrB is not able to form a functional complex with TolC (33). Another potential approach is to target the transcriptional regulators that modulate the expression of these RND pumps. For example, the global regulator CmeR controls the expression of several genes, including *cmeABC* (108). By inhibiting the interactions between the regulator and its substrates, expression of these transporter genes could potentially be blocked. Recently, more examples of RND transporters regulated by two-component systems have been identified, including *E. coli* CusRS, which controls the expression of CusABC (85). The structures of different components of these regulatory systems may potentially allow us to rationally design inhibitors that reduce the level of efflux pump expression.

Acknowledgements

This work is supported by NIH grants R01DK063008 (Q.Z.), R01AI021150 (W.M.S.), R01GM074027 (E.W.Y.), R01GM086431 (E.W.Y.). W. Shafer is supported by a Senior Research Career Scientist Award from the VA Medical Research Service, Y. Zalucki is supported by a C. J. Martin Biomedical Post-doctoral Fellowship from the National Health and Medical Research Council of Australia.

References

1. Barber, M. (1947) Staphylococcal infection due to penicillin-resistant strains, *Br. Med. J.* 29, 863–865.
2. Thompson, K. S., and Smith, M. E. (2000) Version 2000: the new β-lactamases of gram-negative bacteria at the dawn of the new millennium, *Microbes Infect.* 2, 1225–1235.
3. Wright, G. D. (1999) Aminoglycoside-modifying enzymes, *Curr. Opin. Microbiol.* 2, 499–503.

4. Hooper, D. C. (2000) Mechanisms of action and resistance of older and newer fluoroquinolones, *Clin. Infect. Dis. 31*, S24–S28.

5. Piddock, L. J. V. (2006) Clinically relevant bacterial chromosomally encoded multi-drug resistance efflux pumps, *Clin. Microbiol. Rev. 19*, 382–402.

6. Levy, S. (1992) Active efflux mechanisms for antibiotic resistance, *Antimicrob. Agents Chemother. 36*, 695–703.

7. Bolhuis, H., van Veen, H. W., Poolman, B., Driessen, A. J., and Konings, W. N. (1997) Mechanisms of multidrug transporters, *FEMS Microbiol. Rev. 21*, 55–84.

8. Nikaido, H. (1994) Prevention of drug access to bacterial targets: permeability barriers and active efflux, *Science 264*, 382–388.

9. McMurry, L., Petrucci, R. E., Jr., and Levy, S. B. (1980) Active efflux of tetracycline encoded by four genetically different tetracycline resistance determinants in *Escherichia coli*, *Proc. Natl. Acad. Sci. USA 77*, 3974–3977.

10. Higgins, C. F. (1992) ABC transporters: from microorganisms to man, *Annu. Rev. Cell Biol. 8*, 67–113.

11. Tseng, T. T., Gratwick, K. S., Kollman, J., Park, D., Nies, D. H., Goffeau, A., and Saier, M. H., Jr., (1999) The RND permease superfamily: an ancient, ubiquitous and diverse family that includes human disease and development proteins, *J. Mol. Microbiol. Biotechnol. 1*, 107–125.

12. Brown, M. H., Paulsen, L. T., and Skurray, R. A. (1999) The multidrug efflux protein NorM is a prototype of a new family of transporters, *Mol. Microbiol. 31*, 394–395.

13. Morita, Y., Kodama, K., Shiota, S., Mine, T., Kataoka, A., Mizushima, T., and Tsuchiya, T. (1998) NorM, a putative multidrug efflux protein, of *Vibrio parahaemolyticus* and its homolog in *Escherichia coli*, *Antimicrob. Agents Chemother. 42*, 1778–1782.

14. Griffith, J. K., Baker, M. E., Rouch, D. A., Page, M. G., Skurray, R. A., Paulsen, I. T., Chater, K. F., Baldwin, S. A., and Henderson, P. J. (1992) Membrane transport proteins: implications of sequence comparisons, *Curr. Opin. Cell Biol. 4*, 684–695.

15. Marger, M., and Saier, M. H., Jr. (1993) A major superfamily of transmembrane facilitators that can catalyze uniport, symport and antiport, *Trends Biochem. Sci. 18*, 13–20.

16. Pao, S. S., Paulsen, I. T., and Saier, M. H., Jr. (1998) Major facilitator superfamily, *Microbiol. Mol. Biol. Rev. 62*, 1–34.

17. Paulsen, I. T., Skurry, R. A., Tam, R., Saier, M. H., Jr., Turner, R. J., Weiner, J. H., Goldberg, E. B., and Grinius, L. L. (1996) The SMR family: a novel family of multidrug efflux proteins involved with the efflux of lipophilic drugs, *Mol. Microbiol. 19*, 1167–1175.

18. Schweizer, H. P. (2003) Efflux as a mechanism of resistance to antimicrobials in *Pseudomonas aeruginosa* and related bacteria: unanswered questions, *Genet. Mol. Res. 31*, 48–62.

19. Nikaido, H. (1998) Antibiotic resistance caused by gram-negative multidrug efflux pumps, *Clin. Infect. Dis. 27*, S32–S41.

20. Paulsen, I. T., Sliwinski, M. K., and Saier, M. H., Jr. (1998) Microbial genome analyses: global comparisons of transport capabilities based on phylogenies, bioenergetics and substrate specificities, *J. Mol. Biol. 277*, 573–592.

21. Nishino, K., and Yamaguchi, A. (2001) Analysis of a complete library of putative drug transporter genes in *Escherichia coli*, *J. Bacteriol. 183*, 5803–5812.

22. Nikaido, H. (1996) Multidrug efflux pumps of gram-negative bacteria, *J. Bacteriol. 178*, 5853–5859.

23. Zgurskaya, H. I., and Nikaido, H. (2000) Multidrug resistance mechanisms: drug efflux across two membranes, *Mol. Microbiol. 37*, 219–225.

24. Zgurskaya, H. I., and Nikaido, H. (2000) Cross-linked complex between oligomeric periplasmic lipoprotein AcrA and the inner-membrane-associated multidrug efflux pump AcrB from *Escherichia coli*, *J. Bacteriol. 182*, 4264–4267.

25. Mikolosko, J., Bobyk, K., Zgurskaya, H. I., and Ghosh, P. (2006) Conformational flexibility in the multidrug efflux system protein AcrA, *Structure 14*, 577–587.

26. Fralick, J. A. (1996) Evidence that TolC is required for functioning of the Mar/AcrAB efflux pump of *Escherichia coli*, *J. Bacteriol. 178*, 5803–5805.

27. Koronakis, V., Sharff, A., Koronakis, E., Luisi, B., and Hughes, C. (2000) Crystal structure of the bacterial membrane protein TolC central to multidrug efflux and protein export, *Nature 405*, 914–919.

28. Ma, D., Cook, D. N., Alberti, M., Pon, N. G., Nikaido, H., and Hearst, J. E. (1995) Genes *acrA* and *acrB* encode a stress-induced efflux system of *Escherichia coli*, *Mol. Microbiol. 16*, 45–55.

29. (a) Takatsuka, Y., and Nikaido, H. (2006) Threonine-978 in the transmembrane segment of the multidrug efflux pump AcrB of *Escherichia coli* is crucial for drug transporter as a probable component of the proton relay network, *J. Bacteriol. 188*, 7284–7289. (b) Su, C.-C., Li, M., Gu, R., Takatsuka, Y., McDermott, G., Nikaido, H., and Yu, E. W. (2006) Conformation of the AcrB multidrug efflux pump in mutants of the putative proton relay pathway, *J. Bacteriol. 188*, 7290–7296.

30. Zgurskaya, H. I., and Nikaido, H. (1999) Bypassing the periplasm: reconstitution of the AcrAB multidrug efflux pump of *Escherichia coli*, *Proc. Natl. Acad. Sci. USA 96*, 7190–7196.

31. Baucheron, S., Tyler, S., Boyd, D., Mulvey, M. R., Chaslus-Dancla, E., and Cloeckaert, A. (2004) AcrAB–TolC directs efflux-mediated multidrug resistance in *Salmonella enterica* serovar *typhimurium* DT104, *Antimicrob. Agents Chemother. 48*, 3729–3735.

32. Bratu, S., Landman, D., George, A., Salvani, J., and Quale, J. (2009) Correlation of the expression of *acrB* and the regulatory genes *marA*, *soxS* and *ramA* with antimicrobial resistance in clinical isolates of *Klebsiella pneumoniae* endemic to New York City, *J. Antimicrob. Chemother.* May 21, e-pub ahead of print.

33. Okusu, H., Ma, D., and Nikaido, H. (1996) AcrAB efflux pump plays a major role in the antibiotic resistance phenotype of *Escherichia coli* multiple-antibiotic-resistance (Mar) mutants, *J. Bacteriol. 178*, 306–308.

34. Webber, M. A., Talukder, A., and Piddock, L. J. (2005) Contribution of mutation at amino acid 45 of AcrR to acrB expression and ciprofloxacin resistance in clinical and veterinary *Escherichia coli* isolates, *Antimicrob. Agents Chemother. 49*, 4390–4392.

35. Murakami, S., Nakashima, R., Yamashita, E., and Yamaguchi, A. (2002) Crystal structure of bacterial multidrug efflux transporter AcrB, *Nature 419*, 587–593.

36. Yu, E. W., McDermott, G., Zgurskaya, H. I., Nikaido, H., and Koshland, D. E., Jr. (2003) Structural basis of multiple drug-binding capacity of the AcrB multidrug efflux pump, *Science 300*, 976–980.

37. Murakami, S., Nakashima, R., Yamashita, E., Matsumoto, T., and Yamaguchi, A. (2006) Crystal structures of a multidrug transporter reveal a functionally rotating mechanism, *Nature 443*, 173–179.

38. Seeger, M. A., Schiefner, A., Eicher, T., Verrey, F., Diederichs, K., and Pos, K. M. (2006) Structural asymmetry of AcrB trimer suggests a peristaltic pump mechanism, *Science 313*, 1295–1298.

39. Sennhauser, G., Amstutz, P., Briand, C., Storchenegger, O., and Grütter, M. G. (2007) Drug export pathway of multidrug exporter AcrB revealed by DARPin inhibitors, *PLoS Biol. 5*, e7.

40. Mikolosko, J., Bobyk, K., Zgurskaya, H. I., and Ghosh, P. (2006) Conformational flexibility in the multidrug efflux system protein AcrA, *Structure 14*, 577–587.

41. Sennhauser, G., Bukowska, M. A., Briand, C., and Grütter, M. G. (2009) Crystal structure of the multidrug exporter MexB from *Pseudomonas aeruginosa*, *J. Mol. Biol. 389*, 134–145.

42. Akama, H., Kanemaki, M., Yoshimura, M., Tsukihara, T., Kashiwagi, T., and Yoneyama, H., Narita, S., Nakagawa, A., and Nakae, T. (2004) Crystal structure of the drug discharge outer membrane protein, OprM, of *Pseudomonas aeruginosa*: dual modes of membrane anchoring and occluded cavity end, *J. Biol. Chem. 279*, 52816–52819.

43. Higgins, M. K., Bokma, E., Koronakis, E., Hughes, C., and Koronakis, V. (2004) Structure of the periplasmic component of a bacterial drug efflux pump, *Proc. Natl. Acad. Sci. USA 101*, 9994–9999.

44. Akama, H., Matsuura, T., Kashiwagi, S., Yoneyama, H., Narita, S., Tsukihara, T., Nakagawa, A., and Nakae, T. (2004) Crystal structure of the membrane fusion protein, MexA, of the multidrug transporter in *Pseudomonas aeruginosa*, *J. Biol. Chem. 279*, 25939–25942.

45. Symmons, M. F., Bokma, E., Koronakis, E., Hughes, C., and Koronakis, V. (2009) The assembled structure of a complete tripartite bacterial multidrug efflux pump, *Proc. Natl. Acad. Sci. USA 106*, 7173–7178.

46. Elkins, C. A., and Nikaido, H. (2003) Chimeric analysis of AcrA function reveals the importance of its C-terminal domain in its interaction with the AcrB multidrug efflux pump, *J. Bacteriol. 185*, 5349–5356.

47. Su, C.-C., Yang, F., Long, F., Reyon, D., Routh, M. D., Kuo, D. W., Mokhtari, A. K., Van Ornam, J. D., Rabe, K. L., Hoy, J. A., et al. (2009) Crystal structure of the membrane fusion protein CusB from *Escherichia coli*, *J. Mol. Biol. 393*, 342–355.

48. Yu, E. W., Aires, J. R., McDermott, G., and Nikaido, H. (2005) A periplasmic drug-binding site of the AcrB multidrug efflux pump: a crystallographic and site-directed mutagenesis study, *J. Bacteriol. 187*, 6804–6815.

49. Klepsch, M. M., Newstead, S., Flaig, R., De Gier, J. W., Iwata, S., and Beis, K. (2008) The structure of the efflux pump AcrB in complex with bile acid. *Mol. Membr. Biol. 25*, 677–682.

50. Eckert, B., and Beck, C. F. (1989) Overproduction of transposon Tn10-encoded tetracycline resistance protein results in cell death and loss of membrane potential, *J. Bacteriol. 171*, 3557–3559.

51. Kurland, C. G., and Dong, H. (1996) Bacterial growth inhibition by overproduction of protein, *Mol. Microbiol. 21*, 1–4.

52. Lee, S. W., and Edlin, G. (1985) Expression of tetracycline resistance in pBR322 derivatives reduces the reproductive fitness of plasmid-containing *Escherichia coli*, *Gene 39*, 173–180.

53. Rosenberg, E. Y., Bertenthal, D., Nilles, M. L., Bertrand, K. P., and Nikaido, H. (2003) Bile salts and fatty acids induce the expression of *Escherichia coli* AcrAB multidrug efflux pump through their interaction with Rob regulatory protein, *Mol. Microbiol. 48*, 1609–1619.

54. Martin, R. G., and Rosner, J. L. (2003) Analysis of microarray data for the *marA, soxS*, and *rob* regulons of *Escherichia coli*, *Methods Enzymol. 370*, 278–280.

55. Zhang, A., Rosner, J. L., and Martin, R. G. (2008) Transcriptional activation by MarA, SoxS and Rob of two *tolC* promoters using one binding site: a complex promoter configuration for *tolC* in *Escherichia coli*, *Mol. Microbiol. 69*, 1450–1455.

56. Ma, D., Alberti, M., Lynch, C., Nikaido, H. and Hearst, J. E. (1996). The local repressor AcrR plays a moderating role in the regulation of *acrAB* genes of *Escherichia coli* by global stress signals, *Mol. Microbiol. 19*, 101–112.

57. Ramos, J. L., Martinez-Bueno, M., Molina-Henares, A. J., Teran, W., Watanabe, K., Zhang, X. D., Gallegos, M. T., Brennan, R., and Tobes, R. (2005) The TetR family of transcriptional repressors, *Microbiol. Mol. Biol. Rev. 69*, 326–356.

58. Rodionov, D. A., Gelfand, M. S., Mironov, A. A., and Rakhmaninova, A. B. (2001) Comparative approach to analysis of regulation in complete genomes: multidrug resistance systems in gramma-proteobacteria, *J. Mol. Microbiol. Biotechnol. 3*, 319–324.

59. Su, C.-C., Rutherford, D. J., and Yu, E. W. (2007) Characterization of the multidrug efflux regulator AcrR from *Escherichia coli*, *Biochem. Biophys. Res. Commun. 361*, 85–90.

60. Su, C.-C., and Yu, E. W. (2007) Ligand–transporter interaction in the AcrB multidrug pump determined by fluorescence polarization assay, *FEBS Lett. 581*, 4972–4976.

61. Li, M., Gu, R., Su, C.-C., Routh, M. D., Harris, K. C., Jewell, E. S., McDermott, G., and Yu, E. W. (2007) Crystal structure of the transcriptional regulator AcrR from *Escherichia coli*, *J. Mol. Biol. 374*, 591–603.

62. Gu, R., Li, M., Su, C.-C., Long, F., Routh, M. D., Yang, F., McDermott, G., and Yu, E. W. (2008) Conformational change of the AcrR regulator reveals a possible mechanism of induction, *Acta Crystallogr. F64*, 584–588.

63. Hinrichs, W., Kisker, C., Duvel, M., Muller, A., Tovar, K., Hillen, W., and Saenger, W. (1994) Structure of the Tet repressor–tetracycline complex and regulation of antibiotic resistance, *Science 264*, 418–420.

64. Orth, P., Schnappinger, D., Hillen, W., Saenger, W., and Hinrichs, W. (2000) Structural basis of gene regulation by the tetracycline inducible Tet repressor–operator system, *Nat. Struct. Biol. 7*, 215–219.

65. Schumacher, M. A., Miller, M. C., Grkovic, S., Brown, M. H., Skurray, R. A., and Brennan, R. G. (2001) Structural mechanisms of QacR induction and multidrug recognition, *Science 294*, 2158–2163.

66. Schumacher, M. A., Miller, M. C., Grkovic, S., Brown, M. H., Skurray, R. A., and Brennan, R. G. (2002) Structural basis for cooperative DNA binding by two dimers of the multidrug-binding protein QacR, *EMBO J. 21*, 1210–1218.

67. Gu, R., Su, C.-C., Shi, F., Li, M., McDermott, G., Zhang, Q., and Yu, E. W. (2007) Crystal structure of the transcriptional regulator CmeR from *Campylobacter jejuni*, *J. Mol. Biol. 372*, 583–593.

68. Natsume, R., Ohnishi, Y., Senda, T., and Horinouchi, S. (2003) Crystal structure of a γ-butyrolactone autoregulator receptor protein in *Streptomyces coelicolor* A3(2), *J. Mol. Biol. 336*, 409–419.

69. Dover, L. G., Corsino, P. E., Daniels, I. R., Cocklin, S. L., Tatituri, V., Besra, G. S., and Futterer, K. (2004) Crystal structure of the TetR/CamR family repressor *Mycobacterium tuberculosis* EthR implicated in ethionamide resistance, *J. Mol. Biol. 340*, 1095–1105.

70. Frenois, F., Engohang-Ndong, J., Locht, C., Baulard, A. R., and Villeret, V. (2004) Structure of EthR in a ligand bound conformation reveals therapeutic perspectives against tuberculosis, *Mol. Cell. 16*, 301–307.

71. Routh, M. D., Su, C.-C., Zhang, Q., and Yu, E. W. (2009) Structures of AcrR and CmeR: insight into the mechanisms of transcriptional repression and multi-drug recognition in the TetR family of regulators, *Biochim. Biophys. Acta 1794*, 844–851.

72. Nies, D. H. (2003) Efflux-mediated heavy metal resistance in prokaryotes, *FEMS Microbiol. Rev. 27*, 313–339.

73. Zgurskaya, H., and Nikaido, H. (1999) Bypassing the periplasm: reconstitution of the AcrAB multidrug efflux pump of *Escherichia coli*, *Proc. Natl. Acad. Sci. USA 96*, 7190–7195.

74. Aires, J. R., and Nikaido, H. (2005) Aminoglycosides are captured from both periplasm and cytoplasm by the AcrD multidrug efflux transporter of *Escherichia coli*, *J. Bacteriol. 187*, 1923–1929.

75. Lau, S. Y., and Zgurskaya, H. I. (2005) Cell division defects in *Escherichia coli* deficient in the multidrug efflux transporter AcrEF–TolC, *J. Bacteriol. 187*, 7815–7825.

76. Baranova, N., and Nikaido, H. (2002) The BaeSR two-component regulatory system activates transcription of *yegMNOB* (*mdtABCD*) transporter gene cluster in *Escherichia coli* and increases its resistance to novobiocin and deoxycholate, *J. Bacteriol. 184*, 4168–4176.

77. Nagakubo, S., Nishino, K., Hirata, T., and Yamaguchi, A. (2002) The putative response regulator BaeR stimulates multidrug resistance of *Escherichia coli* via a novel multidrug exporter system, MdtABC, *J. Bacteriol. 184*, 4161–4167.

78. Elkins, C. A., and Mullis, L. B. (2006) Mammalian steroid hormones are substrates for the major RND- and MFS-type tripartite multidrug efflux pumps of *Escherichia coli*, *J. Bacteriol. 188*, 1191–1195.

79. Franke, S., Grass, G., and Nies, D. H. (2001) The product of the *ybdE* gene of the *Escherichia coli* chromosome is involved in detoxification of silver ions, *Microbiology 147*, 965–972.

80. Franke, S., Grass, G., Rensing, C., and Nies, D. H. (2003) Molecular analysis of the copper-transporting efflux system CusCFBA of *Escherichia coli*, *J. Bacteriol.* 185, 3804–3812.

81. Loftin, I. R., Franke, S., Roberts, S. A., Weichse, A., Heroux, A., Montfort, W. R., Rensing, C., and McEvoy, M. M. (2005) A novel copper-binding fold for the periplasmic copper resistance protein CusF, *Biochemistry* 44, 10533–10540.

82. Xue, Y., Davis, A. V., Balakrishnan, G., Stasser, J. P., Staehlin, B. M., Focia, P., Spiro, T. G., Penner-Hahn, J. E., and O'Halloran, T. V. (2008) Cu(I) recognition via cation-π and methionine interactions in CusF, *Nat. Chem. Biol.* 4, 107–109.

83. Loftin, I. R., Franke, S., Blackburn, N. J., and McEvoy, M. M. (2007) Unusual Cu(I)/Ag(I) coordination of *Escherichia coli* CusF as revealed by atomic resolution crystallography and x-ray absorption spectroscopy, *Protein Sci.* 16, 2287–2293.

84. Bagai, I., Rensing, C., Blackburn, N. J., and McEvoy, M. M. (2008) Direct metal transfer between periplasmic proteins identifies a bacterial copper chaperone, *Biochemistry* 47, 11408–11414.

85. Munson, G. P., Lam, D. L., Outten, F. W., and O'Halloran, T. V. (2000) Identification of a copper-responsive two-component system on the chromosome of *Escherichia coli* K-12, *J. Bacteriol.* 182, 5864–5871.

86. Borges-Walmsley, M. I., Beauchamp, J., Kelly, S. M., Jumel, K., Candlish, D., Harding, S. E., Price, N. C., and Walmsley, A. R. (2003) Identification of oligomerization and drug-binding domains of the membrane fusion protein EmrA, *J. Biol. Chem.* 278, 12903–12912.

87. Bagai, I., Liu, W., Rensing, C., Blackburn, N. J., and McEvoy, M. M. (2007) Substrate-linked conformational change in the periplasmic component of a Cu(I)/Ag(I) efflux system, *J. Biol. Chem.* 282, 35695–35702.

88. Ruiz-Palacios, G. M. (2007) The health burden of *Campylobacter* infection and the impact of antimicrobial resistance: playing chicken, *Clin. Infect. Dis.* 44, 701–703.

89. van Doorn, P. A., Ruts, L., and Jacobs, B. C. (2008) Clinical features, pathogenesis, and treatment of Guillain–Barré syndrome, *Lancet Neurol.* 7, 939–950.

90. Blaser, M. J., and Engberg, J. (2008) Clinical aspects of *Campylobacter jejuni* and *Campylobacter coli* infections, in *Campylobacter*, 3rd ed. Nachamkin, I., Szymanski, C. M., and Blaser, M. J., Eds., ASM Press, Washington, DC, pp. 99–121.

91. Engberg, J., Aarestrup, F. M., Taylor, D. E., Gerner-Smidt, P., and Nachamkin, I. (2001) Quinolone and macrolide resistance in *Campylobacter jejuni* and *C. coli:* resistance mechanisms and trends in human isolates, *Emerg. Infect. Dis.* 7, 24–34.

92. Gibreel, A., and Taylor, D. E. (2006) Macrolide resistance in *Campylobacter jejuni* and *Campylobacter coli*, *J. Antimicrob. Chemother.* 58, 243–255.

93. Luangtongkum, T., Jeon, B., Han, J., Plummer, P., Logue, C. M., and Zhang, Q. (2009) Antibiotic resistance in *Campylobacter:* emergence, transmission and persistence, *Future Microbiol.* 4, 189–200.

94. Zhang, Q., and Plummer, P. (2008) Mechanisms of antibiotic resistance in *Campylobacter*, in *Campylobacter*, 3rd ed. Nachamkin, I., Szymanski, C. M., and Blaser, M. J., Eds., ASM Press, Washington, DC, pp. 263–276.

95. Lin, J., Michel, L. O., and Zhang, Q. (2002) CmeABC functions as a multidrug efflux system in *Campylobacter jejuni*, *Antimicrob. Agents Chemother. 46*, 2124–2131.

96. Pumbwe, L., and Piddock, L. J. (2002) Identification and molecular characterisation of CmeB, a *Campylobacter jejuni* multidrug efflux pump, *FEMS Microbiol. Lett. 206*, 185–189.

97. Luo, N., Sahin, O., Lin, J., Michel, L. O., and Zhang, Q. (2003) In vivo selection of *Campylobacter* isolates with high levels of fluoroquinolone resistance associated with *gyrA* mutations and the function of the CmeABC efflux pump, *Antimicrob. Agents Chemother. 47*, 390–394.

98. Cagliero, C., Mouline, C., Payot, S., and Cloeckaert, A. (2005) Involvement of the CmeABC efflux pump in the macrolide resistance of *Campylobacter coli*, *J. Antimicrob. Chemother. 56*, 948–950.

99. Mamelli, L., Prouzet-Mauleon, V., Pagès, J. M., Megraud, F., and Bolla, J. M. (2005) Molecular basis of macrolide resistance in *Campylobacter:* role of efflux pumps and target mutations, *J. Antimicrob. Chemother. 56*, 491–497.

100. Ge, B., McDermott, P. F., White, D. G., and Meng, J. (2005) Role of efflux pumps and topoisomerase mutations in fluoroquinolone resistance in *Campylobacter jejuni* and *Campylobacter coli*, *Antimicrob. Agents Chemother. 49*, 3347–3354.

101. Lin, J., Yan, M., Sahin, O., Pereira, S., Chang, Y. J., and Zhang, Q. (2007) Effect of macrolide usage on emergence of erythromycin-resistant *Campylobacter* isolates in chickens, *Antimicrob. Agents Chemother. 51*, 1678–1686.

102. Cagliero, C., Mouline, C., Cloeckaert, A., and Payot, S. (2006) Synergy between efflux pump CmeABC and modifications in ribosomal proteins L4 and L22 in conferring macrolide resistance in *Campylobacter jejuni* and *Campylobacter coli*, *Antimicrob. Agents Chemother. 50*, 3893–3896.

103. Yan, M., Sahin, O., Lin, J., and Zhang, Q. (2006) Role of the CmeABC efflux pump in the emergence of fluoroquinolone-resistant *Campylobacter* under selection pressure, *J. Antimicrob. Chemother. 58*, 1154–1159.

104. Akiba, M., Lin, J., Barton, Y. W., and Zhang, Q. J. (2006) Interaction of CmeABC and CmeDEF in conferring antimicrobial resistance and maintaining cell viability in *Campylobacter jejuni*, *J. Antimicrob. Chemother. 57*, 52–60.

105. Lin, J., Sahin, O., Michel, L. O., and Zhang, Q. (2003) Critical role of multidrug efflux pump CmeABC in bile resistance and in vivo colonization of *Campylobacter jejuni*, *Infect. Immun. 71*, 4250–4259.

106. Lin, J., Akiba, M., Sahin, O., and Zhang, Q. (2005) CmeR functions as a transcriptional repressor for the multidrug efflux pump CmeABC in *Campylobacter jejuni*, *Antimicrob. Agents Chemother. 49*, 1067–1075.

107. Grkovic, S., Brown, M. H., and Skurray, R. A. (2002) Regulation of bacterial drug export systems, *Microbiol. Mol. Biol. Rev. 66*, 671–701.

108. Guo, B., Wang, Y., Shi, F., Barton, Y. W., Plummer, P., Reynolds, D. L., Nettleton, D., Grinnage-Pulley, T., Lin, J., and Zhang, Q. (2008) CmeR functions as a pleiotropic regulator and is required for optimal colonization of *Campylobacter jejuni* in vivo, *J. Bacteriol. 190*, 1879–1890.

109. Lin, J., Cagliero, C., Guo, B., Barton, Y. W., Maurel, M. C., Payot, S., and Zhang, Q. (2005) Bile salts modulate expression of the CmeABC multidrug efflux pump in *Campylobacter jejuni*, *J. Bacteriol. 187*, 7417–7424.

110. Stintzi, A., Marlow, D., Palyada, K., Naikare, H., Panciera, R., Whitworth, L., and Clarke, C. (2005) Use of genome-wide expression profiling and mutagenesis to study the intestinal lifestyle of *Campylobacter jejuni*, *Infect. Immun. 73*, 1797–1810.

111. Rajan, S. S., Yang, X., Shuvalova, L., Collart, F., and Anderson, W. F. (2006) Crystal structure of YfiR, an unusual TetR/CamR-type putative transcriptional regulator from *Bacillus subtilis*, *Proteins: Struct. Funct. Genet. 65*, 255–257.

112. Rouquette-Loughlin, C., Veal, W. L., Lee, E.-H., Zarantonelli, L., Balthazar, J. T., and Shafer, W. M. (2002) Antimicrobial efflux systems possessed by *Neisseria gonorrhoeae* and *Neisseria meningitidis* viewed as virulence factors, in *Microbial Drug Efflux*, Paulsen, I., and Lewis, K., Eds., Horizon Scientific Press, Wymonham, UK, pp. 187–200.

113. Luna, V. A., Cousin, S., Jr., Whittington, W. L. H., and Roberts, M. C. (2000) Identification of the conjugative *mef* gene in clinical *Acinetobacter junii* and *Neisseria gonorrhoeae* isolates, *Antimicrob. Agents Chemother. 44*, 2503–2506.

114. Hagman, K. E., Pan, W., Spratt, B. G., Balthazar, J. T., Judd, R. C., and Shafer, W. M. (1995) Resistance of *Neisseria gonorrhoeae* to antimicrobial hydrophobic agents is modulated by the *mtrCDE* efflux system, *Microbiology 141*, 611–622.

115. Hagman, K. E., Lucas, C. E., Balthazar, J. T., Snyder, L., Nilles, M., Judd, R. C., and Shafer, W. M. (1997) The MtrD protein of *Neisseria gonorrhoeae* is a member of the resistance/nodulation/division protein family constituting part of an efflux system, *Microbiology 143*, 2117–2125.

116. Veal, W. L., Nicholas, R. A., and Shafer, W. M. (2002) Overexpression of the MtrC–MtrD–MtrE efflux pump due to an *mtrR* mutation is required for chromosomally mediated penicillin resistance in *Neisseria gonorrhoeae*, *J. Bacteriol. 184*, 5619–5624.

117. Faruki, H., Kohmescher, R. N., McKinney, W. P., and Sparling, P. F. (1985). A community-based outbreak of infection with penicillin-resistant *Neisseria gonorrhoeae* not producing penicillinase (chromosomally-mediated resistance), *N. Engl. J. Med. 313*, 607–611.

118. Olesky, M., Hobbs, M., and Nicholas, R. A. (2002) Identification and analysis of amino acid mutations in porin IB that mediate intermediate-level resistance to penicillin and tetracycline in *Neisseria gonorrhoeae*, *Antimicrob. Agents Chemother. 46*, 2811–2820.

119. Zarantonelli, L., Borthagary, G., Lee, E. H., Veal, W., and Shafer, W. M. (2001) Decreased susceptibility to azithromycin and erythromycin mediated by a novel *mtrR* promoter mutation in *Neisseria gonorrhoeae*, *J. Antimicrob. Chemother. 47*, 651–654.

120. Johnson, S. R., Sandul, A. L., Parekh, M., Wang, S. A., Knapp, J. S., and Trees, D. L. (2003) Mutations causing in vitro resistance to azithromycin in *Neisseria gonorrhoeae*, *Int. J. Antimicrob. Agents 21*, 414–419.

121. Shafer, W. M., Qu, X.-D., Waring, A. J., and Lehrer, R. I. (1998) Modulation of *Neisseria gonorrhoeae* susceptibility to vertebrate antibacterial peptides due to a member of the resistance/nodulation/division efflux pump family, *Proc. Natl. Acad. Sci. USA 95*, 1829–1833.

122. Lee, E. H., and Shafer, W. M. (1999) The *farAB*-encoded efflux pump mediates resistance of gonococci to long-chained antibacterial fatty acids, *Mol. Microbiol. 33*, 839–845.

123. Jerse, A. E., Sharma, N. D., Bodner, A. N. B., Snyder, L. A., and Shafer, W. M. (2003) A gonococcal efflux pump system enhances bacterial survival in a female mouse model of genital tract infection, *Infect. Immun. 71*, 5576–5582.

124. Warner, D. M., Folster, J. P., Shafer, W. M., and Jerse, A. E. (2007) Regulation of the MtrC–MtrD–MtrE efflux pump system modulates the in vivo fitness of *Neisseria gonorrhoeae*, *J. Infect. Dis. 196*, 1804–1812.

125. Warner, D. M., Shafer, W. M., and Jerse, A. E. (2008) Clinically relevant mutations that cause derepression of the *Neisseria gonorrhoeae* MtrC–MtrD–MtrE efflux pump system confer different levels of antimicrobial resistance and in vivo fitness, *Mol. Microbiol. 70*, 462–478.

126. Maness, M. J., and Sparling, P. F. (1973) Multiple antibiotic resistance due to a single mutation in *Neisseria gonorrhoeae*, *J. Infect. Dis. 128*, 321–330.

127. Guymon, L. F., Walstad, D. L., and Sparling, P. F. (1978) Cell envelope alterations in antibiotic-sensitive and -resistant strains of *Neisseria gonorrhoeae*, *J. Bacteriol. 136*, 391–401.

128. Pan, W., and Spratt, B. G. (1994) Regulation of the permeability of the gonococcal cell envelope by the *mtr* system, *Mol. Microbiol. 11*, 769–775.

129. Delahay, R. M., Robertson, B. D., Balthazar, J. T. Shafer, W. M., and Ison, C. (1997) Involvement of the gonococcal MtrE in the resistance of *Neisseria gonorrhoeae* to toxic hydrophobic compounds, *Microbiology 143*, 2127–2133.

130. Veal, W. L., and Shafer, W. M. (2003) Identification of a cell envelope protein (MtrF) involved in hydrophobic antimicrobial resistance in *Neisseria gonorrhoeae*, *J. Antimicrob. Chemother. 51*, 27–37.

131. Rouquette-Loughlin, C., Stojiljkovic, I., Hrobowski, T., Balthazar, J. T., and Shafer, W. M. (2002) Inducible, but not constitutive resistance of gonococci to hydrophobic agents due to the MtrC–MtrD–MtrE efflux pump requires the TonB–ExbB–ExbD proteins, *Antimicrob. Agents Chemother. 46*, 561–565.

132. Shafer, W. M. (2009) Unpublished work, Emory University, Atlanta, GA.

133. Tzeng, Y.-L., Datta, A., Ambrose, K. A., Davies, J. K., Carlson, R. W., Stephens, D. S., and Kahler, C. M. (2004) The MisR/MisS two-component regulatory system influences inner core structure and immunotype of lipooligosaccharide in *Neisseria meningitides*, *J. Biol. Chem. 279*, 35053–35062.

134. Rouquette-Loughlin, C. E., Balthazar, J. T., Hill, S. A., and Shafer, W. M. (2004) Modulation of the *mtrCDE*-encoded efflux pump gene complex due to a Correia Element insertion sequence, *Mol. Microbiol. 54*, 731–741.

135. Hoffman, K. M., Williams, D., Shafer, W. M., and Brennan, R. G. (2005) Characterization of the multiple transferrable repressor, MtrR, from *Neisseria gonorrhoeae*, *J. Bacteriol. 187*, 5008–5012.

136. Lucas, C. E., Balthazar, J. T., Hagman, K. E., and Shafer, W. M. (1997) The MtrR repressor binds the DNA sequence between the *mtrR* and *mtrC* genes of *Neisseria gonorrhoeae*, *J. Bacteriol. 179*, 4123–4128.

137. Shafer, W. M., Balthazar, J. T., Hagman, K. E., and Morse, S. A. (1995) Missense mutations that alter the DNA-binding domain of the MtrR protein occur frequently in rectal isolates of *Neisseria gonorrhoeae* that are resistant to fecal lipids, *Microbiology 41*, 907–911.

138. Folster, J. P., and Shafer, W. M. (2005) Regulation of *mtrF* expression in *Neisseria gonorrhoeae* and its role in high-level antimicrobial resistance, *J. Bacteriol. 187*, 3713–3720.

139. Ducey, T. F., Carson, J. M., Orvis, B., Stinzi, A. P., and Dyer, D. W. (2005) Identification of the iron-responsive genes of *Neisseria gonorrhoeae* by microarray analysis in defined media, *J. Bacteriol. 187*, 4865–4874.

140. Dyer, D. W., Jackson, L., Mercante, Shafer, W. M. (2009) Unpublished work, Emory University.

141. Dyer, D. (2009) Personal communication, University of Oklahoma, Norman, OK.

142. Lee, E.-H., and Shafer, W. M. (2009) Unpublished work, Emory University, Atlanta, GA.

143. Soncini, F. C., Garcia, E., Vescovi, E. G., Solomon, F., and Groisman, E. A. (1996) Molecular basis of the magnesium deprivation response in *Salmonella:* identification of PhoP-regulated genes, *J. Bacteriol. 178*, 5092–5099.

144. Folster, J. P., Johnson, P. J. T., Jackson, L., Dhulipali, V., Dyer, D. W., and Shafer, W. M. (2009) MtrR modulates *rpoH* expression and levels of antimicrobial resistance in *Neisseria gonorrhoeae*, *J. Bacteriol. 191*, 287–297.

145. Lee, E.-H., Rouquette-Loughlin, C., Folster, J. P., and Shafer, W. M. (2003) FarR regulates the *farAB*-encoded efflux pump of *Neisseria gonorrhoeae* via an MtrR regulatory mechanism, *J. Bacteriol. 185*, 7145–7152.

146. Hagman, K. E., and Shafer, W. M. (1995) Transcriptional control of the *mtr* efflux system of *Neisseria gonorrhoeae*, *J. Bacteriol. 171*, 4162–4165.

147. Warner, D. M., and Shafer, W. M. (2009) Unpublished work, Emory University, Atlanta, GA.

148. Ma, D., Alberti, M., Lynch, C., Nikaido, H., and Hearst, J. E. (1996) The local repressor AcrR plays a modulating role in regulation of the *acrAB* genes of *Escherichia coli* by global stress signals, *Mol. Microbiol. 19*, 101–112.

149. Rouquette, C., Harmon, J. B., and Shafer, W. M. (1999) Induction of the *mtrCDE*-encoded efflux pump system of *Neisseria gonorrhoeae* requires MtrA, an AraC-like protein, *Mol. Microbiol. 33*, 651–658.

150. Shafer, W. M. (2009) Unpublished work, Emory University, Atlanta, GA.

151. Courvalin, P. (1994) Transfer of antibiotic resistance genes between gram-positive and gram-negative bacteria, *Antimicrob. Agents Chemother. 38*, 1447–1451.

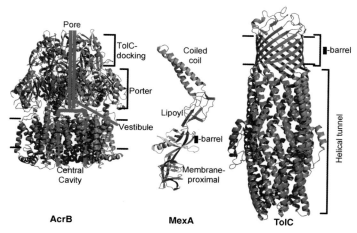

Figure 1.3.   Crystallographic structures of MexA, AcrB, and TolC. (*See text for full caption.*)

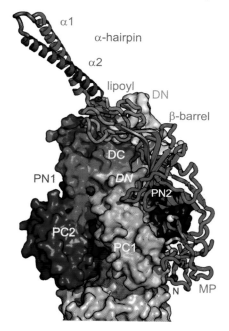

Figure 1.4.   Model of an AcrAB complex based on the recently refined AcrA structure and the cross-linking data. (*See text for full caption.*)

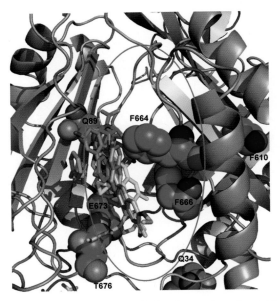

Figure 1.6.   Large cleft of the AcrB periplasmic domain, seen from the outside. (*See text for full caption.*)

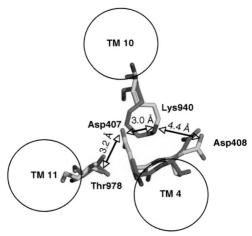

Figure 1.10.   Putative salt bridge/H-bonding network in the transmembrane domain of the wild-type AcrB protomer (yellow stick model) and the Asp407Ala mutant (green stick model), based on PDB files 1IWG and 2HQC. (*See text for full caption.*)

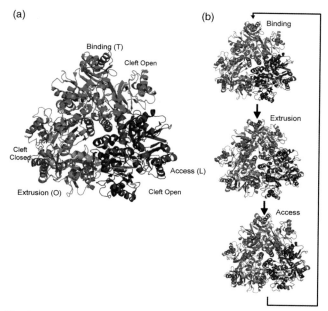

Figure 1.11. Periplasmic domain of the asymmetric crystal structure of AcrB (A) and the functionally rotating mechanism proposed (B). (*See text for full caption.*)

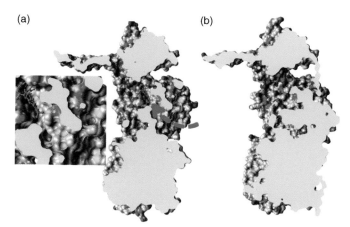

Figure 1.12. Cutout view of the binding protomer with the bound minocycline in a green stick representation (A) and the extruding protomer (B), both from PDB file 2DRD, drawn using the UCSF Chimera package (133). (*See text for full caption.*)

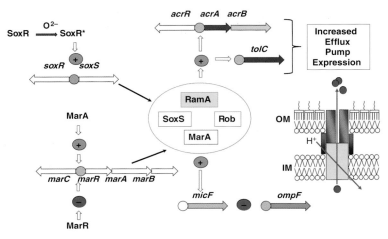

Figure 2.13.   Schematic of part of the network of global regulatory control.

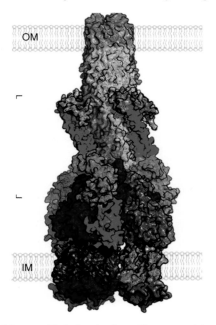

Figure 3.1.   Model of the assembled tripartite drug efflux pump of AcrAB–TolC in the form of TolC$_3$–AcrA$_3$–AcrB$_3$. (*See text for full caption.*)

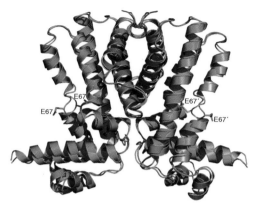

Figure 3.2.   Structural comparison of the $P3_1$ and $P222_1$ structures of AcrR. (*See text for full caption.*)

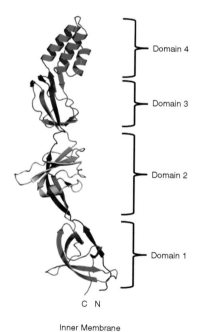

Figure 3.3.   Crystal structure of the CusB membrane fusion protein. The structure can be divided into four distinct domains. (*See text for full caption.*)

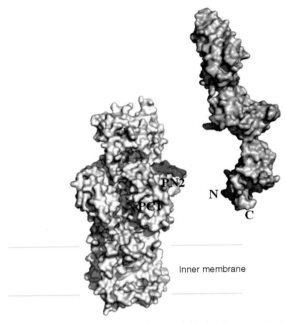

Figure 3.4. Specific interaction between CusA and CusB. (*See text for full caption.*)

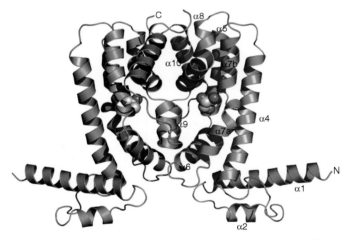

Figure 3.5. Crystal structure of the CmeR regulator. (*See text for full caption.*)

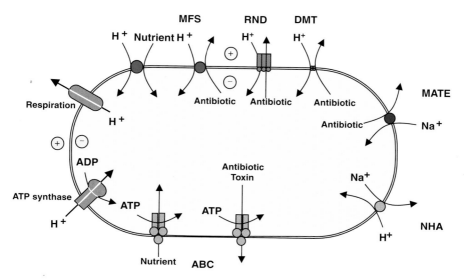

Figure 4.1.   Efflux systems and their energization in bacteria. (*See text for full caption.*)

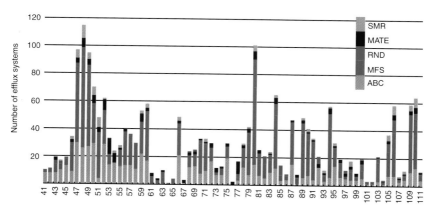

Figure 4.2.   Distribution of efflux systems in different microrganisms. (*See text for full caption.*)

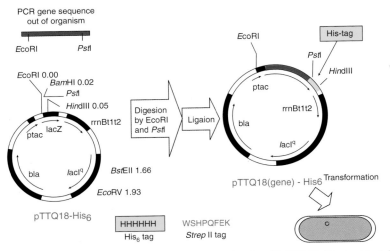

Figure 4.3. Figure 4.3. Cloning strategy for membrane proteins using plasmid pTTQ18. (*See text for full caption.*)

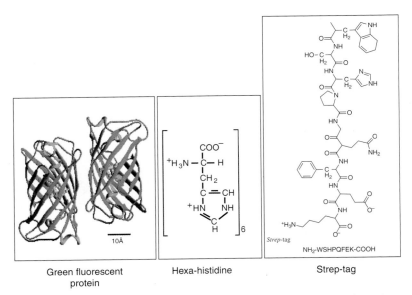

| Green fluorescent protein | Hexa-histidine | Strep-tag |

Figure 4.4. Protein tags used for identification and purification of expressed proteins.

THE MFS EFFLUX PROTEINS OF GRAM-POSITIVE AND GRAM-NEGATIVE BACTERIA

By MASSOUD SAIDIJAM, *Department of Molecular Medicine and Genetics, School of Medicine, Hamedan University of Medical Sciences, Hamedan, Iran,* KIM E. BETTANEY, DONG LENG, PIKYEE MA, ZHIQIANG XU, JEFFREY N. KEEN, NICHOLAS G. RUTHERFORD, ALISON WARD, PETER J. F. HENDERSON*, *Institute for Membrane and Systems Biology, Astbury Centre for Structural Molecular Biology, Faculty of Biological Sciences, University of Leeds, Leeds, LS2 9JT, UK,* GERDA SZAKONYI, *Institute of Pharmaceutical Analysis, Faculty of Pharmacy, University of Szeged, Szeged, Hungary,* QINGHU REN, IAN T. PAULSEN, *Department of Chemistry and Biomolecular Sciences, Macquarie University, Sydney, Australia,* and INGERID NES, JASMIN K. KROEGER, ANNE-BRIT KOLSTO*, *Laboratory for Microbial Dynamics (LaMDa), Department of Pharmaceutical Biosciences, University of Oslo, P.O. Box 1068 Blindern, 0316 Oslo, Norway*

CONTENTS

*Corresponding authors: Peter J. F. Henderson (p.j.f.henderson@leeds.ac.uk) and Anne-Brit Kolsto (a.b.kolsto@farmasi.uio.no)

Advances in Enzymology and Related Areas of Molecular Biology, Volume 77
Edited by Eric J. Toone   Copyright © 2011 John Wiley & Sons, Inc.

I. Introduction

Analyses of the available bacterial genomes (1) predict that membrane transport proteins (2) likely to catalyze drug efflux comprise 2 to 7% of the genetic complement of most bacteria, although they may comprise a much lower percentage of expressed cell protein (2, 3). Their activities are energy linked in bacteria, and this allows transport against the prevailing electrochemical gradient of the solute (Figure 1) (1). Drugs and antibiotics are often hydrophobic molecules that penetrate the cell membrane and are then actively secreted from the cell by such proteins, thereby conferring resistance.

Like the majority of membrane transporters, drug efflux proteins fall predominantly into the two classes (1–3) of the ATP-binding cassette (ABC) superfamily (4, 5) or the major facilitator superfamily (MFS) (6–9), which is energized by the electrochemical gradient of protons (sometimes, sodium ions) in an antiport mechanism (Figure 1). In addition, there are at least three more classes of efflux transport systems (1, 3). The small multidrug resistance family (SMR), itself a subfamily of the drug metabolite transporter (DMT) family of transporters (10, 11) the resistance–nodulation–division (RND) family (1, 11, 12) members of which (e.g., the AcrAB–TolC complex) (13–15) are especially important for the antibiotic resistance of gram-negative pathogens; and the multiantimicrobial extrusion family (MATE), a subfamily of the multidrug/oligosaccharidyl-lipid/polysaccharide (MOP) flippase superfamily (1, 3). The distribution of each of these types of efflux systems varies considerably in different microorganisms, as illustrated in Figure 2 (1, 3). For example, in *Escherichia coli* O157 the ABC/MFS/RND/MATE/DMT ratio predicted is 7 : 32 : 9 : 4 : 6, while in *Bacillus cereus* it is 27 : 56 : 3 : 4 : 5 (3).

While the actual substrates of many efflux systems have yet to be identified, there is a growing awareness that they can contribute to drug and antibiotic resistance in infectious microbes, and that we should be actively seeking new drugs that inhibit their activities (1, 3, 11–13, 16). Many individual efflux proteins have a wide range of substrates and contribute to multidrug resistance (Mdr) (1, 3, 8, 11–16). However, these proteins are very hydrophobic and are generally expressed at too low a level for satisfactory identification, characterization of function, purification, and structural studies.

To overcome these problems, a strategy is reviewed in this chapter that enables the amplified expression and purification of efflux proteins, from

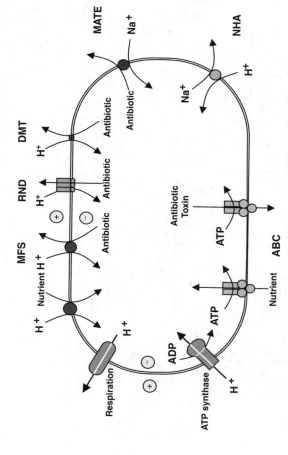

Figure 1. Efflux systems and their energization in bacteria. A transmembrane electrochemical gradient of protons across the bacterial inner membrane is generated by respiration or ATP hydrolysis (both dark gray). This proton/electrical gradient is used to drive ATP synthesis and the MFS proton–nutrient symport and proton–substrate antiport (both red circles) secondary active transport systems. Each MFS transporter is generally a monomer, usually containing 12 to 14 transmembrane helices. Also illustrated are the RND (green), DMT (orange), MATE (deep blue), sodium–proton antiporter (light gray), and ABC (light blue) types of transport systems (see the text and references 1 to 3 for details). MFS, major facilitator superfamily; RND, resistance–nodulation–division; DMT, drug metabolite transporter; MATE, multidrug and toxin extrusion; ABC, ATP-binding cassette; NHA, Na$^+$/H$^+$ antiporter. (See insert for color representation.)

149

150

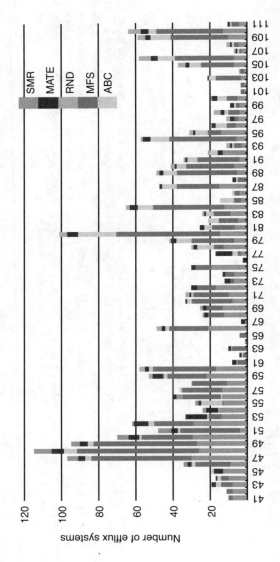

Figure 2. Distribution of efflux systems in different microorganisms. A selection of organisms relevant to this chapter was made from an analysis of 132 bacterial genomes (3). The color code is shown and is the same as in Figure 1. The organisms are as follows: 47 = *Bacillus anthracis* Ames with 66 MFS efflux systems; 49 = *Bacillus cereus* with 56 MFS efflux systems; 51 = *Bacillus subtilis* with 32 MFS efflux systems; 71 = *Staphylococcus aureus* with 22 MFS efflux systems; 82 = *Brucella melitensis* with 12 MFS efflux systems; 94 = *Neisseria meningitides* with four MFS efflux systems; 100 = *Helicobacter pylori* with only two MFS efflux systems; 106 = *Escherichia coli* 0157 with 32 MFS efflux systems; and 109 = *Pseudomonas aeruginosa* with 31 MFS efflux systems. (See insert for color representation.)

many species of bacteria, in amounts required for functional and structural studies. Many of these prokaryote transport proteins are members of the major facilitator superfamily, found in a range of higher organisms, such as protozoan parasites, fungi, plants, and mammals (1–3, 6–9, 17, 18). It is clear, therefore, that detailed comparative studies of the structure–activity relationships of these bacterial proteins are necessary if they are to be useful targets for the discovery of novel antibacterials.

This chapter focuses on efflux proteins of the major facilitator superfamily (MFS), which dominate in many bacteria (1, 3, 8) (Figure 2). From the hydropathy profiles of their amino acid sequences they are generally believed to comprise 12 or 14 transmembrane α-helices (1–3, 8, 17, 18). In the postgenome age, identification of possible efflux proteins through similarities of their amino acid sequences is trivial (1, 6, 8, 17, 18), but their complete biochemical, functional, and structural characterization in purified systems is only just beginning. Although these transport proteins catalyze vectorial, not chemical, events, the activities of the purified proteins can be illuminated by many of the techniques used in classical enzymology. Thus, our focus is on structural genomics—devising a general strategy to move quickly from identification of a gene of interest to having sufficient quantities of pure protein for functional and structural studies.

II. Cloning a Putative Efflux Protein into a Plasmid Expression Vector

The genes encoding selected efflux proteins were amplified from the corresponding samples of genomic DNA using appropriate mutagenic oligonucleotides. These oligonucleotides were generally designed to introduce an *Eco*RI site at the 5′ end and a *Pst*I site at the 3′ end, to promote subsequent ligation with the 4.56-kb pTTQ18/(His)$_6$ fragment (see below). The resulting polymerase chain reaction (PCR) product was isolated from an agarose gel and then digested with *Eco*RI and *Pst*I.

To clone each of the genes into the pTTQ18 plasmid vector (Figure 3) (19–24), the plasmid pNorAH$_6$ (pTTQ18 containing the gene *norAH$_6$*) is isolated from *E. coli* strain BLR and digested with the restriction endonucleases *Eco*RI and *Pst*I to yield two DNA fragments of 4.56 and 1.2 kb. The larger fragment [pTTQ18 with the (His)$_6$-coding DNA sequence] was isolated from an agarose gel.

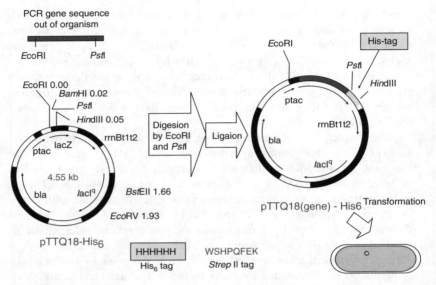

Figure 3.   Cloning strategy for membrane proteins using plasmid pTTQ18. Details are given in references 20 to 25. The target gene is amplified from genomic DNA of the originating organism by the polymerase chain reaction. Both the PCR product from the original gene and the pTTQ18 plasmid are digested by *Eco*RI and *Pst*I restriction enzymes. The digested DNA fragments are ligated and then transformed into XL1 blue or XL10 gold *E. coli* cells for propagation prior to transformation into the expression host *E. coli* BL21(DE3). (See insert for color representation.)

Ligation reactions are performed using the *Eco*RI–*Pst*I digested gene and pTTQ18–(His)$_6$ fragments at various vector/insert molar ratios. The ligated product is subsequently transformed into *E. coli* XL1 Blue Stratagene cells [(*recA1, endA1, gyrA96, thi-I, hsdR17, supE44, relA1, lac* (F′*proAB lacI^qZΔM15, Tn10 (Tet^R*))] or XL10 Gold Stratagene [*recA1, endA1, gyrA96, relA1, lac* (F′*proAB lacI^qZΔM15, Tn10 (Tet^R*) Amp Cam)] and recombinant clones are selected on LB plates containing carbenicillin. Automated DNA sequencing is used to confirm the presence of each gene and the absence of any adventitious base changes. The plasmid is then transformed into *E. coli* strain BL21 (DE3) Novagen [F$^-$*ompT hsdS$_B$(r$_B^-$ m$_B^-$) gal dcm (DE3)*] for expression studies.

This procedure, which is illustrated in Figure 3, can be applied to any gene that does not include *Eco*RI or *Pst*I restriction sites and, using our modified pTTQ18 vector, results in fusion of an (His)$_6$ tag (Figure 4) in the

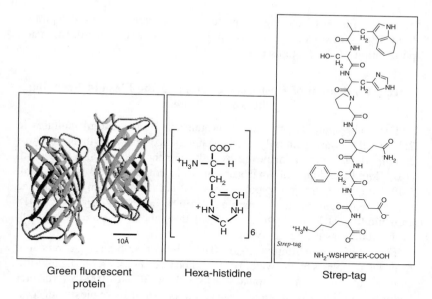

Green fluorescent   Hexa-histidine   Strep-tag
protein

Figure 4. Protein tags used for identification and purification of expressed proteins. (See insert for color representation.)

frame at the C-terminus of the gene inserted (21–25). If these two restriction sites are present in the coding region, *Eco*RI and *Pst*I can still be introduced as flanking sites with partial digestion used to obtain a fragment uncut at the internal site(s), or a two-step procedure can be adopted [see, e.g., the article by Potter et al. (25)]. Alternatively, other flanking restriction sites compatible with the multicloning site in pTTQ18 can be chosen (19): for example, *Bam*HI at the 5′ end and *Hin*dIII at the 3′ end of the gene.

Amplification of expression using the *tac* promoter in plasmid pTTQ18 is usually successful, but there can be some leakage of expression in the absence of inducer [usually, isopropyl-β-D-galactoside (IPTG)]. If this is thought to be a problem, or if there is a need to pursue even higher levels of expression, a construct using the arabinose-inducible *araC* promoter may be helpful (26, 27).

If the C-terminus of the target membrane protein is predicted to lie in the periplasm, a Strep tag (Figure 4) may well be successful if the (His)$_6$ tag is not. Also, fusion of the tagged protein with a green fluorescent protein (GFP, Figure 4) via a protease-susceptible site may assist identification and purification of the proteins (28).

III. Cultivation of Host Cells Containing the Plasmid Encoding the Target Protein

For small-scale investigation of protein expression, 50-mL cultures in 250-mL flasks are generally used. Maintenance and growth of the host *E. coli* BL21 strains is achieved by culturing the bacteria in Luria broth (LB) liquid medium, minimal salts medium containing 20 mM glycerol (M9), or double-strength tryptone-yeast extract (2TY), or on plates of the before-mentioned media containing 1.5% agar. For plasmid pTTQ18 carbenicillin (at least 100 mg/L) is used throughout all stages of growth to maintain plasmid integrity.

For small-scale cultutres (5 to 50 mL) total membranes are prepared from sphaeroplasts by the water lysis method (21). For larger-scale 500-mL cultures in 2-L baffled conical flasks or for 30- to 100-L fermenter cultures (29), inner membrane vesicles are prepared by explosive decompression using a Constant Cell Disruption System (www.constantsystems. com). The inner and outer cell membranes can then be separated by sucrose density centrifugation, followed by washing in buffer to remove the sucrose and EDTA (21).

For both small-scale tests and larger-scale production of inner membranes, the growth of the *E. coli* strains is allowed to continue until the cell density reaches an A_{680} value of approximately 0.6. At this point the expression of the cloned gene is induced by the addition of isopropyl-β-D-thiogalactopyranoside (IPTG) to a final concentration of 0.1 to 2.0 mM (see below). Growth sometimes continues for 3 to 4 h following induction of the *tac* promoter. However, for many of the cloned efflux proteins, the induction of expression attenuates growth and in extreme cases the cells may even lyse (21, 22, 27). Accordingly, it is important to undertake pilot experiments to optimize cell and protein production: for example, in different media with different concentrations of IPTG, using different periods of induction, and possibly, different temperatures for growth. For scaling up production with the IPTG-inducible promoter, economies can be achieved by implementing autoinduction using glucose as an initial growth substrate followed by lactose as both substrate and inducer (30).

IV. Detection of Histidine-Tagged Efflux Proteins in *E. coli* Membrane Preparations

E. coli BL21(DE3) cells harboring each plasmid are generally first cultured in a 2TY medium and expression trials performed with different concentrations of IPTG (0.0 to 2.0 mM). Membrane samples (21, 22) are prepared for analysis of the overexpressed protein by SDS-PAGE (Figure 5). It is common for membrane proteins to migrate at 65 to 75% of their true molecular weight, possibly as a result of their retention of secondary or tertiary structure, accelerating their passage through the gel (21–24). Scanning densitometry analyses of Coomassie Blue-and/or silver-stained gels (examples in Figure 5) showed that the induced proteins were expressed at 6 to 16% of total membrane proteins, indicating that overexpression has occurred. In comparison, protein at the same position in uninduced samples (not shown) is only expressed at <3% of inner membrane protein. The identity of the over-expressed protein in the membranes is confirmed by Western blotting (not shown) (21, 22, 24).

V. Solubilization of the Efflux Proteins with Detergents

The purification of membrane proteins requires the use of detergents to extract the desired membrane protein from the membrane and to keep it in solution (31–34). Detergents used in the solubilization and purification of membrane proteins must maintain the structural integrity of the protein and its activity on reconstitution (21, 31, 33, 34). The physicochemical properties of the detergent provide information regarding its suitability for biophysical studies. These are the critical micelle concentration (CMC), the lowest concentration above which monomers cluster to form micelles, and the aggregation number, the number of monomeric detergent molecules contained in a single micelle (32).

We have explored the use of the following detergents for solubilization and stabilization of bacterial membrane transport proteins: dodecyl-β-D-maltoside, decyl-β-D-maltoside, nonyl-β-D-maltoside, cyclohexyl-β-D-maltoside, Hecameg, Mega-10, Zwittergen, Thesit, Triton X-100, Triton X-114, Mega-8, CHAPS, CHAPSO, sodium cholate, sodium deoxycholate, lauryldimethylamine oxide (LDAO), octyl-β-D-glucoside, and dodecyl-β-D-glucoside. The first four are usually suitable for solubilization, although it is important to explore and establish the optimal detergent

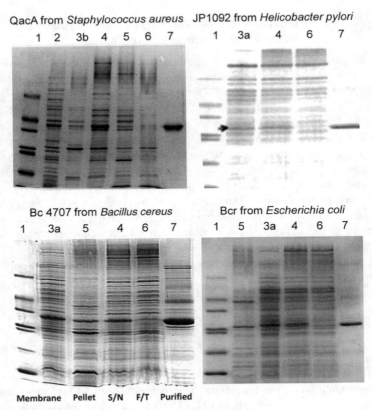

Figure 5. Examples of variable expression and purification of efflux proteins. IPTG-induced *E. coli* BL21(DE3) harboring pTTQ18-containing genes encoding the indicated proteins were generally grown on LB or 2TY with 20 mM glycerol. In a typical experiment, mixed membranes were prepared by water lysis and/or inner membranes by using a constant flow disruptor and sucrose density gradient centrifugation (21). The protein was solubilized from the membrane in DDM (0.5 to 2.0%) and the insoluble and soluble fractions examined. The soluble fraction was exposed to Ni^{2+}-NTA-agarose (IMAC) and a sample of the material that failed to bind was examined. The Ni^{2+}-NTA column containing bound protein was washed with buffer containing a low concentration of imidazole and subsequently eluted with 200 mM imidazole. Following analysis of the samples on a 15% SDS-PAGE gel, the proteins were visualized with Coomassie Brilliant Blue. The gel tracks contained samples as follows: (1) molecular weight standards ordered from the top: 66, 45, 35, 29, 25, and 14 kDa; (2) mixed membrane preparation; (3a) inner membrane preparation; (3b) outer membrane preparation; (4) inner membrane preparation solubilized in DDM; (5) unsolubilized material; (6) flow-through from the washed column using a low imidazole concentration; (7) purified protein eluted with a high imidazole concentration.

concentration, pH, and time of exposure for each protein, always at 0 to 4 °C, to minimize degradation.

The choice of detergent has a particularly critical role in crystallization of membrane proteins. The detergent chosen for solubilizing and purifying proteins may not be the most suitable for crystallization, so Howard et al. (35) recommended the use of detergents that can be readily exchanged.

VI. Purification of Histidine-Tagged Efflux Proteins

Initial purification of (His)$_6$-tagged proteins (Figure 4) is achieved by immobilized metal affinity chromatography [IMAC (22); Figure 5]. The method is based on the affinity of Ni-NTA resin for proteins and peptides that have six consecutive histidine residues at either their N- or C-terminus (hexahistidine affinity tag) (36). The hybrid protein is purified by immobilized metal ion affinity chromatography followed by elution of the immobilized protein (3, 20–22, 24, 25). Within the NTA ligand there are four chelating sites, which can interact with the metal ions; NTA will occupy four of the ligand-binding sites in the coordination sphere of the Ni^{2+} ion; thus, two sites remain to interact with the hexahistidine tag (36, 37).

Since there is a potential for binding background contaminants, low concentrations of imidazole in the lysis and wash buffers (10 to 20 mM) are recommended. The bound-tagged protein can be eluted in a number of ways. The histidine residues in the His$_6$ tag have a pK_a value of approximately 6.0 and will become protonated if the pH is reduced (pH 4.5 to 5.3). Under these conditions the His$_6$-tagged protein can no longer bind to the nickel ions and will dissociate from the Ni-NTA resin. Similarly, if the imidazole concentration is increased above 100 mM, the His$_6$-tagged proteins will also dissociate because they can no longer compete for binding sites on the Ni-NTA resin.

The identity of the protein is confirmed by Western blotting (not shown), using an antibody to the C-terminal (His)$_6$-tag, and by N-terminal sequencing. Some contaminants may be visible, and bands of higher molecular weight may be completely unfolded oligomers of the protein (Figure 5).

The conditions for solubilization and purification (i.e., concentration of dodecyl-β-D-maltoside (DDM) for solubilization and imidazole concentration for IMAC) vary depending on the characteristics of each transport protein. However, the generic conditions described next have proved generally useful. The efflux transporters are initially solubilized in 20 mM Tris pH 8.0, 20 mM imidazole pH 8.0, 300 mM NaCl, 20% glycerol, and

1% DDM. After washing out the unspecifically bound proteins with the same buffer, the (His)$_6$-tagged proteins could generally be eluted by the following buffer: 20 mM Tris pH 8.0, 200 mM imidazole pH 8.0, 150 mM NaCl, 5% glycerol, and 0.05% DDM. Note that a relatively high detergent concentration is needed for solubilization, but can often be substantially lowered and still maintain the protein(s) in solution during purification thereafter.

The four examples shown in Figure 5 have been chosen deliberately from a panel of efflux proteins (Table 1) to illustrate the variability in both levels of expression and homogeneity of product that can be achieved in the one-step purification.

VII.   Confirmation of Integrity and Monodispersity of the Purified Protein

It is important to have rigorous tests for both the identity and integrity of the purified overexpressed proteins before further, often time-consuming studies are undertaken. Also, before crystallization trials are initiated, it is not only important to confirm the integrity of the overexpressed protein but also to test for its monodispersity. Monodispersity is important because aggregated material usually inhibits crystallization. Size-exclusion chromatography is an established technique for the purification of proteins on the basis of their size. In the case of membrane proteins isolated in detergents, the apparent molecular size is increased by the presence of the detergent micelle (33, 35). Even so, this method can be used as both a purification step (in addition to IMAC) and as a means to assess monodispersity.

Measuring the spectrum of absorbances of circularly polarized light [circular dichroism (CD)] in the far-ultraviolet (UV)/UV range of wavelengths is a useful technique for the detection of secondary structure within proteins in solution and can be used as an indicator, for membrane proteins, of α-helical content (38), including efflux proteins (3, 20–22, 39). This is a convenient test in a protein chemistry laboratory for confirming retention of integrity during purification (Figure 6). Confirmation of the N-terminal amino acid sequence of the proteins and a positive Western blot result confirm that the protein is intact, as mass spectrometry can also (40).

Importantly, however, the activity of the isolated protein still requires confirmation. This can be achieved if there is a high-affinity substrate or inhibitor suitable for direct binding studies, or by reconstitution into

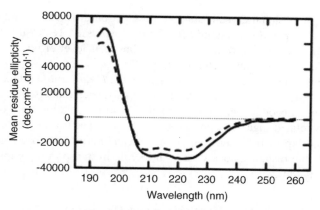

Figure 6.   Circular dichroism analysis of the *Staph. aureus* NorA(His)₆ and the *E. coli* Bcr(His)₆ efflux resistanceproteins for norfloxacin and bicyclomycin, respectively. A sample of each purified protein from the gram positive and gram negative organisms was washed and exchanged into 10 mM KPi (pH 7.6), 0.05% DDM and 1 mM DTT. Following dilution to 25 µg/ml in the above buffer, 300 µl of sample was scanned between 190 and 300 nm using the Jasco J-715 spectropolarimeter with constant nitrogen flushing. Samples were analysed in Hellma quartz cuvettes of 1mm path length. Spectra were recorded with 1nm sampling intervals at a scan rate of 50 nm/min. Each spectrum shown is an average of 20 scans, and is typical of an high content of alpha-helix for both NorAH₆ (dashed line) and BcrH₆ (continuous line). A spectrum of solvent alone has been subtracted.

artificial membranes, where the transport activity of the protein can be measured (41).

VIII.   Crystallization Trials on Purified Proteins

After purification and checks for structural and functional integrity, crystallization trials are set up using sparse-matrix screens from Hampton, Sigma, and Molecular Dimensions. At present, we prefer to retain glycerol in the crystallization drop, as this provides a ready-made cryosolution if and/or when crystals appear. However, hydropathy plots of many efflux protein sequences reveal very little hydrophilic regions between the transmembrane regions, and it is generally thought that these hydrophilic regions facilitate crystal formation as possible protein–protein contact points. Therefore, crystallization trials have also been performed based on the lipidic cubic phase technique (42) using the dedicated screen developed by Emerald Biosystems. These trials are also ongoing.

Two-dimensional crystals suitable for electron diffraction crystallography were prepared successfully for the overexpressed TetA MFS tetracycline transport protein from *E. coli* (43), but their level of resolution was insufficient for structure determination (43).

Recently, a model has been proposed for the *E. coli* EmrD efflux protein based on x-ray diffraction data (44). Previously, such models were derived from gene fusion experiments (45) and tenuous relationships to more distantly related sugar transport proteins (46, 47). The EmrD protein model shows the expected 12-helix structure, with an interior composed mostly of hydrophobic residues, consistent with its role in transporting amphipathic molecules (44).

IX. Conclusions and Discussion

The prime purpose of this chapter is to illustrate experimental procedures for obtaining sufficient quantities of correctly folded efflux protein(s) from a variety of microorganisms, including pathogens, for functional and structural studies. We have achieved this for a number of such proteins from both gram-positive and gram-negative organisms (Table 1).

TABLE 1
Drug Efflux Transport Proteins Expressed

Protein	Organism
JM1092	*Helicobacter pylori*
JM1181	*H. pylori*
QacA	*Staphylococcus aureus*
NorA	*S. aureus*
EF 1078	*Enterococcus faecalis*
EF 2068	*E. faecalis*
BMEI 1555	*Brucella melitensis*
BMEII 0795	*B. melitensis*
LmrB	*Bacillus cereus*
Bc4707	*B. cereus*
Bc5372	*B. cereus*
Bc4568	*B. cereus*
Bc3310	*B. cereus*
Bmr	*B. subtilis*
Blt	*B. subtilis*
YjiO	*Escherichia coli*
Bcr	*E. coli*

Target selection can follow at least two precepts. A single efflux protein may be selected for investigation because its activity is thought to be of importance in the infectivity of an established pathogen. Alternatively, to reduce the risk of failure in determining a three-dimensional structure, a series of orthologs can be selected from the same and different organisms. These orthologs should be related by amino acid sequence (at least 20% identity in aligned sequences) so that their three-dimensional structures can be assumed to be similar. In this second case, a broad-based trial of amplified expression, purification, and stability is then conducted on all targets in parallel. At each stage it is to be expected that some will prove unsuitable, so by a process of elimination, starting from about 40 candidates, one hopes to arrive at a panel of 5 to 10 proteins that may crystallize. Of these, there may be one or two that yield crystals that diffract x-rays to sufficient resolution to allow structure determination.

It is possible to proceed from identification of a gene that encodes a membrane protein in a bacterial genome to the production of milligram quantities of purified protein in a few weeks, and the application of high-throughput methods for cloning and purification will hopefully reduce this time still further (48, 49). The yield and purity of protein may well be increased by further optimization of conditions, especially for detergent extraction of protein from the membrane. Our typical yields from 30- to 100-L fermentations (29) conducted approximately monthly are sufficient to maintain three-dimensional crystallization trials for x-ray crystallography and nuclear magnetic resonance (NMR) studies (50, 51).

In addition, the purified protein can be examined by a variety of biophysical techniques: mass spectrometry for precise M_r and sequence determination; circular dichroism and Fourier transform infrared spectroscopy to measure the content of secondary protein structure; fluorimetry and calorimetry to measure ligand binding; and electron spin resonance spectroscopy to investigate protein–lipid interactions. These can also illuminate structure–activity relationships, especially when performed in conjunction with site-directed mutagenesis and genetic recombination. Expression of genes encoded in the pTTQ18 vector is independent of the E. coli host in our experience so far, and movement between host strains by transformation is easily accomplished in order to optimize, for example, specific labeling with stable isotopes for NMR studies.

The three-dimensional structures of the lactose (LacY, 52) and α-glycerophosphate (GlpT, 53) and possibly the EmrD (efflux) (44) MFS transport proteins of *E. coli* have been determined by x-ray crystallography. These structures can now be used to make better models of the structures of related proteins, including those responsible for efflux (45, 46). Our understanding of the molecular origin of the promiscuity of many multi-drug substrates (54) is based currently on the structure of the ligand-binding site of the nonmembrane QacR protein (55, 56). Future elucidation of the actual structures of efflux proteins from many pathogenic microorganisms may very well uncover new targets for preventing or treating bacterial infections, an increasingly urgent requirement, as "superbugs" acquire apparently unconquerable resistances (3, 12, 16, 57, 58).

Acknowledgements

The work in the P.J.F.H. laboratory was funded by European Membrane Protein Consortium contract LSGH-CT-2004-504601, BBSRC, and by equipment grants from the Wellcome Trust and BBSRC. M.S. is grateful to the Iranian government for financial support. MRC provided a studentship for K.E.B. The work in the A-B.K. laboratory was funded by the Norwegian Functional Genomics and Consortium for Advanced Microbial Sciences and Technologies platform of the Research Council of Norway. Collaborations between the laboratories are funded by the EU BacMT, EDICT and ATENS initiatives. P.J.F.H. is grateful for a Leverhulme Research Fellowship. G.S. is grateful for the Bolyai Fellowship of the Hungarian Academy of Sciences and the financial support of Norway Grants and the Hungarian Research Fund (OTKA NNF 78930).

References

1. Paulsen, I. T., Chen, J., Nelson, K. E., et al. (2002) Comparative genomics of microbial drug efflux systems, in *Microbial Multidrug Efflux*, Paulsen, I. T., and Lewis, K. Eds., Horizon Press, Norwich, UK, pp. 5–20.

2. Henderson, P. J. F. (1998) Function and structure of membrane transport proteins, in *The Transporter FactsBook*, Griffith, J. K., and Sansom, C. E. Eds., Academic Press, London, pp. 3–29.

3. Saidijam, M., Benedetti, G., Ren, Q., et al. (2006) Microbial drug efflux proteins of the major facilitator superfamily, *Curr. Drug Targets 7*, 793–811.

4. Ponte-Sucre, A., Ed. (2009) *ABC Transporters in Microorganisms*, Caister Academic Press, Great Yarmouth.

5. Schmitt, L., and Tampe, R. (2002) Structure and mechanism of ABC transporters, *Curr. Opin. Struct. Biol. 12*, 754–760.

6. Henderson, P. J. F., and Maiden, M. C. (1987) Homologous sugar transport systems in *Escherichia coli* and their relatives in both prokaryotes and eukaryotes, *Philos. Trans. R. Soc. Lond. B 326*, 391–410.

7. Griffith, J. K., Baker, M. E., Rouch, D. A., et al. (1992) Evolution of transmembrane transport: relationships between transport proteins for sugars, carboxylate compounds, antibiotics and antiseptics, *Curr. Top. Cell Biol. 4*, 684–695.

8. Griffith, J. K., and Sansom, C. E. (1998) H^+-dependent antiporters, in *The Transporter FactsBook*, Griffith, J. K., and Sansom, C. E. Eds., Academic Press, London, pp. 335–370.

9. Pao, S. S., Paulsen, I. T., and Saier, M. H., Jr., (1998) Major facilitator superfamily, *Microbiol. Mol. Biol. Rev. 62*, 1–34.

10. Brown, M. H., Paulsen, I. T., and Skurray, R. A. (1999) The multidrug efflux protein NorM is a prototype of a new family of transporters, *Mol. Microbiol. 31*, 394–395.

11. Paulsen, I. T. (2003) Multidrug efflux pumps and resistance: regulation and evolution, *Curr. Opin. Microbiol. 6*, 446–451.

12. Li, X. Z., and Nikaido, H. (2004) Efflux-mediated drug resistance in bacteria, *Drugs 64*, 159–204.

13. Nikaido, H., and Takatsuka, Y. (2009) Mechanisms of RND multidrug efflux pumps, *Biochim. Biophys. Acta 1794*, 769–781.

14. Pos, K. M. (2009) Drug transport mechanism of the AcrB efflux pump, *Biochim. Biophys. Acta 1794*, 782–793.

15. Tseng, T. T., Gratwick, K. S., Kollman, J., et al. (1999) The RND permease superfamily: an ancient, ubiquitous and diverse family that includes human disease and development proteins, *J. Mol. Microbiol. Biotechnol. 1*, 107–125.

16. Lomovskaya, O., and Watkins, W. (2002) Inhibition of efflux pumps as a novel approach to combat multidrug resistance in bacteria, in *Microbial Multidrug Efflux*, Paulsen, I. T., and Lewis, K. Eds., Horizon Press, Norwich, UK, pp. 201–231.

17. Saier, M. H., Jr., (2000) A functional-phylogenetic classification system for transmembrane solute transporters, *Microbiol. Mol. Biol. Rev. 64*, 354–411.

18. Ren, Q., and Paulsen, I. T. (2007) Large-scale comparative genomic analyses of membrane transport systems in prokaryotes, *J. Mol. Microbiol. Biotechnol. 12*, 165–179.

19. Stark, M. J. (1987) Multicopy expression vectors carrying the *lac* repressor gene for regulated high-level expression of genes in *Escherichia coli*, *Gene 51*, 255–267.

20. Ward, A., Hoyle, C., Palmer, S., et al. (2001) Prokaryote multidrug efflux proteins of the major facilitator superfamily: amplified expression, purification and characterisation, *J Mol. Microbiol. Biotechnol. 3*, 193–200.

21. Ward, A., Sanderson, N. J., O'Reilly, J., et al. (2000) The amplified expression identification, purification, and properties of hexahistidine-tagged bacterial membrane transport proteins, in *Membrane Transport: A Practical Approach*, Baldwin, S. A. Ed., Oxford University Press, New York, pp. 141–166.

22. Saidijam, M., Psakis, G., Clough, J. L., et al. (2003) Collection and characterisation of bacterial membrane proteins, *FEBS Lett.* *555*, 170–175.

23. Liang, W.-J., Wilson, K., Xie, H., Knol, J., Suzuki, S., Rutherford, N. G., Henderson, P. J. F., and Jefferson, R. (2005) The *gusBC* genes of *Escherichia coli* encode a glucuronide transport system, *J. Bacteriol.* *187*, 2377–2385.

24. Suzuki, S., and Henderson, P. J. F. (2006) The hydantoin transport protein from *Microbacterium liquefaciens*, *J. Bacteriol.* *188*, 3329–3336.

25. Potter, C. A., Ward, A., Laguri, C., et al. (2002) Expression, purification and characterisation of full-length histidine protein kinase RegB from *Rhodobacter sphaeroides*, *J. Mol. Biol.* *320*, 201–213.

26. Guzman, L. M., Belin, D., Carson, M. J., and Beckwith, J. (1995) Tight regulation, modulation, and high-level expression by vectors containing the arabinose PBAD promoter, *J. Bacteriol.* *177*, 4121–4130.

27. Hassan, K. A., Xu, Z., Watkins, R. E., Brennan, R. G., Skurray, R. A., and Brown, M. H. (2009) Optimized production and analysis of the staphylococcal multidrug efflux protein QacA, *Protein Expr. Purif.* *64*, 118–124.

28. D'Ulisse, V., Fagioli, M., Ghelardini, P., and Paolozzi, L. (2007) Three functional subdomains of the *Escherichia coli* FtsQ protein are involved in its interaction with the other division proteins, *Microbiology 153*, 124–138.

29. Roach, P. C. J., O'Reilly, J., Norbertczak, H., et al. (2008) Equipping a research scale fermentation laboratory for production of membrane proteins, in *Practical Fermentation Technology*, McNeil, B., and Harvey, L. M., Eds., Wiley, Chichester, UK, pp. 37–67.

30. Deacon, S. E., Roach, P. C. J., Postis, V. L. G., et al. (2008) Reliable scale-up of membrane protein over-expression by bacterial auto-induction: from microwell plates to pilot scale fermentations, *Mol. Membr. Biol.* *25*, 588–598.

31. Grisshamer, R., and Tate, C. G. (1995) Overexpression of integral membrane proteins for structural studies, *Q. Rev. Biophys.* *28*, 315–422.

32. Bhairi, S. M. (1997) *Detergents: A Guide to the Properties and Uses of Detergents in Biological Systems*, Calbiochem-Novabiochem Corporation, Nottingham, UK.

33. Keyes, M. H., Gray, D. N., Kreh, K. E., et al. (2003) Solubilizing detergents for membrane proteins, in *Methods and Results in Crystallization of Membrane Proteins*, Iwata, S., Ed., International University Line, La Jolla, CA, pp. 15–33.

34. Gutman, D. A. P., Mizohata, E., Newstead, S., et al. (2007) A high-throughput method for membrane protein solubility screening: the ultracentrifugation dispersity sedimentation dispersion assay, *Protein Sci.* *16*, 1422–1428.

35. Howard, T. D., McAuley-Hecht, K. E., and Cogdell, R. J. (2000) Crystallisation of membrane proteins, in *Membrane Transport: A Practical Approach*, Baldwin, SA, Ed., Oxford University Press, New York.

36. Hochuli, E., Bannwarth, W., Döbeli, H., et al. (1988) Genetic approach to facilitate purification of recombinant proteins with a novel metal chelate adsorbent, *Bio/Technology 6*, 1321–1325.

37. Quiagen (2003) *The QIA Expressionist: A Handbook for High Level Expression and Purification of 6xHis-Tagged Proteins*, QIAGEN Gmbh, Basel.

38. Wallace, B. A., Lees, J. G., Orry, A. J., et al. (2003) Analyses of circular dichroism spectra of membrane proteins, *Protein Sci. 12,* 875–884.

39. Morrison, S., Ward, A., Hoyle, C. J., and Henderson, P. J. F. (2003) Cloning, expression, purification and properties of a putative multidrug resistance efflux protein from *Helicobacter pylori, Int. J. Antimicrob. Agents 22*(3), 242–249.

40. Venter, H., Ashcroft, A. E., Keen, J. N., Henderson, P. J. F., and Herbert, R. B. (2002) Molecular dissection of membrane transport proteins: mass spectrometry and sequence determination of the galactose-proton symport protein, GalP, of *Escherichia coli* and quantitative assay of the incorporation of [ring-2-^{13}C]histidine and ^{15}NH$_3$, *Biochem. J. 363,* 243–252.

41. Yu, J. L., Grinius, L., and Hooper, D. C. (2002) NorA functions as a multidrug efflux protein in both cytoplasmic membrane vesicles and reconstituted proteoliposomes, *J. Bacteriol. 184,* 1370–1377.

42. Caffrey, M. (2000) A lipid's eye view of membrane protein crystallization in mesophases, *Curr. Opin. Struct. Biol. 10,* 486–497.

43. Yin, C. C., Aldema-Ramos, M. L., Borges-Walmsley, M. I., et al. (2000) The quarternary molecular architecture of TetA, a secondary tetracycline transporter from *Escherichia coli, Mol. Microbiol. 38,* 482–492.

44. Yin, Y., He, X., Szewczyk, P., et al. (2006) Structure of the multidrug transporter EmrD from *Escherichia coli, Science 312,* 741–744.

45. Allard, J. D., and Bertrand, K. P. (1992) Membrane topology of the pBR322 tetracycline resistance protein, *J. Biol. Chem. 267,* 17809–17819.

46. Vardy, E., Arkin, I. T., Gottschalk, K. E., et al. (2004) Structural conservation in the major facilitator superfamily as revealed by comparative modeling, *Protein Sci. 13,* 1832–1840.

47. Hirai, T., Heymann, J. A., Maloney, P. C., et al. (2003) Structural model for 12-helix transporters belonging to the major facilitator superfamily. *J. Bacteriol. 185,* 1712–1718.

48. Cogdell, R. J., Ed. (2008) Special issue on the membrane protein structure initiative of the BBSRC, *Mol. Membr. Biol. 25,* 585–701.

49. Geertsma, E., and Poolman, B. (2007) High-throughput cloning and expression in recalcitrant bacteria, *Nat. Methods 4,* 705–707.

50. Langton, K. P., Henderson, P. J. F., and Herbert, R. B. (2005) Antibiotic resistance: multidrug efflux proteins, a common transport mechanism? *Nat. Prod. Rep. 22,* 439–451.

51. Basting, D., Lehner, I., Lorch, M., et al. (2006) Investigating transport proteins by solid state NMR, *Naunyn-Schmiedeberg's Arch. Pharmacol. 372,* 451–464.

52. Abramson, J., Smirnova, I., Kasho, V., et al. (2003) Structure and mechanism of the lactose permease of *Escherichia coli, Science 301,* 610–615.

53. Huang, Y., Lemieux, M. J., Song, J., et al. (2003) Structure and mechanism of the glycerol-3-phosphate transporter from *Escherichia coli, Science 301,* 616–620.

54. Neyfakh, A. A. (2002) The ostensible paradox of multidrug recognition, in *Microbial Multidrug Efflux,* Paulsen, I. T., and Lewis, K. Eds., Horizon Press, Norwich, UK, pp. 21–30.

55. Murray, D. S., Schumacher, M. A., and Brennan, R. G. (2004) Crystal structures of QacR–diamidine complexes reveal additional multidrug-binding modes and a novel mechanism of drug charge neutralization, *J. Biol. Chem. 279,* 14635–14371.

56. Newberry, K. J., Huffman, J. L., Miller, M. C., Vazquez-Laslop, N., Neyfakh, A. A., and Brennan, R. G. (2008) Structures of BmrR–drug complexes reveal a rigid multidrug binding pocket and transcription activation through tyrosine expulsion, *J. Biol. Chem. 283*, 26795–26804.

57. Stavri, M., Piddock, L. J., and Gibbons, S. (2007) Bacterial efflux pump inhibitors from natural sources, *J. Antimicrob. Chemother. 59*, 1247–1260.

58. Chopra, I. (2005) New drugs for the superbugs, *Microbiol. Today 27*, 4–6.

EFFLUX PUMPS AS AN IMPORTANT MECHANISM FOR QUINOLONE RESISTANCE

By JORDI VILA, ANNA FÀBREGA,
IGNASI ROCA, *Departamento de Microbiología Clínica,*
Centro de Diagnóstico Biomédico, Hospital Clínic, Facultad
de Medicina, Universidad de Barcelona, Barcelona, Spain, and
ALVARO HERNÁNDEZ, JOSÉ LUIS MARTÍNEZ,
Departamento de Biotecnología Microbiana, Centro Nacional de
Biotecnología-CSIC and CIBERESP, Madrid, Spain

CONTENTS

Advances in Enzymology and Related Areas of Molecular Biology, Volume 77
Edited by Eric J. Toone Copyright © 2011 John Wiley & Sons, Inc.

I. Background

Fluoroquinolones are wide-spectrum antibacterial agents. Nalidixic acid was the first quinolone to show antibacterial activity and was found as a by-product of antimalarial research. Since then, different changes in the basic structure of nalidixic acid have increased the spectrum and potency of these antibacterial agents (1). Currently, quinolones can be classified into three generations. The first-generation quinolones, such as nalidixic acid and pipemidic acid, lack the fluorine atom at position 6. These quinolones reach high concentrations in the urinary tract and hence have been used for treatment of urinary tract infections caused by gram-negative bacilli, except for *Pseudomonas aeruginosa*. They lack activity against gram-positive cocci and anaerobes. The second-generation quinolones (norflox-acin, ciprofloxacin, and ofloxacin) incorporated a cyclic diamine at position 7 and a fluorine atom at position 6 in the quinolone nucleus. This generation showed excellent activity against gram-negative bacteria. However, despite presenting enhanced antibacterial activity in comparison to the first gener-ation, even against gram-positive cocci, they showed moderate activity against *Staphylococcus aureus* and *Streptococcus pneumoniae* and no activity against anaerobes. The third generation (levofloxacin, moxiflox-acin, and gatifloxacin) has greater activity against gram-positive cocci, and some are also active against anaerobes. The main difference in the structure of the third with respect to the second-generation quinolones is found in the substituent located at position 8 of the quinolone nucleus (1).

Quinolones act by inhibiting the DNA gyrase and topoisomerase IV, two enzymes essential for the viability of bacteria. The DNA gyrase is a tetrameric enzyme with two A subunits and two B subunits. The A subunit catalyzes the breakage and reunion of the DNA, whereas the B subunit shows ATPase activity, thereby generating energy from the hydrolysis of the ATP, which is needed for the reaction to take place. The A subunit is encoded in the *gyrA* gene, whereas the B subunit is encoded in the *gyrB* gene. This enzyme catalyzes the negative supercoiling of the DNA, thereby playing an important role in both the replication and transcription of the DNA. The topoisomerase IV is also a tetrameric enzyme with two A subunits and two B subunits, encoded in the *parC* and *parE* genes, respectively. It is attached to the inner membrane of the cell and catalyzes the decatenation of the two daughter nucleoids (2).

Acquisition of quinolone resistance is due to two mechanisms:

1. Mutations in the genes encoding protein targets and mutations related to the decreased accumulation of the drug, either by a decreased permeability linked to the lack of expression of certain porins in gram-negative bacteria or by increased efflux associated with the overexpression of some efflux pumps in both gram-negative and gram-positive bacteria.

2. Acquisition of genetic elements carrying quinolone-resistant determinants. Among these determinants the following have been described: (a) the *qnr* gene encoding a protein that protects protein targets from the binding of the quinolones; (b) the *aac(6′)Icr* gene encoding an aminoglycoside-modifying enzyme which can acetylate the nitrogen of the piperacyline ring found in most fluoroquinolones, and (c) plasmid-encoded efflux pumps such as that encoded in the *qepA* gene and that encoded in the *oqxAB* gene, both pumping quinolones out of the cell (2).

Efflux pumps are expressed in all living cells, being their number for each bacterial species proportional to the genome size for their expression. According to the substrate specificity, efflux pumps can be single and multidrug efflux pumps. Drug-specific efflux pumps can extrude only one type of antimicrobial agent, such as TetA specific for tetracyclines (3) and CraA specific for chloramphenicol (4). The multidrug efflux pumps can extrude different classes of antimicrobial agents. The multidrug efflux systems have been grouped into six families: the ATP-binding cassette (ABC) family, the major facilitator superfamily (MFS), the resistance–nodulation–division (RND) family, the multidrug and toxic compound extrusion (MATE) family, the small multidrug-resistance (SMR) family, and the drug/metabolite transporter (DMT) superfamily (5).

ABC-type efflux pumps are ATP-dependent multidrug transporters and use ATP as a source of energy to expel the antimicrobial agents from the cell. The members of this family are rarely involved in the acquisition of resistance to antimicrobial agents in gram-negative bacteria, but they have a relevant role in gram-positive bacteria. The other types of efflux pumps are drug–proton or drug–Na^+ antiporters. Antimicrobial efflux is accomplished utilizing the proton-motive force.

MDR efflux pumps contribute to antibiotic resistance in bacteria in two ways: (1) intrinsic resistance to one or several antimicrobial drugs, and (2) resistance conferred by overexpression of an efflux pump. Overexpressed efflux pumps can be mediated by mutations in (1) the local regulator gene (often a repressor), (2) a global regulatory gene, (3) the promoter region of the transporter gene, or (4) insertion elements upstream from the transporter gene.

This chapter is focused on the current knowledge of efflux pumps affecting quinolones in a plethora of bacteria.

II. Efflux Pumps in Enterobacteriaceae

Intrinsic resistance has been reported in Enterobacteriaceae, as in most gram-negative bacteria, to account for decreased outer membrane permeability as well as constitutive expression of multidrug efflux transporters that can export a wide range of substrates. These types of pumps belong primarily to the RND family and protect bacteria by providing immediate response to structurally diverse antimicrobial agents (Table 1). Furthermore, the increasingly more prevalent overexpression of such systems is leading to decreased susceptibility or higher levels of resistance to several unrelated antibiotics (6, 7).

A. *Escherichia coli*

Several studies have focused on the characterization of particular genes within the *Escherichia coli* genome selected on the basis of sequence similarities of their open reading frames with previously known drug transporter proteins. Only four efflux pumps have been reported to be clearly involved in quinolone efflux in in vitro experiments: (1) AcrAB and (2) AcrEF belong to the RND family and confer eight- and fourfold

TABLE 1

Fluoroquinolone Efflux Pumps in Enterobacteriaceae

Microorganism	Efflux Pump[a]	MDR Family	Quinolones Extruded	Other Compounds Extruded[b]	Refs.
Citrobacter freundii	AcrAB[ESR]	RND	Ciprofloxacin	Chloramphenicol	(69)
Enterobacter aerogenes	AcrAB	RND	Norfloxacin, ciprofloxacin, nalidixic acid	Chloramphenicol, tetracycline, ceftazidime, cefepime, novobiocin, mitomycin, acriflavine, SDS	(54, 56)
E. cloacae	AcrAB	RND	Norfloxacin, ciprofloxacin, nalidixic acid	Chloramphenicol, tetracycline, tigecycline, telithromycin, piperacillin, imipenem, meropenem, aztreonam, oxacillin, erythromycin, clindamycin, tobramycin, gentamicin, amikacin, linezolid, cotrimoxazol, novobiocin, rifampin	(57)
Escherichia coli	AcrAB	RND	Norfloxacin, ciprofloxacin, nalidixic acid, ofloxacin, enoxacin	Chloramphenicol, tetracycline, minocycline, ampicillin, oxacillin, cefuroxime, erythromycin, puromycin, clarithromycin, azithromycin, clindamycin, linezolid, trimethoprim, rifampin, doxorubicin, novobiocin, acriflavine, crystal violet, EtBr, rhodamine 6G, TPP, benzalkonium, SDS, deoxycholate	(9, 17)
	AcrEF	RND	Ciprofloxacin, ofloxacin	Chloramphenicol, tetracycline, trimethoprim, erythromycin, clarthromycin, azithromycin, clindamycin, linezolid, cefuroxime, oxacillin doxorubicin, acriflavine, EtBr, rhodamine 6G, SDS, deoxycholate	(9, 17)

(continued)

171

TABLE 1 (*Continued*)

Microorganism	Efflux Pump[a]	MDR Family	Quinolones Extruded	Other Compounds Extruded[b]	Refs.
	YdhE	MATE	Norfloxacin, ciprofloxacin, enoxacin	Chloramphenicol, fosfomycin, doxorubicin, trimethoprim, EtBr, TPP, benzalkonium, deoxycholate	(9, 10)
	MdfA	MFS	Norfloxacin, ciprofloxacin	Chloramphenicol, tetracycline, doxorubicin, trimethoprim, kanamycin, neomycin, erythromycin, puromycin, daunomycin, rifampin, acriflavine, EtBr, TPP, benzalkonium, rhodamine 6G	(8, 9)
Klebsiella penumoniae	AcrAB	RND	Norfloxacin, ciprofloxacin, nalidixic acid	Chloramphenicol, tetracycline, tigecycline, minocycline, erythromycin, trimethoprim, novobiocin, acriflavine, EtBr	(48, 50)
	KmrA	MFS	Norfloxacin, ofloxacin	Kanamycin, gentamicin, erythromycin, acriflavine, CTAB, EtBr, H33342, mehyl viologen, TPP, DAPI	(51)
	KdeA	MFS	Norfloxacin, ciprofloxacin	Chroamphenicol, puromycin, daunomycin, acriflavine, EtBr, TPP	(52)
Proteus mirabilis	AcrAB	RND	Ciprofloxacin	Chloramphenicol, tigecycline, minocycline, ampicillin, novobiocin, erythromicin, trimethoprim, EtBr, acriflavine, SDS	(63)
P. vulgaris	Unknown	Unknown	Norfloxacin, ciprofloxacin, ofloxacin	Chloramphenicol, tetracycline, gentamicin, ceftazidime, imipenem	(66)

Salmonella enterica	AcrAB	RND	Ciprofloxacin, nalidixic acid, levofloxacin, enrofloxacin, gatifloxacin, sarafloxacin, danofloxacin, marbofloxacin, difloxacin, orbifloxacin, flumequine	Ciprofloxacin, nalidixic acid, levofloxacin, enrofloxacin, gatifloxacin, sarafloxacin, danofloxacin, marbofloxacin, difloxacin, orbifloxacin, flumequine	(32, 33)
	AcrEF	RND	Ciprofloxacin, enrofloxacin, marbofloxacin	Florfenicol, erythromycin	(35)
Serratia marcescens	AcrA-like[ESR]	RND	Ciprofloxacin, ofloxacin	EtBr	(59)
	SdeXY	RND	Norfloxacin	Tetracycline, erythromycin, ampicillin, benzalkonium chloride, EtBr, acriflavine, rhodamine 6G, H33342, triclosan, clorhexidine gluconate	(60)
	SmfY	MFS	Norfloxacin, ofloxacin	Acriflavine, DAPI, benzalkonium chloride, EtBr, TPP, methyl viologen, H33342	(61)
	SmdAB	ABC	Norfloxacin, ciprofloxacin	Tetracycline, DAPI, H33342, TPP	(62)

[a]ESR, EPI suggested results (data based on MIC increment after EPI addition).

[b]EtBr, ethidium bromide; TPP, tetraphenylphosphonium; DAPI, 4′,6-diamidino-2-phenylindole; H33324, Hoechst 33342; SDS, sodium dodecyl sulfate.

increases in the minimal inhibitory concentrations (MICs) of norfloxacin and nalidixic acid, respectively; (3) MdfA, a member of the MFS superfamily, and (4) YdhE (also called NorE), a MATE member, both confer eightfold increases in the MIC of norfloxacin, although they do not impair the susceptibility levels of nalidixic acid (8–10). Similar, albeit irreproducible experiments performed with AcrD (RND), YegO (RND), YjiO (40% homolog to MdfA), YceL (MFS), and YdgFE (SMR) have reported a limited ability to extrude quinolones in vitro (9, 11). The combined expression of three out of the four most influent transporters, AcrAB, YdhE, and MdfA, has been studied by Yang et al. (10). Simultaneous overexpression of AcrAB with either YdhE or MdfA led to 7- to 11-fold increases in the MICs of ciprofloxacin and norfloxacin, respectively. These values showed a synergistic effect since they were significantly higher than those supported by the overexpression of AcrAB by itself. However, the existence of an upper limit of approximately 10-fold in the increase in drug resistance has been proposed, either due to the overexpression of a single efflux pump or to the combined increased expression of several systems, since the correlation reported between increased expression of the pump and increased antibiotic MICs is dismissed over this value (10, 12).

Despite the high diversity of genes encoding efflux pumps, AcrAB–TolC is the most important multidrug transporter characterized so far which plays a major role in conferring quinolone resistance not only in *E. coli* clinical isolates but also in multidrug-resistant (MDR) mutants selected in vitro (13–15). Sulavik et al. (16) showed that inactivation of the *mdfA*, *yegMNO*, *acrD*, or *acrEF* genes in wild-type *E. coli* strains did not trigger any change in fluoroquinolone susceptibility, and therefore a putative role in intrinsic resistance was ruled out. Contrarily, the inactivation of either *acrB* or *tolC* has been reported to be the only system capable of increasing the susceptibility of quinolones in wild-type strains and increasing internal drug accumulation in AcrAB-overexpressing mutants. Thus, the role played by this pump has been well characterized in contributing to the intrinsic levels of resistance due to its constitutive expression and in mediating a MDR phenotype by means of its overexpression (13, 14, 16).

The overexpression of an efflux pump other than AcrAB has been detected in particular situations. Jellen-Ritter and Kern (17) reported the selection of a Δ*acrAB* fluoroquinolone-resistant mutant showing enhanced efflux. The results showed that the increased AcrEF expression accounted for the phenotype observed. Kobayashi et al. (18) also found an AcrEF-overexpressing strain in which *acrB* had previously been inactivated. In this situation, the authors selected the strain based on organic solvent tolerance, which had previously been attributed to increased AcrAB expression, also responsible for the MDR phenotype, detected prior to the *acrB* disruption (19). In both studies, AcrEF overexpression was due to the insertional activation of IS elements upstream from the operon. However, fluoroquinolone-resistant strains, either clinical isolates or in vitro selected mutants, with a wild-type *acrAB* locus and overexpressing an efflux system other than AcrAB have not emerged.

The regulatory mechanisms that govern the AcrAB expression levels have been extensively characterized in *E. coli* strains [see the review by Fabrega et al. (2)]. There are three transcriptional activators that belong to the AraC/XylS family (20): MarA (21), SoxS (22), and Rob (23). These three proteins bind to specific sequences, called *marbox*, detected within the promoter of each gene that belongs to their regulons (24). Since regulons are highly overlapping and affect a significant number of genes, they are usually referred to as the *mar regulon*. Both the *acrAB* operon and the *tolC* gene are members of the *mar regulon* (25). Furthermore, the *acrR* gene is the local repressor divergently encoded upstream from the *acrAB* operon, which negatively regulates expression of the efflux pump (12, 26). Chromosomal-encoded mutations acquired within *acrR* that lead to a loss of repressor activity (12, 17), or within the genes that govern *marA* and *soxS* and lead to an overexpression of both proteins (14, 27, 28), give rise to an MDR phenotype by means of AcrAB–TolC. However, no relevant clinical information on Rob and AcrAB overexpression has been published.

B. *Salmonella enterica*

There are two serovars among *Salmonella enterica* isolates which are predominantly isolated: the serovars *typhimurium* and *enteritidis* (29, 30). The efflux pumps contributing to quinolone extrusion have been studied particularly in *typhimurium*. AcrAB has also been reported to play a main role in fluoroquinolone-resistance acquisition in *Salmonella* strains, both isolated from animals and humans (31), and its contribution to the intrinsic resistance phenotype has been demonstrated (32). Moreover, several studies have proposed that the contribution of AcrAB overexpression to the fluoroquinolone-resistance phenotype is greater than that of target gene mutations, and in addition, it is probably the first step acquired in the selection process (31, 32).

The inactivation of either *acrB* or *tolC* in multidrug-resistant isolates leads to a 4- to 64-fold decrease in the MICs of these compounds (32, 33). Furthermore, the lack of mutants selected in vitro in the presence of ciprofloxacin from *acrB*-deleted fluoroquinolone-susceptible strains reinforces the main role that AcrAB plays in quinolone resistance (34). Only Olliver et al. (35) reported the selection of fluoroquinolone-resistant mutants from *acrB*-deleted strains. However, the parent strains 902SA92 and 102SA00 already showed a high level of fluoroquinolone resistance due to target gene mutations in addition to AcrAB overexpression. The resistant mutants showed an overexpression of AcrEF, due to the insertion of IS elements upstream from the operon, presumably activating the transcription of the pump. These results suggest that the background of the *acrB*-deleted mutants probably plays a key role in allowing the emergence of other efflux systems.

Apart from AcrAB, other efflux pump genes have been detected in the chromosome and have been studied in *Salmonella typhimurium*. Chen et al. (32) studied known and putative multidrug transporters among laboratory-selected fluoroquinolone-resistant *Salmonella typhimurium* strains. They reported a ≥2-fold increase in the AcrA, AcrB, AcrE, AcrF (80% homolog to AcrB), EmrB, EmrD, or MdlB expression levels, whereas TolC, MdtB, MdtC, or EmrA showed an increase of only ≤2-fold. Further characterization of these pumps putatively involved in fluoroquinolone resitance was carried out: the inactivation of *acrEF*, *mdtABC*, *emrAB*, *mdlAB*, or *acrD* either in a wild-type background or in fluoroquinolone-resistant strains did not lead to any significant change in the MICs of fluoroquinolones. These results suggest that the transporters evaluated so far play a limited role or no role in quinolone resistance (32, 34, 35).

The regulatory proteins described in *E. coli* capable of governing the AcrAB levels have also been studied in *S. enterica* serovars *typhimurium*, *enteritidis*, and *choleraesius*; homologs of *soxS* (36), *marA* (37), and *acrR* (38) have been characterized, and the same types of mutations leading to their overexpression or repression, respectively, have been detected (39–41). However, a truncated form of the AcrR protein was detected by Chu et al. (42) among both quinolone-susceptible and quinolone-resistant *S. chloreaesius* clinical isolates suggesting that *acrR* may have evolved to become a pseudogene in most *S. choleraesius*. In addition to these known AraC regulators, there is a MarA homolog not found in *E. coli* but detected among the genus *Salmonella* as well as in other Enterobacteriaceae, named RamA, which is also capable of inducing an increased expression of AcrAB and the MDR phenotype (43). Mutations acquired within the local repressor, RamR, account for the increased levels of RamA transcription (40, 44).

C. *Klebsiella pneumoniae*

The studies on the role that efflux plays in conferring quinolone resistance in *Klebsiella pneumoniae* were initiated in 1997 by Deguchi et al. (45) and rely on experiments performed with fluoroquinolone-resistant strains in which a decrease in the MIC of several quinolones has been reported in the presence of phenylalanyl arginyl β-naphthylamide (PAβN). Other experiments have shown a decrease in the internal accumulation of ciprofloxacin upon the addition of CCCP (carbonyl cyanide *m*-chlorophenylhydrazone) (45, 46). The homologous loci of *acrAB* and *acrR* have been sequenced in *K. pneumoniae*. Mazzariol et al. (47) were the first to demonstrate a correlation between efflux-mediated norfloxacin resistance and increased levels of AcrA expression. Such experiments were repeated later, with efflux being attributed to an overexpression of AcrAB (48). In terms of regulation, AcrR, MarA, SoxS, and RamA, the latter of which was first described in this pathogen (49), have been studied. Overexpression of AcrA has been associated with strains displaying mutations within *acrR* and strains showing increased levels of RamA (48, 50). However, increased expression of neither MarA nor SoxS has been associated with quinolone resistance.

Two newly described efflux pumps, KmrA (51) and KdeA (52), have been characterized for their ability to export norfloxacin. Both proteins belong to the MFS family and show similarity with EmrB and MdfA of *E. coli*, respectively. The expression of these proteins in a hypersusceptible *E. coli* strain resulted in an eightfold increase in the MIC of norfloxacin for KmrA, whereas KdeA only accounted for a fourfold increase. No other change in the MIC of nalidixic acid or ciprofloxacin was detected. A constitutive expression of the *kdeA* gene has been proposed in *K. pneumoniae* ATCC 10031, whereas the same cannot be extended to the *kmrA* gene (51, 52). However, the role played by these pumps in the acquisition of quinolone resistance in clinical isolates remains to be clarified.

D. *Enterobacter aerogenes*

Efflux was evidenced among quinolone-resistant *Enterobacter aerogenes* clinical isolates in 1998 by Mallea et al. (53). They reported a significant decrease in the internal accumulation of norfloxacin in the presence of CCCP. Furthermore, the homolog sequences of the *acrRAB* locus and the *tolC* gene have been characterized (54). Disruption of the *acrA* gene led to a four- and twofold decrease in the MICs of norfloxacin and ciprofloxacin, whereas disruption of the *tolC* gene was associated with a 16- and eightfold decrease in the MICs of the same antibiotics. These results suggested that other efflux systems, probably acting in conjunction with TolC, are involved in quinolone efflux. Furthermore, the MDR phenotype observed in an *E. aerogenes* isolate was associated with a mutation within *acrR* leading to the repression of the efflux pump. Concerning the other regulators, the *marRAB* operon (55) and the *ramA* gene (56) have been detected. Both activators, MarA and RamA, lead to an MDR phenotype in *E. aerogenes*, associated with a decreased expression of porins and decreased internal accumulation of norfloxacin related to increasing AcrA levels. These accumulation values can revert in the presence of CCCP (55, 56). Furthermore, it has been demonstrated that RamA from *E. aerogenes* can *marA*-independently and *marA*-dependently induce the MDR phenotype in *E. coli* since this protein can increase expression of the *marRAB* operon (56).

E. *Enterobacter cloacae*

Other studies have focused on quinolone resistance in *Enterobacter cloacae* strains. Proteins similar to MarR, AcrAB, RamA, and Rob have been detected. The MDR phenotype has been associated with increased expression of *marR*, and therefore *marA* (since both genes are cotranscribed) and *acrB* (57, 58). No homolog of the *soxS* gene or the *soxRS* region has been found (58). These results obtained from *E. cloacae* strains may be extensive for *E. aerogenes*.

F. *Serratia marcescens*

In 2002, quinolone efflux was detected among clinical isolates of *Serratia marcescens* by Kumar and Worobec (59). Efflux of ciprofloxacin and ofloxacin was detected and attributed, at least partially, to a proton gradient-dependent efflux pump since the internal accumulation of both antibiotics increased upon addition of CCCP. Furthermore, at least two AcrA-like proteins were detected using immunoblot assays, suggesting that a putative RND efflux pump may be involved in the intrinsic resistance to fluoroquinolones. However, no increase was detected in the accumulation of norfloxacin.

More recently, other studies have associated quinolone resistance with the overexpression of specific efflux pump mechanisms. SdeXY was the first pump characterized in this pathogen as an RND multidrug efflux system. The overexpression of either SdeXY or SdeY alone in a *E. coli* KAM32 strain (Δ*acrB* and Δ*ydhE*) led to a fourfold increase in norfloxacin resistance. These results demonstrated the role played by SdeXY as a new quinolone efflux pump and suggested an interaction between SdeY and AcrA and TolC in order to constitute an active system (60). The second efflux system, SmfY, was characterized as a putative MFS member. Its overproduction in vitro conferred a four- and twofold increase in the resistance to norfloxacin and ofloxacin, respectively (61). Finally, SmdAB is the third efflux system characterized so far. It has been proposed to belong to the ABC superfamily. It has been reported to increase the norfloxacin and ciprofloxacin resistance by eight- and twofold, respectively. Furthermore, this system represents the first heterodimeric ABC-type multidrug efflux pump that has been reported in gram-negative bacteria (62).

G. *Proteus mirabilis*

The first description of the AcrAB homolog locus in a *Proteus mirabilis* clinical isolate was made by Visalli et al. in 2003 (63). They noticed a significant decrease in the MICs of several antibiotics after disruption of the *acrB* gene, including a four- to eightfold decrease in the MIC of ciprofloxacin. Later, nine clinical isolates were studied for their resistance to fluoroquinolones, with only three reporting high levels of fluoroquinolone resistance while simultaneously overexpressing the *acrB* gene. These results suggested a synergistic effect between efflux and the target gene mutations acquired by these strains in conferring the high level of fluoroquinolone resistance observed (64).

H. *Proteus vulgaris*

The evidence of efflux as a mechanism involved in quinolone resistance was detected earlier in *Proteus vulgaris* than in *P. mirabilis*. CCCP was used by Ishii et al. in 1991 (65) in a *P. vulgaris* fluoroquinolone-resistant clinical isolate and revealed efflux as one of the mechanisms of ofloxacin resistance. However, less information is available concerning efflux and quinolones. There is only one other study in which a regulator capable of increasing the levels of resistance to several quinolones was characterized.

The protein was dubbed PqrA and showed significant amino acid sequence similarity to SoxS and MarA. Thus, it may presumably confer the MDR phenotype in a manner similar to that exerted by the known regulators: via overexpression of AcrAB. Although this pump has not been reported in *P. vulgaris*, it has been studied in *P. mirabilis* (66).

I. *Citrobacter freundii*

Fluoroquinolone resistance has been studied in *Citrobacter freundii* among a set of quinolone-susceptible and quinolone-resistant clinical isolates (67) and among 10 fluoroquinolone-resistant mutants selected in vitro (68). Beside the acquisition of target gene mutations, both studies revealed efflux as a mechanism contributing to fluoroquinolone resistance, since either norfloxacin or ciprofloxacin accumulation, respectively, was lower among resistant isolates and increased upon addition of CCCP. In addition, a parallel increase in the MICs of chloramphenicol (67) and tetracycline (68) was shown in the resistant strains, suggesting the presence of quinolone-resistance mechanisms other than target gene mutations.

In a more recent study (69) a partial characterization of the *acrAB* locus was reported in two fluoroquinolone-resistant *C. freundii* clinical isolates. Both isolates showed the same target gene mutations but had a fourfold difference in the MIC of ciprofloxacin related to a decreased ciprofloxacin accumulation in the more resistant strain. Microarray analysis revealed an overexpression of the acrA and acrB genes in the strain showing higher levels of resistance, suggesting that efflux may play a role in modulating the final levels of fluoroquinolone resistance.

III. Efflux Pumps in Nonfermentative Gram-Negative Bacilli

Until recently, nonfermentative gram-negative bacilli have been considered as commensals or environmental bacteria with little clinical importance. However, recent studies have shown that these organisms account for almost 15% of all the isolations in clinical microbiology laboratories, associated primarily with nosocomial infections. This indicates that nonfermentative gram-negative bacilli constitute an important group of opportunistic pathogens (70). From a clinical point of view, the most important species within this group are *Pseudomonas aeruginosa*, *Acinetobacter baumannii*, *Stenotrophomonas maltophilia*, and *Burkholderia* spp. (71) (Table 2). One of the most worrisome characteristics of

TABLE 2
Fluoroquinolone Efflux Pumps in Nonfermentative Gram-Negative Bacilli

Microorganism	Efflux Pump[a]	MDR Family	Quinolones Extruded[b]	Other Compounds Extruded[c]	Refs.
Acinetobacter baumannii	AdeIJK	RND	Levofloxacin, pefloxacin	β-Lactams, chloramphenicol, tetracycline, erythromycin, lincosamides, fusidic acid, novobiocin, rifampin, trimethoprim, acridine, safranin, pyronine, SDS	(117)
	AdeABC	RND	Sparfloxacin, ofloxacin, norfloxacin, pefloxacin, levofloxacin	Kanamycin, genistein, tobramycin, amikacin, cefotaxime, tetracycline, erythromycin, chloramphenicol, trimethoprim, tigecycline	(260)
	AdeDE	RND	Ciprofloxacin, ofloxacin[DT], nalidixic acid[DT]	Amikacin, ceftazidime, tetracycline, chloramphenicol, erythromycin, meropenem, rifampin, EtBr	(123)
	AbeM	MATE	Ciprofloxacin, norfloxacin, ofloxacin	Gentamicin, kanamycin, erythromycin, chloramphenicol, trimethoprim, daunomycin, doxorubicin, triclosan, acriflavine, DAPI, EtBr, H33342, rhodamine 6G, TPP	(125)
Burkholderia cepacia	BcrA	MFS	Nalidixic acid	Tetracycline	(152)
B. pseudomallei	Unknown[ESR]	RND	Levofloxacin	—	(150)
	BPSS1119/ BMAA1045	RND	Pefloxacin, ofloxacin, ciprofloxacin	Ceftazidime, ampicillin, genistein, erythromycin, chloramphenicol	(261)

(continued)

181

TABLE 2 (*Continued*)

Microorganism	Efflux Pump[a]	MDR Family	Quinolones Extruded[b]	Other Compounds Extruded[c]	Refs.
Pseudomonas aeruginosa	MexAB–OprM	RND	Nalidixic acid	β-Lactams, chloramphenicol, carbenicillin, β-lactams, novobiocin, trimethoprim, tigecycline, sulphonamides, EtBr, acriflavine, crystal violet, SDS, aromatic hydrocarbons, homoserine lactones, cerulenin, thiolactomycin, triclosan, TPP, rhodamine, lincomycin, mureidomycin	(262, 263)
	MexCD–OprJ	RND	Trovafloxacin	Erythromycin, cephems, β-lactams, chloramphenicol, novobiocin, trimethoprim, tetracycline, macrolides, crystal violet, EtBr, acriflavine, SDS, aromatic hydrocarbons, cerulenin, triclosan, tigecycline, lincomycin, TPP, rhodamine, benzalkonium, chorhexidine, mureidomycin	(79, 262)
	MexEF–OprN	RND	Ciprofloxacin	Chloramphenicol, trimethoprim, aromatic hydrocarbons, triclosan	(80, 262)

Organism	Pump	Family		Substrates	Ref.
	MexXY	RND	Norfloxacin, ofloxacin	Aminoglycosides, erythromycin, tetracycline, β-lactams, aminoglycosides, erythromycin, tetracycline, tigecycline, mureidomycin, acriflavine, EtBr, erythromycin	(82)
	PmpM	MATE	Ciprofloxacin, norfloxacin	Fradiomycin, chlorhexidine, acriflavine, EtBr, rhodamine 6G, TPP, benzalkonium chloride	(264)
	MexVW	RND	Norfloxacin, ofloxacin	Chloramphenicol, cefpirome, tetracycline, EtBr, acriflavine, erythromycin	(84)
P. fluorescens	EmhABC	RND	Nalidixic acid, ciprofloxacin	Toluene, chloramphenicol, rhodamine 6G, dequalinium, polycyclic aromatic hydrocarbons	(265, 266)
P. putida	TtgABC	RND	Nalidixic acid	Ampicillin, carbenicillin, tetracycline, chloramphenicol, aromatic hydrocarbons	(126, 267)
P. stutzeri	TbtABM	RND	Nalidixic acid	Tributyltin, chloramphenicol, sulfamethoxazole, *n*-hexane	(113)
Stenotrophomonas maltophilia	SmeDEF	RND	Ciprofloxacin, nalidixic acid, norfloxacin, ofloxacin	Tetracycline, chloramphenicol, erythromycin, EtBr, triclosan	(129)
	SmeIJK	RND	Ciprofloxacin	Amikacin, tetracycline, minocycline	(126)

[a]ESR, EPI suggested results (data based on MIC increment after EPI addition).
[b]DT, Data from disk diffusion tests.
[c]EtBr, ethidium bromide; TPP, tetraphenylphosphonium; DAPI, 4′,6-diamidino-2-phenylindole; H33324, Hoechst 33342; SDS, sodium dodecyl sulfate.

nonfermentative gram-negative bacilli consists in their elevated intrinsic drug resistance and their capability to increase this resistance level further either by mutation or by horizontal gene transfer (72). This low-level susceptibility to antibiotics includes quinolones. Nonfermentative gram-negative bacilli are usually free-living bacteria presenting large genomes, which allows their adaptation to very different habitats. It has been suggested that at least for some of these bacterial species, the clinical isolates are not a subset of strains presenting specific virulence and resistance properties (73, 74). This indicates that their characteristic phenotype of low susceptibility to quinolones, shared by all members, has been acquired in their natural, nonclinical habitats, long before the invention of these antibiotics (75, 76). The fact that *P. aeruginosa* environmental strains isolated before the use of quinolones in the clinical setting can extrude these synthetic antibiotics (73) strongly suggests that the determinants capable of extruding quinolones are, in fact, relevant elements for bacterial physiology and can shape the intrinsic low susceptibility to quinolones of nonfermentative gram-negative bacilli as a side effect of their primary functional role (77).

A. *Pseudomonas aeruginosa*

In its chromosome, *Pseudomonas aeruginosa* contains several putative MDR efflux pump-encoding genes belonging to the RND family of bacterial transporters. The best studied are MexAB–OprM (78), MexCD–OprJ (79), MexEF–OprN (80), and MexXY (81, 82). Other MDR pumps probably contributing to antibiotic resistance in *P. aeruginosa* are MexJK (83), MexVW (84), MexPQ–OpmE, and MexMN (85). Among them, it has been described that quinolones are substrates of MexAB–OprM, MexCD–OprJ, MexEF–OprN, MexXY, and MexVW. A different issue concerns the MDR efflux pump MexGHI–OpmD (86). In contrast to the situation for other MDR pumps, mutations in *mexGHI* reduce the susceptibility of *P. aeruginosa* to quinolones. This result means that MexGHI–OpmD does not extrude quinolones like the aforementioned MDR pumps. However, it can affect the susceptibility of *P. aeruginosa* to these drugs (and also bacterial virulence) by mechanisms that are not completely understood (86).

MexAB–OprM and MexXY are the only pumps that are constitutively expressed and thus contribute to *P. aeruginosa* intrinsic resistance to quinolones under standard growth conditions (87). The contribution of the other pumps to intrinsic resistance is probably negligible unless the appropriate induction conditions are given. Expression of MDR pumps is strongly down-regulated by transcriptional factors that are frequently upstream from the operon encoding for the pump (88). Mutations in the elements regulating expression of MDR pumps (88) render a constitutive expression of these pumps as well as a concomitant phenotype of low-level acquired resistance to several antibiotics, quinolones included.

A survey on the mechanisms of quinolone resistance in a collection of 20 quinolone-resistant *P. aeruginosa* isolates detected that 17 of such isolates (85%) overproduced at least one of the MDR efflux pumps MexAB–OprM, MexCD–OprJ, MexEF–OprN, or MexXY (89). This indicates that overproduction of MDR efflux pumps is highly prevalent among *P. aeruginosa* quinolone-resistant isolates. It has been shown that different types and concentrations of quinolones select mutants overexpressing one or another of the *P. aeruginosa* MDR pumps, at least under laboratory conditions (90). This specificity in mutant selection depending on the quinolone used was also observed using an experimental model of *P. aeruginosa* pneumonia in rats. For ciprofloxacin-treated rats, the most predominant cause of resistance was overproduction of MexEF–OprN, whereas trovafloxacin selected mainly MexCD–OprJ overproducers (91). Quite surprisingly, mutants overproducing these pumps emerged in this model with higher frequencies than predicted (91).

It has been described that mutations in topoisomerases genes are the predominant mechanisms of quinolone resistance for *P. aeruginosa* isolates from wounds and urine, whereas overproduction of the MDR efflux systems MexEF–OprN and MexCD–OprJ are common in quinolone-resistant isolates from the lungs of cystic fibrosis (CF) patients (91). Furthermore, overexpression of the MexCD–OprJ pump has been associated with in vivo acquired antibiotic resistance to quinolones in a mouse model of *P. aeruginosa* lung infection (92). Some studies have shown that overproduction of MDR efflux pumps diminishes the virulence of *P. aeruginosa* (93–95). Some of these MDR efflux pumps can extrude quorum-sensing (QS) autoinducers, so that mutants overexpressing them display low-level expression of QS-regulated virulence determinants (96, 97), which might explain the lower virulence of *P. aeruginosa* mutants overexpressing MDR pumps. Despite these results, the finding of MDR *P. aeruginosa* overproducing isolates in human patients indicates that the physiological burden associated with the constitutive expression of these pumps does not fully compromise the capability of *P. aeruginosa* for producing infections (98, 99).

The lifestyle of *P. aeruginosa* is different in chronic and acute infections (100). Isolates from chronic infection present an overall lower level of expression of QS-regulated virulence determinants and type III secretion and are capable of forming thicker biofilms (101–103). Notably, it has been described that overproduction of either MexCD–OprJ or MeEF–OprN (but not MexAB–OprM or MexXY) is associated with a reduction in type III secretion (104). Whether or not this side effect of efflux-driven quinolone resistance favors the persistence of MexCD–OprJ and MexEF–OprN overproducers in the lung of the chronically infected patient remains to be established. In fact, MexCD–OprJ overproducers are not often isolated from treated patients, although they can be isolated after single therapy with quinolones (98). On the other hand, a retrospective analysis of nonredundant *P. aeruginosa* strains recovered sequentially from CF patients has shown that 28% of the samples presented defects on MexAB–OprM, whereas overexpression of MexXY was common among these isolates (105).

It should be considered that although the different effects that overproduction of one efflux pump or another may have on *P. aeruginosa* physiology might be important for the persistence of some specific mutants in the absence of further selection, the type of antibiotics used for therapy is probably more important (106). For example, a recent report (107) indicated that the most frequent mechanisms of quinolone resistance in a collection of multiresistant *P. aeruginosa* strains are mutations in *gyrA* (91%) and high-level expression of MexXY (82%). Given that MexXY is the most relevant aminoglycoside-efflux determinant in *P. aeruginosa* (81, 105), treatment with aminoglycosides might help to select these mutants that also display a phenotype of low-level resistance to quinolones. This is an example of second-order selection, a selective mechanism that is very important in the case of determinants, such as the MDR efflux pumps, which can confer simultaneous resistance to different antibiotics (cross-resistance).

Although overexpression of MDR efflux pumps can be associated with low-level resistance to quinolones in *P. aeruginosa*, combinations with other mechanisms of resistance have been reported. For example, a multiresistant *P. aeruginosa* strain producing PER-1 β-lactamase (108) presents mutations in *parC* and *gyrA*, together with overproduction of the MDR efflux pump MexXY that account for high-level quinolone resistance (108).

B. NON-*aeruginosa Pseudomonas*

P. aeruginosa is the most important pathogen belonging to the *Pseudomonas* genus. Nevertheless, some cases of infections by other species belonging to the same genus, and with an environmental origin such as *P. aeruginosa*, have been reported.

1. *Pseudomonas fluorescens and Pseudomonas putida*

These bacterial species produce infections in immunocompromized patients (109) and have been reported to be responsible for some bloodstream infections produced by contaminated blood products (110). Although they are different species, *P. fluorescens* and *P. putida* isolates are sometimes grouped in the same complex because of the strong similarity between them. The chromosomes of some *P. fluorescens* and *P. putida* isolates have been sequenced and annotated (http://v2.pseudomonas.com/and http://cmr.tigr.org/tigrscripts/CMR/shared/Genomes.cgi). Sequence analysis of these genomes demonstrates that they contain several genes encoding MDR efflux pumps, some of which, such as the TtgABC efflux pump (111), extrude antibiotics, including quinolones, ampicillin, carbenicillin, tetracycline, and chloramphenicol.

2. *Pseudomonas stutzeri*

Pseudomonas stutzeri is an unusual cause of human infection, although it has been recovered from wounds, bloodstream, respiratory and urinary tracts, and skin (112). A search for genes involved in resistance to tributyltin (TBT), a toxic agent used in antifouling paints, demonstrated the presence of a novel efflux pump, TbtABM (113), in *P. stutzeri*. This pump effluxes TBT, but also antibiotics, including quinolones, chloramphenicol, and sulfamethoxazole. Differing from other chromosomally encoded MDR efflux pumps, the *tbtABM* operon was present in only half of the *P. stutzeri* strains analyzed, suggesting that the presence/absence of *tbtABM* might be the consequence of subspecific diversification in *P. stutzeri*.

C. *Acinetobacter baumannii*

The role of efflux determinants in the resistance to quinolones of *Acinetobacter baumannii* has been estimated using inhibitors of MDR efflux pumps. A survey using PAβN demonstrated that in the presence of this efflux pump inhibitor, the MIC of nalidixic acid decreased at least eightfold in 45% of the strains from a collection of epidemiologically unrelated clinical isolates (114). Another survey, using reserpine as an efflux pump inhibitor, demonstrated that the MIC of ciprofloxacin decreased at least fourfold (in the presence of the inhibitor) in 33% of the *A. baumannii* clinical isolates (115). This indicates that efflux pumps contributing to quinolone resistance are widely distributed in *A. baumannii*. Notably, the reduction in ciprofloxacin MICs observed with reserpine was not parallel with an increased susceptibility to nalidixic acid. Altogether, these results indicate that different efflux pumps, with different substrate specificities and belonging to different structural families are involved in quinolone resistance in *A. baumannii*.

The sequence of the genomes of two *A. baumannii* isolates, one of which (AYE) is highly resistant and another (SDF) which presents a wild-type antibiotic susceptibility profile, gives new clues as to the mechanisms involved in quinolone efflux. As expected, both strains harbored a large number of genes encoding putative MDR efflux pumps (116). However, the activity of these determinants in extruding quinolones has been demonstrated in only a few cases.

Both strains contain genes encoding the efflux pump AdeIJK (117). This MDR determinant belongs to the RND family of transporters and can extrude a variety of amphiphilic compounds, fluoroquinolones among them. The MIC of levofloxacin is reduced upon deletion of *adeIJK*, which indicates that this efflux is involved in intrinsic resistance to quinolones in *A. baumannii* (117). Stable overexpression of plasmid-encoded AdeIJK inhibits bacterial growth, both in *E. coli* and in *Acinetobacter*, suggesting that mutants overexpressing this system will rarely be selected, given the associated fitness costs (117). A survey of the presence of *adeIJK* demonstrated the presence of this determinant in the genomes of all *A. baumannii* strains tested. Moreover, very similar MDR elements were also found in the chromosomes of *Acinetobacter* genomic DNA group 3 as well as in *Acinetobacter baylyi* (118, 119). Together with the strong burden that its overexpression produces, this suggests that as in other MDR efflux pumps (77), AdeIJK is an ancient element with an important role in the physiology of *Acinetobacter*.

The multidrug-resistant isolate AYE hosts another RND efflux pump, which is not present in the genome of the SDF strain. The genes coding for this determinant, named *adeABC*, are present in a resistance genomic island in *A. baumannii* AYE (116). AdeABC was first described as an aminoglycoside efflux pump present in the chromosome of *Acinetobacter baumannii* BM4454 (120). This MDR efflux pump is expressed at a low level and is not likely to contribute to intrinsic resistance to fluoroquinolones in a wild-type strain. However, *adeABC* is expressed in the clinical isolate BM4454, and its deletion increases the susceptibility to quinolones in this strain (121). Furthermore, the simultaneous deletion of *adeABC* and *adeIJK* in the strain BM4454 increases *A. baumannii* susceptibility to quinolones dramatically (117). Expression of *adeABC* is regulated by the AdeRS two-component system. Specific mutations in both *adeR* and *adeS* account for the constitutive expression of *adeABC* in BM4454 (121). Although *adeABC* is present in the resistance island of the *A. baumannii* AYE strain, it is widely distributed among other *A. baumannii* isolates, being present in more than 80% of the strains of this bacterial species (122). Analysis of the molecular basis of quinolone resistance in different isolates of an outbreak of *A. baumannii* demonstrated that mutations in *gyrA* and *parC* paralleled the constitutive high-level expression of *adeB*, indicating that this system is probably necessary for the development of high-level quinolone resistance in the large majority of *A. baumannii* isolates (122).

Another efflux pump that contributes to intrinsic resistance to quinolones in *Acinetobacter* is AdeDE, which was first described in an isolate belonging to the *Acinetobacter* genomic DNA group 3 (123). Further work demonstrated that, indeed, *adeDE* genes are only present in those clinical isolates of the *A. baumannii–Acinetobacter calcoaceticus* complex that do not present *adeABC/adeIJK*. It is thus likely that these isolates belong to the *Acinetobacter* genomospecies 3 (124).

The MATE family of transporters might also contribute to the extrusion of quinolones in *Acinetobacter*. The AbeM transporter belonging to this family confers resistance to quinolones upon its expression in an *E. coli* strain lacking the major multidrug efflux pumps AcrAB and YdhE (125). The *abeM* gene has been found to be widely distributed among clinical isolates of *A. baumannii*. However, the contribution of this determinant to intrinsic or acquired resistance to quinolones of *A. baumannii* remains to be studied in detail.

D. *Stenotrophomonas maltophilia*

The recent release of the full-genome sequence of two different isolates of *Stenotrophomonas maltophilia* has shown that as in other free-living organisms, this bacterial species hosts several efflux pumps (126, 127). Most of these pumps are present in both genomes, although a few are strain-specific (127). At present, just two of these MDR efflux pumps, Sme-DEF (128–132) and SmeABC (133, 134), have been characterized. Whereas SmeDEF contributes to both intrinsic and acquired resistance to quinolones in *S. maltophilia* (132), SmeAB is not involved in antibiotic resistance, although overexpression of the porin of this tripartite efflux pump, SmeC, reduces the susceptibility of *S. maltophilia* to quinolones, most probably as the consequence of the coupling of this porin to another MDR efflux pump (134). Among the other efflux pumps annotated in the genome of *S. maltophilia*, only SmeIJK seems to be involved in intrinsic resistance to fluoroquinolones (126).

Expression of SmeDEF is regulated by the local transcriptional repressor SmeT, which binds a pseudopalindromic sequence present in the intergenic region between *smeT* and *smeD* (135, 136). However, the fact that some mutants overexpressing SmeDEF do not present relevant changes either in the SmeT protein or in its operator sequence (137) indicates that the regulation of *smeDEF* expression involves more regulators than SmeT.

One important aspect of SmeDEF is that differing from other MDR efflux pumps, overexpression of this determinant confers high-level resistance to quinolones (128–130). Given that overexpression of SmeDEF strongly diminishes *S. maltophilia* virulence (131), it could be thought that usually, these mutants should not be found among the isolates of this bacterial species. Nevertheless, the analysis of a collection of *S. malto-philia* isolates has shown that SmeDEF is overexpressed by a high percentage of isolates and that its expression levels correlate with the level of susceptibility to quinolones in *S. maltophilia* (130, 137). This suggests that SmeDEF might play a primary role in the acquisition of resistance to quinolones in *S. maltophilia*. In fact, it has been demonstrated that *S. maltophilia* quinolone-resistant clinical isolates do not carry mutations in their topoisomerase genes (138, 139), a situation in sharp contrast to what has been observed in other organisms. The presence of a gene encoding a Qnr protein (Sm*qnr*) in the genome of *S. maltophilia* has recently been reported (140, 141). The Qnr family is a group of proteins involved in plasmid-encoded resistance mainly in Enterobacteriaceae (142, 143). Whether or not this element, which is chromosomally encoded in *S. maltophilia*, can contribute to acquired resistance to quinolones, either alone or in combination with SmeDEF overexpression, remains to be explored. This potential synergy between two chromosomally encoded quinolone resistance elements might explain why mutations in the topo-isomerase genes are not the major cause of quinolone resistance in *S. maltophilia*.

As in other MDR pumps, SmeDEF can extrude different antibiotics, and thus its overexpression confers resistance not just to quinolones but to other drugs as well. The other side of the coin is that selection of mutants resistant to any other SmeDEF substrate can also confer high-level resistance to quinolones in *S. maltophilia*. Since the widely used biocide triclosan can select mutants that overexpress SmeDEF (144), the likelihood of selecting quinolone-resistant *S. maltophilia* mutants in environments containing biocides should be evaluated carefully.

E. *Burkholderia*

The genus *Burkholderia* is formed by more than 30 different species that inhabit diverse ecological niches, from soil to humans. This genus contains primary pathogens of animals and humans such as *Burkholderia mallei* and *Burkholderia pseudomallei* and opportunistic pathogens such as the *Burkholderia cepacia* complex (145).

The genome sequence of a *B. pseudomallei* isolate (146) has demonstrated the presence in its genome of at least 10 operons encoding putative efflux pumps belonging to the RND family of transporters. An analysis of the expression levels of the seven MDR pumps most likely to be involved in resistance in this bacterial pathogen has shown that more than 80% of the isolates tested expressed all of these efflux pumps, suggesting that they might participate in intrinsic resistance (147). The best-studied efflux pumps in *B. pseudomallei* are AmrAB–OprA, BpeAB–OprB (148), and BpeEF–OprC (149). None has been described to extrude quinolones. However, the possibility that any of these pumps can extrude quinolones cannot be ruled out formally, given that the number of quinolones used in the antibiotic susceptibility tests in the aforementioned studies was very low (either one or none). The existence of efflux pumps contributing to quinolone resistance in *B. pseudomallei* is further supported by the increased susceptibility to levofloxacin displayed by this bacterial species in the presence of agents such as phenothiazines or omeprazol, which interfere with the proton gradient at the inner membrane and impede the activity of RND/MDR efflux pumps (150).

In the case of *B. cenocepacia*, the genome sequence of one isolate allowed the identification of 14 putative RND-type efflux pumps encoded in its chromosome. One such determinant conferred low-level resistance to quinolones when expressed in an *E. coli* strain lacking its major MDR efflux pump (151). However, the impact of this MDR pump on intrinsic resistance and clinically acquired resistance of *B. cenocepacia* remains to be established.

A different situation concerns BcrA. This protein was isolated as an immunodominant antigen present in cystic fibrosis patients infected with *B. cepacia* (152). Further sequencing and functional analysis demonstrated that BcrA belongs to the MFS family of transporters and is able to extrude tetracycline and nalidixic acid. Since BcrA is recognized by immune systems, this means that it is expressed during infection and may participate in the establishment of *B. cepacia* in the lung of a CF patient, thus highlighting the dual role that MDR pumps may have in both resistance and bacterial infection (77, 153).

IV. Efflux Pumps in Gram-Positive Bacteria

It was not until the advent of fluoroquinolones in the 1980s (154) that quinolone-derived agents were used in combating gram-positive bacteria, and since then, quinolone-resistant isolates have been reported extensively. Quinolone resistance mechanisms in gram-positive bacteria parallel those in gram-negative microorganisms, with sequential mutations in the *gyrA* and *parC* genes giving rise to high-level quinolone resistance and expression of efflux pumps originating in low-level resistant clones.

Compared to gram-negative bacteria, however, fluoroquinolone efflux transporters in gram-positive microorganisms are far less well characterized and have been reported in only a few species, primarily *Staphylococcus aureus*, *Streptococcus pneumoniae*, *Enterococcus faecalis*, *Bacillus subtilis*, *Listeria monocytogenes*, and *Mycobacterium* spp. (Table 3). Among these, MFS efflux pumps are predominant, although ABC and MATE transporters have also been found (see below), all being of chromosomal origin. It is worth mentioning that these pumps are not quinolone-specific but are capable of transporting several distinct substrates, most of which, however, are not clinically relevant.

A. *Staphylococcus aureus*

S. aureus is by far the best-characterized gram-positive bacterium when it comes to efflux-mediated quinolone resistance, with up to six different pumps having been described so far. In 1990, NorA became the first fluoroquinolone efflux transporter to be described in gram-positive bacteria (155), although homologs in *S. pneumoniae* and *B. subtilis* were rapidly identified (156, 157). It is predicted to contain 12 transmembrane hydrophobic domains and has been classified among the MFS family of transport proteins, whose activity is dependent on the proton-motive force.

Constitutive expression levels of NorA are kept to a minimum in wild-type *S. aureus*, but increased expression in selected mutants confers resistance to hydrophilic fluoroquinolones (i.e., norfloxacin and ciprofloxacin) as well as to various compounds with no apparent structural similarity, such as rhodamine 6G, ethidium bromide, acriflavine, and tetraphenylphosphonium (158). NorA-conferred multidrug resistance can be reversed upon addition of the plant alkaloid reserpine as well as the membrane photonophore CCCP.

Increased expression of NorA was first described as the result of single nucleotide mutations within the 5′ untranslated region (5′ UTR) of *norA* mRNA, which apparently increases mRNA stability and the half-life of the transcript (the *flqB* mutation) (159). Subsequent studies, however, indicate that in most cases the *flqB* mutation results only in augmented *norA* transcription, and increased mRNA stability may just be an additive factor in some of the mutants (160).

Another mechanism leading to increased NorA expression is associated with an 18-kDa protein that recognizes nucleotide repeats consisting of the consensus sequence TTAATT upstream from the *norA* promoter. Initially termed NorR, this protein has been shown to up-regulate *norA* expression in an *arlS*-negative background as well as in an *arlS*⁺ background if overexpressed from an uncharacterized plasmid-based promoter (161). ArlR–ArlS constitutes a two-component regulatory system that plays a role in adhesion, autolysis, and proteolytic activity in *S. aureus* (162), but it is not clear how a mutation in *arlS* modifies the effect of NorR on the *norA* promoter.

NorR was later identified as a global regulator involved in the transcriptional control of several known autolytic regulators, including ArlR–ArlS (163), and it has been given several different names, although all authors working with this regulator have finally agreed to designate its gene and protein as *mgrA* and MgrA (multiple gene regulator), respectively (164). More recently it has been shown that overexpression of MgrA from its own promoter results in a significant reduction in *norA* expression (160), suggesting that MgrA may act as a negative regulator of *norA* rather than as an activator, but this is still an open debate.

TABLE 3
Fluoroquinolone Efflux Pumps in Gram-Positive Bacteria

Microorganism	Efflux Pump	MDR Family	Quinolones Extruded	Other Compounds Extruded[a]	Refs.
Bacillus subtilis	Bmr	MFS	Norfloxacin, ciprofloxacin, temafloxacin, nalidixic acid, oxolinic acid	Chloramphenicol, puromycin, EtBr, TPP, rhodamine 6G, acridine orange, netropsin, doxorubicin	(157)
	Blt	MFS	Norfloxacin	EtBr, rhodamine 6G, TPP	(197)
	Bmr3	MFS	Norfloxacin, tosufloxacin	Puromycin, daunomycin, EtBr, acriflavine, TPP	(201)
Enterococcus faecalis	EmeA	MFS	Norfloxacin, ciprofloxacin	EtBr, acriflavine, DAPI, TPP, H33342, benzalkonium	(191, 192)
	EfrAB	ABC	Norfloxacin, ciprofloxacin	Doxycycline, arbekacin, novobiocin, daunorubicin, doxorubbicin, acriflavine, DAPI, TPP, EtBr, safranin O	(193)
E. faecium	EfmA	MFS	Norfloxacin, ciprofloxacin	Erythromycin, oleandomycin, EtBr, DAPI, TPP	(195)
Listeria monocytogenes	Lde	MFS	Norfloxacin, ciprofloxacin	EtBr, acridine orange	(203)
Staphylococcus aureus	NorA	MFS	Norfloxacin, ciprofloxacin	Rhodamine 6G, EtBr, acriflavine, acridine orange, TPP, chloramphenicol, puromycin	(158)
	NorB	MFS	Norfloxacin, ciprofloxacin, sparfloxacin, moxifloxacin	EtBr, cetrimide, TPP	(168)

	Transporter	Family	Antibiotics	Other substrates	References
	NorC	MFS	Norfloxacin, ciprofloxacin, sparfloxacin, moxifloxacin, garefloxacin, premafloxacin	—	(169)
	SdrM	MFS	Norfloxacin	EtBr, acriflavine	(170)
	MdeA	MFS	Ciprofloxacin	Doxorubicin, daunorubicin, virginiamycin, novobiocin, mupirocin, fusidic acid, EtBr, TPP, acriflavine, rhodamine 6G, benzalkonium chloride	(171, 172)
	MepA	MATE	Norfloxacin, ciprofloxacin	Tigecycline, acriflavine, benzalkonium chloride, cetrimide, crystal violet, EtBr, rhodamine, TMA-DPH, chlorhexidine, DAPI, dequalinium, H33342, Pentamidine	(173, 174)
Streptococcus pneumoniae	Pmr	MFS	Norfloxacin, ciprofloxacin	EtBr, acriflavine	(156)
	PatAB	ABC	Norfloxacin, ciprofloxacin, levofloxacin	EtBr, acriflavine, cetrimide, reserpine	(179, 182, 183)

[a]EtBr, ethidium bromide; TPP, tetraphenylphosphonium; DAPI, 4',6-diamidino-2-phenylindole; H33324, Hoechst 33342.

195

Finally, in 2007, Hooper and co-workers identified a novel transcriptional regulator, designated NorG, that has been shown to bind to its own promoter as well as to the *norA* promoter sequence, although no effect on *norA* transcriptional levels could be detected (165). NorG is, however, involved in the direct up-regulation of NorA-related pumps in *S. aureus* (see below), indicating its participation in quinolone resistance.

After the initial discovery and characterization of NorA, several investigations have shown the presence of non-NorA transporters in quinolone-resistant mutants of S. *aureus* (166, 167), but it took several years for Hooper and co-workers to finally identified a novel efflux transporter in S. *aureus* conferring resistance to quinolones (168). This novel efflux protein was detected on a microarray screening for up-regulated genes in an MgrA-negative background, and it has been designated NorB, due to its relatedness to other quinolone efflux transporters, including NorA (30% similarity). The deduced amino acid sequence of NorB indicates the presence of the characteristic 12 transmembrane domains that classify this protein as a member of the MFS group.

Wild-type expression levels of NorB, mimicking those of NorA, are not sufficient to affect drug susceptibility, but when overexpressed from a plasmid-derived copy, NorB causes a significant increase in the MICs of norfloxacin and ciprofloxacin, which are also NorA substrates, as well as of sparfloxacin and moxifloxacin, which are non-NorA substrates. The effect of NorA and NorB on the common substrates is not additive, however, and there is a limit to the extent of resistance due to the overlapping activity of these pumps (168).

Both NorG and MgrA regulate NorB expression. NorG acts as a direct activator of *norB* transcription, whereas MgrA modulates *norB* negatively by binding directly to a single MgrA-binding motif in the *norB* promoter and repressing *norB* indirectly by binding and blocking expression from the *norG* promoter (165). The results obtained by Hooper and co-workers seem, however, to indicate that the increased expression of *norB* and *norA* does not fully account for the quinolone-resistant phenotype observed in some mutants and that additional mechanisms must therefore coexist.

Shortly after the characterization of NorB, Truoung-Bolduc et al. reported the identification of a third quinolone-related MFS pump in S. *aureus*, which was thus designated NorC (169). NorC displays 61% identity with NorB and, similar to what has been described for both NorA and NorB, is poorly expressed in wild-type S. *aureus*, although constitutive expression causes low-level resistance to sparfloxacin and moxifloxacin.

Nevertheless, NorC overexpression promotes a significant increase in the MICs of norfloxacin, ciprofloxacin, garenoxacin, premafloxacin, sparfloxacin, and moxifloxacin.

Up-regulation of NorC, also similar to that of NorB, is negatively regulated by MgrA, but MgrA probably does not interact with the *norB* promoter, and other factors probably act as direct regulators of *norC* expression (169). NorG has also been shown to interact with the *norC* promoter and to induce a twofold transcriptional increase, as in *norB*, suggesting a common regulation for both NorB and NorC, whose substrate profiles are also very much alike (165).

SdrM constitutes the fourth MFS pump described in *S. aureus* so far, and it was identified initially by randomly cloning and expressing putative multidrug efflux pump genes from the genome of *S. aureus* (170). Wild-type levels of SdrM are too low to be detected, but overexpression of a plasmid-derived SdrM leads to a twofold increase in the MICs of norfloxacin and ethidium bromide as well as an eightfold increase in the MIC of acriflavine. An MgrA-binding motif has been located in the upstream region from the putative promoter of *sdrM*, but the role of MgrA or any other regulator in the expression of this pump has not yet been elucidated.

A fifth MFS pump, designated MdeA, has also been suggested to play a role in quinolone extrusion in *S. aureus* (171). MdeA expression levels are similar to those of NorA in wild-type strains, and only when it is expressed from a multicopy plasmid does it cause a twofold increase in the MIC of ciprofloxacin. However, other studies suggest that fluoroquinolones are not a substrate for this pump (172). The precise involvement of MdeA in fluoroquinolone resistance requires further analysis, but the evidence available indicates that fluoroquinolones are, at best, moderate substrates of this transporter.

Besides MFS efflux pumps, *S. aureus* contains at least one quinolone transporter from the MATE family, designated MepA. MepA was identified simultaneously in two different studies (173, 174) as a novel MDR pump encoded within the *mepRAB* operon. The *mepB* gene encodes a predicted soluble protein with no homology to proteins of known function. *mepA* encodes a transporter protein with distant homology to the MATE proteins NorM, from *V. parahaemolyticus,* and CdeA, from *C. difficile* (21% and 26% identity, respectively). Overexpression of MepA resulted in two- to 40-fold increases in the MICs of the fluoroquinolones norfloxacin and ciprofloxacin, the glycylcycline tygecycline and several monovalent (acriflavine, ethidium bromide, cetrimide, rhodamine, tetraphenylphospho-nium) and divalent (chlorhexidine, DAPI, pentamidine) cations (174). All of these substrates, however, share a common positive charge or a nitrogen atom that can easily be protonated. *mepR*, on the other hand, encodes a transcriptional regulator shown to tightly repress the transcriptional expression of *mepRAB*.

Indeed, identification of MepA in both studies was accomplished after the isolation of *S. aureus* mutants that presented a truncated copy of *mepR* and therefore were able to express MepA and generate a MDR phenotype. Nevertheless, MepA augmented expression in the presence of an intact MepR protein has also been described (174), indicating the probable existence of additional regulatory mechanisms for this system. More recently, Kaatz et al. (175) have shown that MepR behaves as a substrate-responsive regulator, meaning that upon binding to MepA substrates, MepR dissociates from the *mepRAB* promoter and releases its repressive effect.

In view of the current knowledge concerning quinolone transport proteins in *S. aureus*, it is clear that their contribution to quinolone resistance in overexpressing mutants is small compared with resistance associated with target mutations. However, low-level resistance might have a specific role in lowering intracellular quinolone concentrations to allow persistence in the hospital environment and favoring the emergence of high-level resistant mutants by means of target mutations. In this respect it has been shown that the prevalence of efflux-overexpressing mutants in clinical isolates is nearly 50%, with NorA and NorB being the most predominant pumps, although increased expression of MepA, MdeA, and NorC has also been detected (176). In addition, dyes and biocides are also substrates for most of these transporters, meaning that MIC increases for these compounds may enhance survival in otherwise "decontaminated" environments, augmenting the risk for nosocomial infections.

B. *Streptococcus pneumoniae*

Compared to that of *S. aureus*, our knowledge of the quinolone-resis-tance mechanisms mediated by efflux proteins in *Streptococci* is far scarcer. To date only two efflux mechanisms have been described. In 1999, Gill et al. used homology search-based methods to identify an MFS pump in *S. pneumoniae* that was substantially similar to NorA (23% identity) (156). Cloning and overexpression of this protein in *S. pneumo-niae*, thereafter designated Pmr (pneumococcal multidrug-resistance protein), caused an increased resistant phenotype to norfloxacin, ciprofloxacin, acriflavine, and ethidium bromide, a phenotype that was partially reversed in the presence of the efflux inhibitor reserpine.

Although the initial work by Gill et al. in assessing the role of PmrA in quinolone efflux seemed straightforward, there is some evidence that questions the actual contribution of this pump to quinolone resistance in clinical isolates of *S. pneumoniae*. In 2002, Piddock et al. analyzed 50 different *S. pneumoniae* isolates and failed to establish a correlation between *pmrA* expression and resistance levels to either norfloxacin or ciprofloxacin (177). In addition, several studies reported no modification of the MIC values of fluoroquinolones when working with *S. pneumoniae* strains lacking a functional *pmrA* gene (178–181), although increased susceptibility was detected upon addition of the efflux inhibitor reserpine, thus demonstrating the presence of an efflux protein other than PmrA. Indeed, two subsequent studies led to the identification of two juxtaposed genes encoding a pair of proteins homologous to components of ABC transporters (179, 182). These genes were designated PatA and PatB, respectively, and were shown to extrude ciprofloxacin, norfloxacin, and levofloxacin, as well as acriflavine, ethidium bromide, cetrimide, and reserpine, but not moxifloxacin, ofloxacin, sparfloxacin, or garenoxacin, probably due to the differences in their hydrophilic/hydrophobic balance (183).

Expression of both *patA* and *patB* can be induced by ciprofloxacin concentrations above 80 μg/mL (182). However, they also display different substrate specificity and insertional mutants of *patA* and *patB*, while conferring hypersusceptibility to fluoroquinolones, apparently pointing to a more relevant role for PatA in the growth of *S. pneumoniae*. The differences observed concerning PatA and PatB may indicate that they constitute two independent systems under similar regulatory mechanisms, rather than forming a single two-component heterodimeric efflux pump (184).

More recently, Avrain et al. demonstrated that exposure of *S. pneumoniae* strains to ciprofloxacin selects for resistant mutants with increased efflux, due to overexpression of *patA* and *patB* but not *pmrA*, thus supporting the proposal that PmrA is not the main factor responsible for quinolone efflux in *S. pneumoniae* (185). Moxifloxacin, on the other hand, selects for resistant mutants with targeted mutations in the QRDR. This work also points out the observation that efflux-mediated resistance to quinolones in *S. pneumoniae* can reach levels similar to those of targeted mutations; thus, efflux-mediated resistance in this microorganism is no longer "low level," and more attention should be paid to the clinical significance of this finding.

Interestingly, Jones et al. have also provided evidence for the presence of efflux transporters involved in quinolone resistance in the closely related *S. pyogenes* (186). This putative efflux mechanism is not affected by any of the common inhibitors of MFS pumps, such as reserpine, rescinnamine, or CCCP, and therefore the efflux phenotype is likely to be caused by an ABC-class transporter such as that of *S. pneumoniae* (182).

C. *Enterococcus faecalis*

Although bacteria from the genus *Enterococcus* are normal dwellers of the human intestine, they are currently emerging as one of the leading causes of life-threatening nosocomial infections, due mainly to the rapid acquisition or up-regulation of multidrug efflux transporters conferring resistance to the most commonly used agents (187–189). In addition, fluoroquinolones show a weak activity toward enterococci, which is due mainly to the presence of a chromosomal *qnr*-like gene that seems to be responsible for the intrinsic resistance of this genus to modern quinolones (190).

In 2001, Jonas et al. identified and characterized the *emeA* gene in *E. faecalis*, an MFS transporter showing 31% identity and 61% similarity to NorA of *S. aureus* (191). EmeA is responsible for a modest decrease in susceptibility to norfloxacin, ciprofloxacin, ethidium bromide, DAPI, acriflavine, and tetraphenylphosphonium chloride, and this effect can be reversed upon addition of efflux inhibitors such as lansoprazole or verapamil. These results were corroborated further by Lee et al. in 2003 (192).

In a subsequent study, Lee et al. also identified two juxtaposed genes that conferred decreased susceptibility to norfloxacin, ciprofloxacin, doxycycline, acriflavine, DAPI, and tetraphenylphosphonium when expressed in susceptible *E. coli* from a multicopy plasmid and thus were termed *efrA* and *efrB*, for *E. faecalis* multidrug resistance (193). Interestingly, neither *efrA* nor *efrB* is able to confer resistance when expressed alone, indicating that both proteins are needed for efflux activity. Hydropathy analysis of these proteins shows that they contain specific features common to ABC transporters, suggesting that they might constitute a heterodimeric two-component ABC efflux transporter, a view that was supported by bioinformatic analysis revealing significant homology with the ABC transporters PatA and PatB from *S. pneumoniae* (184).

Despite the identification of two different efflux transporters involved in quinolone-decreased susceptibility in *E. faecalis*, there is no information available concerning the contribution of these systems to quinolone resistance in clinical isolates. Nevertheless, in the closely related microorganism *E. faecium* it has been shown that fluoroquinolone efflux leading to decreased intracellular accumulation is not common among clinical isolates and that most of the resistance observed originates from the alteration of target enzymes (194). More recently, however, a novel MFS pump (termed EfmA) conferring low-level resistance to norfloxacin and ciprofloxacin, among others, has been described in *E. faecium* (195).

D. *Bacillus subtilis*

To date, only three quinolone-related efflux pumps have been described in *B. subtilis*, all three belonging to the MFS family of multidrug transporters. Bmr was first described in 1992 as a distant homolog of NorA from *S. aureus* (44% identity) (157), and its overexpression confers increased MICs of norfloxacin, ciprofloxacin, chloramphenicol, puromycin, ethidium bromide, and tetraphenylphosphonium. The activity of Bmr is also reversed by addition of reserpine. Furthermore, expression of *bmr* is regulated by the product of a gene located immediately downstream from *bmr*, designated *bmrR*, that encodes a MerR-type transcriptional regulator capable of binding the *bmr* promoter and enhancing its transcription upon interacting with compounds that are also actively effluxed by Bmr (such as rhodamine or tetraphenylphosphonium, but not fluoroquinolones), indicating that it constitutes a substrate-responsive regulator (196).

Blt, the second MFS pump of *B. subtilis*, was identified by Ahmed et al. in 1995, and its substrate specificity is very much like that of Bmr (norfloxacin, ethidium bromide, rhodamine, and tetraphenylphosphonium) (197). Sequence alignment demonstrates strong homology between Bmr and Blt (51% identity), but unlike Bmr, *blt* is very poorly expressed under standard laboratory conditions, and its disruption has no effect on drug susceptibility. Expression of *blt* seems to be controlled by the product of *bltR*, a gene located upstream from *blt* in inverted orientation and presenting sequence homology to BmrR [except for the C-terminal end of the protein, which is thought to be involved in the specific binding of inducers in MerR-type regulators (198, 199)]. BltR, however, does not respond to any of the known substrates of Blt, indicating that despite the multiple similarities between Bmr–BmrR and Blt–BltR, they may have different physiological roles and are thus governed by distinct and specific mechanisms.

Both Bmr and Blt were also found to be regulated by a third MerR-type transcriptional regulator, designated Mta (multidrug transporter activator) (200). Baranova et al. demonstrated that an Mta mutant missing its C-terminal domain was capable of inducing the expression of both *bmr* and *blt* in a BmrR- or BltR-independent fashion. It was suggested that the C-terminal tail of MerR regulators blocks its DNA-binding activity, and only upon binding of the appropriate inducer is the blocking relaxed. The lack of the entire domain may allow for constitutive binding and enhanced expression of target genes. The natural inducer of Mta, however, has not been found, and the clinical significance of this regulation remains to be elucidated.

The third MFS pump of *B. subtilis* was termed Bmr3 and showed significant similarity to Bmr (49%) and, to a lesser extent, to Blt (17%) (201). Its substrate specificity, however, mimicked that of the two other MFS pumps described above. In a subsequent work, Ohki et al. identified point mutations within the 5'UTR region of *bmr3* that enhanced its mRNA stability and resulted in the MDR phenotype (202).

E. *Listeria monocytogenes*

Since fluoroquinolones are not commonly used to treat listeriosis, very little attention has been paid to efflux-mediated fluoroquinolone resistance in this microorganism. In 2003, however, Godreuil et al. described the presence of an MFS efflux pump (designated Lde) responsible for increased resistance to norfloxacin, ciprofloxacin, ethidium bromide, and acridine orange (but not to more hydrophobic fluoroquinolones) in clinical isolates of *L. monocytogenes* (203). Lde shows substantial similarity to Bmr, NorA, EmeA, and PmrA, and similar to other MFS pumps described in this chapter, its activity is inhibited by reserpine.

V. Efflux Pumps in Other Bacterial Species with Clinical Relevance

In this section we discuss the mechanisms of fluoroquinolone efflux in bacterial microorganisms of acknowledged clinical relevance that do not fit in any of the previous categories (Table 4).

TABLE 4
Fluoroquinolone Efflux Pumps in Other Bacterial Species

Microorganism	Efflux Pump[a]	MDR Family	Quinolones Extruded	Other Compounds Extruded[b]	Refs.
Aeromonas salmonicida	Unknown[ESR]	—	Flumequine, oxolinic acid, ciprofloxacin	—	(214)
Bacteroides fragilis	BmeABC 1–16	RNDs	Ciprofloxacin, norfloxacin, garenoxacin, levofloxacin and moxifloxacin	β-Lactams, cephems (cephalothin, cefazolin, ceftizoxime, and cephalexin), polypeptide antibiotics (polymyxin-B, colistin, bacitracin, and vancomycin), fusidic acid, novobiocin, puromycin, EtBr, SDS, triclosan, ampicillin, cefoperazone, cefoxitin, tetracycline	(225, 226)
B. thetaiotaomicron	BexA	MATE	Ciprofloxacin, norfloxacin	EtBr	(217, 229)
Clostridium difficile	CdeA	MATE	Ciprofloxacin, norfloxacin	Acriflavine, EtBr	(217)
C. hathewayi	CmpAB	ABC	Norfloxacin, ciprofloxacin, garifloxacin, moxifloxacin, sparfloxacin, levofloxacin	EtBr	(232)
Campylobacter jejuni	CmeABC	RND	Ciprofloxacin, norfloxacin and nalidixic acid	β-Lactams (cefotaxime and ampicillin), erythromycin, rifampin, tetracycline, chloramphenicol, gentamicin, triclosan	(208)

(continued)

TABLE 4 (*Continued*)

Microorganism	Efflux Pump[a]	MDR Family	Quinolones Extruded	Other Compounds Extruded[b]	Refs.
Haemophilus influenzae	HmrM	MATE	Norfloxacin	Streptomycin, trimethoprim, daunomycin, doxorubicin, acriflavine, DAPI, EtBr, H33342, TPP, berberine, cholate, deoxycholate	(212)
Mycobacterium smegmatis	LfrA	MFS	Norfloxacin, ciprofloxacin, ofloxacin, levofloxacin and enoxacin	EtBr, acriflavine	(233, 234)
	Pst	ABC	Ciprofloxacin, ofloxacin, sparfloxacin	EtBr, acriflavine	(239, 240)
M. tuberculosis	Rv1634	MFS	Norfloxacin, ciprofloxacin, ofloxacin, lomefloxacin	—	(238)
	Rv2686c–2687c–2688c	ABC	Norfloxacin, ciprofloxacin, moxifloxacin, sparfloxacin	—	(241)
Porphyromonas gingivalis	XepCAB	RND	Norfloxacin, ofloxacin and ciprofloxacin	EtBr, puromycin, rifampin, SDS, berberine, acriflavine, tetracycline, minocycline	(230)
Vibrio cholerae	VcmA	MATE	Ciprofloxacin, norfloxacin, ofloxacin	Kanamycin, streptomycin, daunomycin, doxorubicin, acriflavine, DAPI, EtBr, H33342	(268)

VcmB	MATE	Ciprofloxacin, norfloxacin, ofloxacin	Kanamycin, streptomycin, EtBr, H33342	(217, 269)
VcmD	MATE	Ciprofloxacin, norfloxacin, ofloxacin	EtBr, H33342	(217, 269)
VcmH	MATE	Ciprofloxacin, norfloxacin, ofloxacin	Kanamycin, streptomycin, EtBr, H33342	(217, 269)
VcmN	MATE	Ciprofloxacin, norfloxacin, ofloxacin	Kanamycin, streptomycin, EtBr, H33342	(217, 269)
VceCAB	MFS	Nalidixic acid	Chloramphenicol, erythromycin, deoxycholate	(219)
VcaM	ABC	Norfloxacin and ciprofloxacin	Tetracycline, doxorubicin, daunomycin, DAPI, H33324	(220)
V. parahae-molyticus NorM	MATE	Ciprofloxacin, norfloxacin	Kanamycin, streptomycin, EtBr, berberine	(221)

[a]ESR, EPI suggested results (data based on MIC increment after EPI addition).

[b]EtBr, ethidium bromide; TPP, tetraphenylphosphonium; DAPI, 4',6-diamidino-2-phenylindole; H33324, Hoechst 33342; SDS, sodium dodecyl sulfate.

A. *Campylobacter jejuni*

Campylobacter spp. are the most common bacterial enteropathogens in developed countries (204). These organisms are transmitted to humans primarily from contaminated water or food (205). The increasing number of isolates resistant to ciprofloxacin is of considerable concern. This resistance, which persists even in the absence of selective pressure (206), has been associated with point mutations in *gyrA*, the gene encoding for the DNA gyrase A topoisomerase, and mutations that result in the overexpression of efflux pumps (207).

CmeABC is the first MDR system described as responsible for the resistance to fluoroquinolones (ciprofloxacin, norfloxacin, and nalidixic acid) mediated by efflux pumps in *Campylobacter jejuni* (208). CmeABC has been suggested to be a member of the RND efflux pump family and predicted to form a tripartite structure consisting of an RND transporter (CmeB), a membrane fusion protein (CmeA), and an outer membrane protein (CmeC). These three proteins are encoded in the same operon, *cmeABC*, and their expression is controlled by the local transcriptional repressor CmeR (209). *cmeR*, the gene coding for the regulator, is located upstream from the *cmeABC* operon, and the intergenic region *cmeR–cmeA* contains the promoter region of the efflux pump genes.

CmeR represses efflux pump expression through interaction with an inverted repeat sequence localized in the *cmeR–cmeA* intergenic region. Mutations affecting CmeR or the intergenic region *cmeR–cmeA* have been described to be associated with fluoroquinolones resistance (209). Both types of mutations result in the inability of CmeR to bind to its cognate DNA and therefore in its incapability to repress the constitutive expression of the efflux pump. These mutations have been proven, in vivo, to appear after treating poultry with fluoroquinolones such as enrofloxacin (206) and cause not just quinolone resistance but also resistance to β-lactams (cefotaxime and ampicillin), erythromycin, rifampin, tetracycline, chloramphenicol, gentamicin, and other antimicrobials, since all these molecules can be pumped out by CmeABC (208). Besides CmeABC, *C. jejuni* possesses at least nine other putative efflux pumps, including CmeDEF (210), but none have proven to be able to expel fluoroquinolones (207).

B. *Haemophilus influenzae*

Haemophilus influenzae type b, or Hib, is responsible for millions of serious, sometimes deadly cases of meningitis and pneumonia around the world. In developed countries where Hib vaccine is widely used, nontypable *H. influenzae* strains have appeared. Growing resistance of this bacterium to several antibiotics has been reported.

Fluoroquinolone resistance in *H. influenzae* occurs mainly due to mutations in the genes encoding for topoisomerases I and II (211). Porines and increased efflux have also been suggested to contribute to reduced levels of quinolone susceptibility (211). To date, only one efflux pump, HmrR, has been proved to be involved in fluoroquinolone resistance in *H. influenzae* (212).

HmrR (*H. influenzae* multidrug resistance) is a 51-kDa Na^+/drug antiporter that belongs to the MATE family of multidrug efflux pumps. It shares a high degree of homology with other proteins of this family, such as YdhE of *E. coli* (47% identity and 88% similarity) and NorM of *V. parahaemolyticus* (43% identity and 85% similarity). As proposed for other MATE family proteins (213), its 12 transmembrane domains are arranged in the bacterial inner membrane, forming a channel able to pump a variety of molecules to the periplasm. Through this mechanism HmrR confers resistance to quinolones (norfloxacin) and other drugs, including acriflavine, doxorubicin, ethidium bromide, tetraphenylphosphonium chloride, daunomycin, berberine, or sodium deoxycholate (212).

C. *Aeromonas* spp.

Aeromonas are ubiquitous gram-negative bacteria with a preference for an aquatic environment. Even though their major hosts are fishes, several species have been implicated in human infections, including *A. hydrophila*, *A. veronii* biovar *sobria*, *A. jandaei*, *A. schubertii*, *A. caviae*, and *A. salmonicida* (214, 215). Increasing levels of resistance have been reported for both clinical and environmental strains.

The use of fluoroquinolones, added directly into the water or as a part of feeding in aquaculture antibiotherapy, may select quinolone-resistant *Aeromonas* isolates (214). Resistant bacteria are transmitted to fish farm workers through open wounds, contaminated water, and infected fish consumption or handling. Although the main reasons for this acquired resistance are mutations in *gyrA*, differences in the level of resistance between isolates with identical *gyrA* and *parC* genes have been related to efflux activity (214). In *A. salmonicida*, the presence of the efflux pump inhibitor PAβN increases the bacterial susceptibility to flumequine and oxolinic acid. The reduced MIC to the aforementioned quinolones is related to a lower accumulation of flumequine for strains with a higher reduction in MICs in the presence of the efflux pump inhibitor. Actually, the levels of susceptibility in the presence of an efflux pump inhibitor are the same as for strains that do not carry a mutation in the *gyrA* and *parC* genes. Based on these data, Giraud et al. suggested that efflux pump up-regulation might be responsible for significant levels of quinolone resistance in *A. salmonicida* (214).

The efflux pumps responsible for the extrusion of quinolones in *Aeromonas* spp. have not been identified and could be of interest to study whether there is a synergic effect between efflux of quinolones and the activity of the recently described Qnr-like proteins (216) encoded in the genomes of *A. hydrophila* and *A. salmonicida*.

D.   *Vibrio* spp.

In vibrios, MDR efflux pumps belonging to the MFS, MATE and the ABC type of MDR transporters have been involved in resistance to quinolones. The MATE family plays the most important role, with six pumps (NorM, VcmA, VcmB, VcmD, VcmH, and VcmN) contributing to fluoroquinolone resistance (217). For the other two families, only one quinolone efflux system has been described for each: VceCAB (MFS) (218, 219) and VcaM (ABC) (220).

The norfloxacin efflux protein, NorM, was the first reported Na^+-coupled multidrug efflux antiporter (221). NorM is a 49-kDa protein that shares a 12-transmembrane helices topology with the members of the MFS family. Since the sequence similarity with the aforementioned family is low, NorM and the *E. coli* transporter YdhE were categorized in a new family called MATE (multidrug and toxic compound extrusion) (222). Morita et al. (221) demonstrated that expression of the *norM* gene, from the halophilic marine bacterium *V. parahaemolyticus*, into *E. coli* KAM3, a strain that lacks the multidrug efflux transporter AcrB, resulted in MIC increases for norfloxacin and ciprofloxacin. Furthermore, since the accumulation of norfloxacin was lower in the transformant than in the host cell, together with the fact that both levels became almost equal when the H^+ uncoupler CCCP was added, NorM was confirmed as an efflux pump whose activity depends on the proton membrane gradient. Besides quinolones, NorM extrudes other compounds, such as ethidium bromide, kanamycin, and streptomycin. Evidence that the extrusion of ethidium was stimulated by the addition of Na^+ proved that NorM has a Na^+-driven efflux mechanism (223). Homologs to NorM are found in other vibrios, such as *V. vulnificus*, *V. cholerae*, *V. splendidus*, and *V. fischeri*. NorM from *V. cholerae* has been characterized, and its sequence, substrate, and efflux mechanism are rather similar to those of *V. parahaemolyticus*.

In non-O1 *V. cholerae*, five transporters of the MATE family are involved in quinolone efflux. Expression of the MATE transporters VcmA, VcmB, VcmD, VcmH, and VcmN (*vcmA*, *vcmB*, *vcmD*, *vcmH*, and *vcmN*, respectively) in *E. coli* KAM32 cells, deficient in the AcrAB and YdhE efflux pumps, resulted in increased MICs for ciprofloxacin, norfloxacin, ofloxacin, and other compounds, such as ethidium bromide. In the *E. coli* KAM32 strains that carried *vcmA*, *vcmB*, or *vcmD*, the ethidium bromide efflux was enhanced by the presence of NaCl, whereas for cells containing *vcmN* the efflux seemed to be independent of Na^+. In cells expressing VcmH, efflux of Hoechst 33342 also proved to be Na^+ coupled.

The MFS system VceCAB, formerly VceAB, contributes to the nalidixic acid resistance in *V. cholerae* (218, 219). The three proteins are arranged forming the typical tripartite pump structure, in which VceC is the OMP, VceA the MFP, and VceB the MFS translocase. The pump is encoded by the *vceCAB* operon, whose expression is regulated by the TetR family transcriptional repressor VceR. *vceR* is located 232 bp upstream from *vceC*, and in between these genes lies the 28-bp inverted repeat with the operator sequence, in which VceR binds, thereby impeding expression of the pump (219). Along with quinolones, VceCAB pumps out chloramphenicol, erythromycin, and deoxycholate, among others (218).

So far, the only ABC multidrug efflux pump type with a confirmed fluoroquinolone resistance function in vibrios is VcaM (*Vibrio cholerae* ABC multidrug-resistance pump) (220). VcaM is a 69-kDa protein composed of two domains: an N-terminal hydrophobic domain, predicted to form six transmembrane segments, and a C-terminal hydrophilic region, probably cytoplasmic, that constitutes the ATP-binding domain. Its importance in multidrug resistance was confirmed by transforming *E. coli* KAM32 with a plasmid that carried the *vcaM* gene from non-O1 *V. cholerae*. The cells harboring the plasmid showed increased levels of resistance to the quinolones norfloxacin and ciprofloxacin, as well as to other drugs, such as tetracycline, doxorubicin, daunomycin, DAPI, and Hoechst 33342. The addition of increasing concentrations of the ATPase inhibitor sodium *ortho*-vanadate to these transformed cells led to a progressive reduction of Hoechst 33342 efflux, corroborating that VcaM is an ATP-dependent pump.

E. *Anaerobes*

In anaerobes, as for other bacteria, resistance to quinolones is due primarily to mutations in gyrase or topoisomerase genes and to increased efflux of the drugs. Normally, those mutations affecting target enzymes involve only *gyrA* (mutations in *gyrB*, *parC*, and *parE* are exceptional), and a combination of *gyrA* mutations and elevated efflux appears in the most resistant strains. MDR systems capable of expelling quinolones have been described in several species (224).

Bacteroides fragilis and *Bacteroides thetaiotaomicron* are the two most clinically relevant species of bacteroides. The number of resistant strains to fluoroquinolones is increasing in both species. *B. fragilis* harbors 16 tripartite RND pumps, named BmeABC 1 to 16, in which BmeA is the MFP, BmeB the RND transporter, and BmeC the OMP (225). These pumps are expressed constitutively and coordinately, and thus deletion of some of these genes is compensated by overexpression of others (226). Pumbwe et al. (227) suggested that this regulation can be mediated by a global regulator as occurs in *E. coli* with MarRAB. This compensatory response makes it difficult to assign specific substrates to every pump.

Experiments performed in laboratory mutants demonstrated that BmeB3 is responsible for higher MIC levels of quinolones, but these results have not been verified in the clinical environment, with none of the isolates studied so far overexpressing *bmeB3* (226, 227). Nevertheless, the essential role of the BmeABC pumps in the efflux of fluoroquinolones (ciprofloxacin, norfloxacin, garenoxacin, levofloxavin, and moxifloxacin) by *B. fragilis* has been ruled out. Besides the RND pumps, a protein of the MFS family has been suggested to efflux fluoroquinolones (norfloxacin) in *B. fragilis* (228).

In *B. thetaiotaomicron* a MATE-type protein BexA (*Bacteroides exporter A*) extrudes fluoroquinolones (229). Cloning and expression on *E. coli* AG102AX of a plasmid containing *bexA* increased the MICs for norfloxacin (eightfold) and ciprofloxacin (fourfold). The addition of CCCP enhanced the intracellular concentration of norfloxacin, in agreement with an efflux inhibition.

In *Porphyromonas gingivalis*, a RND multidrug efflux pump has been characterized. This MDR determinant, named XepCAB (xenobiotic exporters of *Porphyromonas*), is an RND-type efflux pump suggested to form the typical tripartite complex: XepC (RND), XepA (MFP), and XepB (OMP) (230). It has been reported that this pump is involved in the intrinsic resistance of *P. gingivalis* to norfloxacin, ofloxacin, and ciprofloxacin.

In *C. difficile* a MATE-type Na^+-coupled efflux pump, CdeA, has been proven to extrude fluoroquinolones (norfloxacin and ciprofloxacin) (231). However, significant levels of resistance are achieved only if *cdeA* is overexpressed, and quinolones do not induce its gene expression. Dridi et al. suggested that changes in *cdeA* or in regulatory regions increase quinolone efflux (231).

The first ABC-type transporter from the genus *Clostridium* has recently been described (232). In CmpAB, as for other members of the ABC family, the ATPase component (CmpA) consists of two ATP-binding domains, and the membrane protein (CmpB) serves as a transmembrane transport channel. Cloning and expression of *cmpAB* genes from *C. hathewayi* into *E. coli* and *C. perfringens* resulted in higher resistance to the fluoroquinolones norfloxacin, ciprofloxacin, garifloxacin, moxifloxacin, sparfloxacin, and slightly higher resistance to levofloxacin. Transformed cells accumulated and were unable to extrude ethidium bromide when CCCP was added.

F. *Mycobacterium* spp.

Despite the strong clinical relevance of *M. tuberculosis* MDR (which remains as one of the leading causes of mortality worldwide), very few studies have characterized efflux transporters involved in fluoroquinolone resistance within this microorganism. This lack of information may be attributed to the fact that *M. tuberculosis* is slow growing and difficult to work with, which has prompted use of the fast-growing *M. smegmatis* as a model genetic system.

To date, only two MFS pumps mediating quinolone resistance have been described in mycobacteria as well as two ABC-class transporters. The MFS efflux pump LfrA was described in *M. smegmatis* in 1996 and was shown to confer low-level resistance to hydrophilic quinolones as well as to acriflavine and ethidium bromide when it is expressed from a multicopy plasmid (233, 234). Inactivation of chromosomal LfrA, however, seems to have a very subtle effect on quinolone susceptibility, suggesting that LfrA-mediated efflux does not occur unless LfrA is overexpressed (235, 236).

LfrA is regulated by the product of *lfrR*, a gene located immediately upstream from *lfrA* and in the same orientation (235). LfrR shows substantial homology to several TetR-like transcriptional regulators, and its inactivation causes an increase in the MICs of fluoroquinolones, ethidium bromide, and acriflavine. Both *lfrR* and *lfrA* have been shown to constitute an operon, which is negatively regulated by LfrR but is inducible upon addition of some substrates of LfrA. Substrate induction is caused by the interaction between LfrR and the inducer molecule, which impairs the ability of LfrR to bind the *lfrRA* promoter sequence (237). Inducing compounds include ethidium bromide rhodamine and acriflavine but not ciprofloxacin. This apparent paradox is reminiscent of that found for the Bmr–BmrR MFS pump of *B. subtilis* and might indicate that fluoroquinolones are not preferred substrates for these pumps (196).

Although no homolog of the LfrRA system has been found in *M. tuberculosis*, De Rossi et al. screened the *M. tuberculosis* genome for the presence of MFS-efflux proteins and identified one candidate that confers low-level resistance to norfloxacin, ciprofloxacin, ofloxacin, and lomefloxacin when expressed in a heterologous system such as *M. smegmatis* (238). A more thorough characterization of this protein is expected to assess the clinical significance of this finding.

In addition to the MFS-efflux systems described above, two different ABC-type transporters have also been found in mycobacteria, one in *M. smegmatis* and the second in *M. tuberculosis*. In 2000, Bhatt et al. identified a novel function in fluoroquinolone extrusion for a known phosphate specific transporter (Pst) in *M. smegmatis*. Pst is an ATP-binding cassette permease (ABC-type efflux protein) involved in phosphate transport whose transcriptional levels were highly increased in an in vitro–selected *M. smegmatis* mutant resistant to ciprofloxacin (239). Inactivation of its gene results in hypersusceptibility to ciprofloxacin, ofloxacin, sparfloxacin, acriflavine, and ethidium bromide (240). Homologs to Pst can be found in *M. tuberculosis*, but its involvement in fluoroquinolone resistance remains to be elucidated.

In *M. tuberculosis*, however, Pasca et al. identified a novel ABC transporter leading to fluoroquinolone resistance when expressed in *M. smegmatis* from a multicopy plasmid (241). This novel ABC efflux pump is encoded by three genes arranged in an operonic structure (Rv2686c–Rv2687c–Rv2688c) and heterologous expression of the entire operon results in increased MICs of ciprofloxacin, norfloxacin, moxifloxacin, and sparfloxacin, but not to any other antibiotic or unrelated compounds. Expression of each component by itself, however, does not increase the MICs of any of the substrates tested except for Rv2686, which shows increased MICs of ciprofloxacin and norfloxacin. The authors attributed this observation to possible interactions of this protein with ABC-class homologs in *M. smegmatis*, thus generating a functional transporter (241). This is the only example of an ABC transporter involved in single fluoroquinolone efflux in *M. tuberculosis*.

VI. Plasmid-Encoded Quinolone Efflux Pumps

Since quinolones are synthetic drugs, it was early suggested that the existence of quinolone-resistance genes would be unlikely and that resistance to these drugs would be driven by mutation and not by the acquisition of resistance genes by horizontal gene transfer. Nevertheless, the existence of plasmid-encoded quinolone resistance was predicted initially (242–244) and found soon after (143) (Table 5).

The first mechanism described involved protection of the target from quinolone activity [Qnr determinants (142)]. Later, another plasmid-encoded quinolone-resistance mechanism based on the modification of the quinolones was described (245).

More recently, a plasmid-encoded efflux pump (QepA) has been found in plasmids from Enterobacteriaceae (246). QepA presents similarities to transporters belonging to the 14-transmembrane-segment MFS of MDR pumps, mainly with transporters from environmental actinomycetes, which suggests that these microorganisms might be the origin of the plasmid-encoded *qepA* gene (246). QepA confers resistance to the hydrophilic fluoroquinolones norfloxacin, ciprofloxacin, and enrofloxacin and differs from chromosomally encoded quinolone efflux pumps, which frequently have a wider range of substrates. It has not been reported that QepA can efflux drugs belonging to other structural families. Although the number of studies on the prevalence of *qepA*-encoding plasmids is still low, all reports indicate that this gene has only very recently emerged in the population of bacterial human pathogens, and its prevalence in these pathogens is still low (247, 248).

Nevertheless, the situation with bacterial isolates from animals is different. It has been reported that the prevalence of this gene in plasmids from Enterobacteriaceae isolated from pigs is high (249, 250). This different prevalence strongly suggests that as in other antibiotic resistance genes acquired by HGT (76, 251), animal farms may be where *qepA* was first acquired by pathogenic bacteria and are currently a reservoir for the spread of this gene among different bacteria. Whether this high prevalence in bacteria of animal origin represents a higher risk for the dissemination of

TABLE 5
Plasmid-Encoded Quinolone Efflux Pumps

Efflux Pump	MDR Family	Quinolones Extruded	Other Compounds Extruded[a]	Ref.
QepA	MFS	Norfloxacin, ciprofloxacin, enrofloxacin	—	(246)
OqxAB	RND	Olaquindox, ciprofloxacin, nalidixic acid, norfloxacin, flumequine	Benzalkonium chloride, carbadox, cetrimide, chloramphenicol, SDS, triclosan, trimethoprim, mitomycin C, chlorhexidine, acriflavine, tetracycline	(255)

[a]SDS, sodium dodecyl sulfate.

qepA-encoding plasmids in the population of human pathogens remains to be established. An important feature of this potential dissemination is the association of *qepA* with other resistance genes within the same plasmid (co-resistance). The first report of *qepA* described that this element was present in *E. coli* in an IncFII plasmid conferring resistance to ampicillin, erythromycin, and kanamycin (246). The association of *qepA* with the *rmtB* gene, which encodes a 16S rRNA methyltransferase responsible for high-level resistance to aminoglycosides, was described early (246, 252). Both *qepA* and *rmtB* genes were present in a conjugative plasmid inside a putative transposable element flanked by two copies of IS26 (246), and it has been described that 58.3% of *E. coli* isolates from pigs containing the *rmtB* gene also harbored *qepA*. Recent work has reported the presence of a *qepA* derivative (*qepA2*) in a mobilizable plasmid which does not contain the *rmtB* gene (253). Although the structure of the region surrounding *qepA* is similar to that observed in other plasmids, the *qepA2* gene presents two nucleotide substitutions, leading to Ala99Gly and Val134Ile (253), and it is present between two copies of an IS*CR*-like element, not being associated with the IS26 elements present in other *qepA*-encoding plasmids (248, 252). This suggests that the *qepA* gene is under a process of diversification after being transferred to a new host through HGT.

A recent study has reported the presence of *qepA* together with two other plasmid-encoded quinolone resistance genes, *qnrS2* and *aac(6′)-Ib-cr*, in an *E. coli* isolate from pigs (249). The presence of different plasmid-encoded quinolone-resistance determinants in the same host might facilitate their spread among bacterial populations.

Another plasmid-encoded quinolone efflux pump is OqxAB. This determinant was first described as conferring resistance to the swine growth enhancer olaquindox (254). Further work demonstrated that OqxAB is indeed a multidrug efflux pump, belonging to the RND family of bacterial transporters and capable of effluxing quinolones among other drugs (255). Very few surveillance studies on the prevalence of OqxAB have been published so far. One showed that 10 out of 556 (1.8%) *E. coli* strains isolated from pigs were resistant to olaquindox, and in nine, the *oqxA* gene was detected and was probably plasmid-encoded (256). The presence of *oqxA* in human bacterial pathogens on a plasmid in *E. coli* and on the chromosome of *Klebsiella pneumoniae* has also been reported recently. The presence of IS26-like sequences flanking the plasmid-encoded *oqxAB* genes suggests that they have been acquired as part of a composite transposon (257). Given the scarcity of data available on their distribution in human pathogens, the impact of *oqxA* genes in quinolone resistance remains to be established. However, the fact that this determinant was probably selected in animals because it confers resistance on a growth promoter illustrates the risk of cross-selection mechanisms in the acquisition and spread of this type of quinolone-resistance genes.

VII.   Concluding Remarks

Fluoroquinolone resistance has been increasing steadily in the past decade. Although the molecular bases of fluoroquinolone resistance are associated mainly with mutations in the genes encoding protein targets, the basal expression of efflux pumps contributes to the intrinsic resistance to fluoroquinolones mainly in nonfermentative bacilli. In addition, the over-expression of efflux pumps can complement the mutations in protein target genes and generate a high level of resistance. Studies showing that highly fluoroquinolone-resistant protein target mutants cannot be selected in vitro from efflux-deficient mutants of *E. coli* have been reported and highlight the significant contribution of efflux to high-level fluoroquinolone resistance (13). The efflux pumps affecting quinolones have been studied in depth from the functional and structural points of view. However, the number of studies analyzing the clinical impact of the overexpression of the efflux pumps is scarce, but the wide prevalence and specific characteristics of these pumps make them excellent targets for the search of novel drugs under an antiresistance approach. In fact, different efflux pump inhibitors have been developed (258). In addition, the overexpression of efflux pump(s) could contribute to increased resistance in bacteria forming biofilm (259); therefore, the development of efflux pump inhibitors could also be used in combination with a specific antibiotic as a strategy to treat infections caused by biofilm-producing bacteria.

Acknowledgements

Research in the laboratory of J.V. is funded by the Spanish Ministry of Health (FIS 05/0068), 2009 SGR 1256 from the Departament d'Innovació, Universitats I Empresa de la Generalitat de Catalunya, and by the Ministerio de Sanidad y Consumo, Instituto de Salud Carlos II, Spanish Network for the Research in Infectious Diseases (REIPI RE06/0008). This work has also been supported by funding from the European Community (AntiPathoGN contract HEALTH-F3-2008-223101 and TROCAR contract HEALTH-F3-2008-223031). Research in the laboratory of J.L.M. is funded by BIO2008-00090 from the Spanish Ministerio de Ciencia e Innovación, and LSHM-CT-2005-518152 and LSHM-CT-2005-018705 and BIOHYPO from the European Union.

References

1. Vila, J. (2005) Fluoroquinolone resistance, in *Frontiers in Antimicrobial Resistance: A Tribute to Stuart B. Levy*, White, D. G., Alekshun, M. N., and MacDermott, P. F., Eds., ASM Press, Washington, DC, pp. 41–52.

2. Fabrega, A., Madurga, S., Giralt, E., and Vila, J. (2009) Mechanism of action of and resistance to quinolones, *Microb. Biotechnol. 2*, 40–61.

3. Levy, S. B. (1992) Active efflux mechanisms for antimicrobial resistance, *Antimicrob. Agents Chemother. 36*, 695–703.

4. Roca, I., Marti, S., Espinal, P., Martinez, P., Gibert, I., and Vila, J. (2009) CraA: an MFS efflux pump associated with chloramphenicol resistance in *Acinetobacter baumannii*, *Antimicrob. Agents Chemother. 53*, 4013–4014.

5. Piddock, L. J. (2006) Clinically relevant chromosomally encoded multidrug resistance efflux pumps in bacteria, *Clin. Microbiol. Rev. 19*, 382–402.

6. Nikaido, H. (1996) Multidrug efflux pumps of gram-negative bacteria, *J. Bacteriol. 178*, 5853–5859.

7. Nikaido, H. (1998) Antibiotic resistance caused by gram-negative multidrug efflux pumps, *Clin. Infect. Dis. 27* (Suppl. 1), S32–S41.

8. Edgar, R., and Bibi, E. (1997) MdfA, an *Escherichia coli* multidrug resistance protein with an extraordinarily broad spectrum of drug recognition, *J. Bacteriol. 179*, 2274–2280.

9. Nishino, K., and Yamaguchi, A. (2001) Analysis of a complete library of putative drug transporter genes in *Escherichia coli*, *J. Bacteriol. 183*, 5803–5812.

10. Yang, S., Clayton, S. R., and Zechiedrich, E. L. (2003) Relative contributions of the AcrAB, MdfA and NorE efflux pumps to quinolone resistance in *Escherichia coli*, *J. Antimicrob. Chemother. 51*, 545–556.

11. Rosenberg, E. Y., Ma, D., and Nikaido, H. (2000) AcrD of *Escherichia coli* is an aminoglycoside efflux pump, *J. Bacteriol. 182*, 1754–1756.

12. Webber, M. A., and Piddock, L. J. (2001) Absence of mutations in *marRAB* or *soxRS* in *acrB*-overexpressing fluoroquinolone-resistant clinical and veterinary isolates of *Escherichia coli*, *Antimicrob. Agents Chemother. 45*, 1550–1552.

13. Oethinger, M., Kern, W. V., Jellen-Ritter, A. S., McMurry, L. M., and Levy, S. B. (2000) Ineffectiveness of topoisomerase mutations in mediating clinically significant fluoro-quinolone resistance in *Escherichia coli* in the absence of the AcrAB efflux pump, *Antimicrob. Agents Chemother. 44*, 10–13.

14. Okusu, H., Ma, D., and Nikaido, H. (1996) AcrAB efflux pump plays a major role in the antibiotic resistance phenotype of *Escherichia coli* multiple-antibiotic-resistance (Mar) mutants, *J. Bacteriol. 178*, 306–308.

15. Fralick, J. A. (1996) Evidence that TolC is required for functioning of the Mar/AcrAB efflux pump of *Escherichia coli*, *J. Bacteriol. 178*, 5803–5805.

16. Sulavik, M. C., Houseweart, C., Cramer, C., Jiwani, N., Murgolo, N., Greene, J., DiDomenico, B., Shaw, K. J., Miller, G. H., Hare, R., and Shimer, G. (2001) Antibiotic susceptibility profiles of *Escherichia coli* strains lacking multidrug efflux pump genes, *Antimicrob. Agents Chemother. 45*, 1126–1136.

17. Jellen-Ritter, A. S., and Kern, W. V. (2001) Enhanced expression of the multidrug efflux pumps AcrAB and AcrEF associated with insertion element transposition in *Escherichia coli* mutants selected with a fluoroquinolone, *Antimicrob. Agents Chemother. 45*, 1467–1472.

18. Kobayashi, K., Tsukagoshi, N., and Aono, R. (2001) Suppression of hypersensitivity of *Escherichia coli acrB* mutant to organic solvents by integrational activation of the *acrEF* operon with the IS1 or IS2 element, *J. Bacteriol. 183*, 2646–2653.

19. White, D. G., Goldman, J. D., Demple, B., and Levy, S. B. (1997) Role of the *acrAB* locus in organic solvent tolerance mediated by expression of *marA*, *soxS*, or *robA* in *Escherichia coli*, *J. Bacteriol. 179*, 6122–6126.

20. Gallegos, M. T., Schleif, R., Bairoch, A., Hofmann, K., and Ramos, J. L. (1997) Arac/XylS family of transcriptional regulators, *Microbiol. Mol. Biol. Rev. 61*, 393–410.

21. George, A. M., and Levy, S. B. (1983) Gene in the major cotransduction gap of the *Escherichia coli* K-12 linkage map required for the expression of chromosomal resistance to tetracycline and other antibiotics, *J. Bacteriol. 155*, 541–548.

22. Wu, J., and Weiss, B. (1991) Two divergently transcribed genes, *soxR* and *soxS*, control a superoxide response regulon of *Escherichia coli*, *J. Bacteriol. 173*, 2864–2871.

23. Skarstad, K., Thony, B., Hwang, D. S., and Kornberg, A. (1993) A novel binding protein of the origin of the *Escherichia coli* chromosome, *J. Biol. Chem. 268*, 5365–5370.

24. Martin, R. G., Gillette, W. K., Rhee, S., and Rosner, J. L. (1999) Structural requirements for marbox function in transcriptional activation of *mar/sox/rob* regulon promoters in

Escherichia coli: sequence, orientation and spatial relationship to the core promoter, *Mol. Microbiol. 34,* 431–441.

25. Martin, R. G., and Rosner, J. L. (2002) Genomics of the *marA/soxS/rob* regulon of *Escherichia coli:* identification of directly activated promoters by application of molecular genetics and informatics to microarray data, *Mol. Microbiol. 44,* 1611–1624.

26. Ma, D., Alberti, M., Lynch, C., Nikaido, H., and Hearst, J. E. (1996) The local repressor AcrR plays a modulating role in the regulation of *acrAB* genes of *Escherichia coli* by global stress signals, *Mol. Microbiol. 19,* 101–112.

27. Koutsolioutsou, A., Peña-Llopis, S., and Demple, B. (2005) Constitutive *soxR* mutations contribute to multiple-antibiotic resistance in clinical *Escherichia coli* isolates, *Antimicrob. Agents Chemother. 49,* 2746–2752.

28. Oethinger, M., Podglajen, I., Kern, W. V., and Levy, S. B. (1998) Overexpression of the *marA* or *soxS* regulatory gene in clinical topoisomerase mutants of *Escherichia coli, Antimicrob. Agents Chemother. 42,* 2089–2094.

29. Coburn, B., Grassl, G. A., and Finlay, B. B. (2007) *Salmonella,* the host and disease: a brief review, *Immunol. Cell Biol. 85,* 112–118.

30. Hohmann, E. L. (2001) Nontyphoidal salmonellosis, *Clin. Infect. Dis. 32,* 263–269.

31. Baucheron, S., Imberechts, H., Chaslus-Dancla, E., and Cloeckaert, A. (2002) The AcrB multidrug transporter plays a major role in high-level fluoroquinolone resistance in *Salmonella enterica* serovar *typhimurium* phage type DT204, *Microb. Drug Resist. 8,* 281–289.

32. Chen, S., Cui, S., McDermott, P. F., Zhao, S., White, D. G., Paulsen, I., and Meng, J. (2007) Contribution of target gene mutations and efflux to decreased susceptibility of *Salmonella enterica* serovar *typhimurium* to fluoroquinolones and other antimicrobials, *Antimicrob. Agents Chemother. 51,* 535–542.

33. Baucheron, S., Tyler, S., Boyd, D., Mulvey, M. R., Chaslus-Dancla, E., and Cloeckaert, A. (2004) AcrAB–TolC directs efflux-mediated multidrug resistance in *Salmonella enterica* serovar *typhimurium* DT104, *Antimicrob. Agents Chemother. 48,* 3729–3735.

34. Ricci, V., Tzakas, P., Buckley, A., and Piddock, L. J. (2006) Ciprofloxacin-resistant *Salmonella enterica* serovar *typhimurium* strains are difficult to select in the absence of AcrB and TolC, *Antimicrob. Agents Chemother. 50,* 38–42.

35. Olliver, A., Valle, M., Chaslus-Dancla, E., and Cloeckaert, A. (2005) Overexpression of the multidrug efflux operon *acrEF* by insertional activation with IS1 or IS10 elements in *Salmonella enterica* serovar *typhimurium* DT204 *acrB* mutants selected with fluoroquinolones, *Antimicrob. Agents Chemother. 49,* 289–301.

36. Pomposiello, P. J., and Demple, B. (2000) Identification of SoxS-regulated genes in *Salmonella enterica* serovar *typhimurium, J. Bacteriol. 182,* 23–29.

37. Sulavik, M. C., Dazer, M., and Miller, P. F. (1997) The *Salmonella typhimurium mar* locus: molecular and genetic analyses and assessment of its role in virulence, *J. Bacteriol. 179,* 1857–1866.

38. Piddock, L. J., White, D. G., Gensberg, K., Pumbwe, L., and Griggs, D. J. (2000) Evidence for an efflux pump mediating multiple antibiotic resistance in *Salmonella enterica* serovar *typhimurium, Antimicrob. Agents Chemother. 44,* 3118–3121.

39. Koutsolioutsou, A., Martins, E. A., White, D. G., Levy, S. B., and Demple, B. (2001) A *soxRS*-constitutive mutation contributing to antibiotic resistance in a clinical isolate of *Salmonella enterica* (serovar *typhimurium*), *Antimicrob. Agents Chemother. 45*, 38–43.

40. O'Regan, E., Quinn, T., Pagès, J. M., McCusker, M., Piddock, L., and Fanning, S. (2009) Multiple regulatory pathways associated with high-level ciprofloxacin and multi-drug resistance in *Salmonella enterica* serovar *enteritidis*: involvement of RamA and other global regulators, *Antimicrob. Agents Chemother. 53*, 1080–1087.

41. Olliver, A., Valle, M., Chaslus-Dancla, E., and Cloeckaert, A. (2004) Role of an *acrR* mutation in multidrug resistance of in vitro-selected fluoroquinolone-resistant mutants of *Salmonella enterica* serovar *typhimurium*, *FEMS Microbiol. Lett. 238*, 267–272.

42. Chu, C., Su, L. H., Chu, C. H., Baucheron, S., Cloeckaert, A., and Chiu, C. H. (2005) Resistance to fluoroquinolones linked to *gyrA* and *parC* mutations and overexpression of *acrAB* efflux pump in *Salmonella enterica* serotype *choleraesuis*, *Microb. Drug Resist. 11*, 248–253.

43. van der Straaten, T., Janssen, R., Mevius, D. J., and van Dissel, J. T. (2004) *Salmonella* gene *rma* (*ramA*) and multiple-drug-resistant *Salmonella enterica* serovar *typhimurium*, *Antimicrob. Agents Chemother. 48*, 2292–2294.

44. Abouzeed, Y. M., Baucheron, S., and Cloeckaert, A. (2008) *ramR* mutations involved in efflux-mediated multidrug resistance in *Salmonella enterica* serovar *typhimurium*, *Antimicrob. Agents Chemother. 52*, 2428–2434.

45. Deguchi, T., Kawamura, T., Yasuda, M., Nakano, M., Fukuda, H., Kato, H., Kato, N., Okano, Y., and Kawada, Y. (1997) In vivo selection of *Klebsiella pneumoniae* strains with enhanced quinolone resistance during fluoroquinolone treatment of urinary tract infections, *Antimicrob. Agents Chemother. 41*, 1609–1611.

46. Martínez-Martínez, L., Garcia, I., Ballesta, S., Benedi, V. J., Hernandez-Alles, S., and Pascual, A. (1998) Energy-dependent accumulation of fluoroquinolones in quinolone-resistant *Klebsiella pneumoniae* strains, *Antimicrob. Agents Chemother. 42*, 1850–1852.

47. Mazzariol, A., Zuliani, J., Cornaglia, G., Rossolini, G. M., and Fontana, R. (2002) AcrAB efflux system: expression and contribution to fluoroquinolone resistance in *Klebsiella* spp., *Antimicrob. Agents Chemother. 46*, 3984–3986.

48. Schneiders, T., Amyes, S. G., and Levy, S. B. (2003) Role of AcrR and *ramA* in fluoroquinolone resistance in clinical *Klebsiella pneumoniae* isolates from Singapore, *Antimicrob. Agents Chemother. 47*, 2831–2837.

49. George, A. M., Hall, R. M., and Stokes, H. W. (1995) Multidrug resistance in *Klebsiella pneumoniae*: a novel gene, *ramA*, confers a multidrug resistance phenotype in *Escherichia coli*, *Microbiology 141*(Pt. 8), 1909–1920.

50. Ruzin, A., Visalli, M. A., Keeney, D., and Bradford, P. A. (2005) Influence of transcriptional activator RamA on expression of multidrug efflux pump AcrAB and tigecycline susceptibility in *Klebsiella pneumoniae*, *Antimicrob. Agents Chemother. 49*, 1017–1022.

51. Ogawa, W., Koterasawa, M., Kuroda, T., and Tsuchiya, T. (2006) KmrA multidrug efflux pump from *Klebsiella pneumoniae*, *Biol. Pharm. Bull. 29*, 550–553.

52. Ping, Y., Ogawa, W., Kuroda, T., and Tsuchiya, T. (2007) Gene cloning and characterization of KdeA, a multidrug efflux pump from *Klebsiella pneumoniae*, *Biol. Pharm. Bull. 30*, 1962–1964.

53. Mallea, M., Chevalier, J., Bornet, C., Eyraud, A., vin-Regli, A., Bollet, C., and Pagès, J. M. (1998) Porin alteration and active efflux: two in vivo drug resistance strategies used by *Enterobacter aerogenes*, *Microbiology 144*(Pt. 11), 3003–3009.

54. Pradel, E., and Pagès, J. M. (2002) The AcrAB-TolC efflux pump contributes to multidrug resistance in the nosocomial pathogen *Enterobacter aerogenes*, *Antimicrob. Agents Chemother. 46*, 2640–2643.

55. Chollet, R., Bollet, C., Chevalier, J., Mallea, M., Pagès, J. M., and vin-Regli, A. (2002) *mar* operon involved in multidrug resistance of *Enterobacter aerogenes*, *Antimicrob. Agents Chemother. 46*, 1093–1097.

56. Chollet, R., Chevalier, J., Bollet, C., Pagès, J. M., and vin-Regli, A. (2004) RamA is an alternate activator of the multidrug resistance cascade in *Enterobacter aerogenes*, *Antimicrob. Agents Chemother. 48*, 2518–2523.

57. Perez, A., Canle, D., Latasa, C., Poza, M., Beceiro, A., Tomas, M. M., Fernandez, A., Mallo, S., Perez, S., Molina, F., et al. (2007) Cloning, nucleotide sequencing, and analysis of the AcrAB–TolC efflux pump of *Enterobacter cloacae* and determination of its involvement in antibiotic resistance in a clinical isolate, *Antimicrob. Agents Chemother. 51*, 3247–3253.

58. Linde, H. J., Notka, F., Irtenkauf, C., Decker, J., Wild, J., Niller, H. H., Heisig, P., and Lehn, N. (2002) Increase in MICs of ciprofloxacin in vivo in two closely related clinical isolates of *Enterobacter cloacae*, *J. Antimicrob. Chemother. 49*, 625–630.

59. Kumar, A., and Worobec, E. A. (2002) Fluoroquinolone resistance of *Serratia marcescens*: involvement of a proton gradient–dependent efflux pump, *J. Antimicrob. Chemother. 50*, 593–596.

60. Chen, J., Kuroda, T., Huda, M. N., Mizushima, T., and Tsuchiya, T. (2003) An RND-type multidrug efflux pump SdeXY from *Serratia marcescens*, *J. Antimicrob. Chemother. 52*, 176–179.

61. Shahcheraghi, F., Minato, Y., Chen, J., Mizushima, T., Ogawa, W., Kuroda, T., and Tsuchiya, T. (2007) Molecular cloning and characterization of a multidrug efflux pump, SmfY, from *Serratia marcescens*, *Biol. Pharm. Bull. 30*, 798–800.

62. Matsuo, T., Chen, J., Minato, Y., Ogawa, W., Mizushima, T., Kuroda, T., and Tsuchiya, T. (2008) SmdAB, a heterodimeric ABC-type multidrug efflux pump, in *Serratia marcescens*, *J. Bacteriol. 190*, 648–654.

63. Visalli, M. A., Murphy, E., Projan, S. J., and Bradford, P. A. (2003) AcrAB multidrug efflux pump is associated with reduced levels of susceptibility to tigecycline (GAR-936) in *Proteus mirabilis*, *Antimicrob. Agents Chemother. 47*, 665–669.

64. Saito, R., Sato, K., Kumita, W., Inami, N., Nishiyama, H., Okamura, N., Moriya, K., and Koike, K. (2006) Role of type II topoisomerase mutations and AcrAB efflux pump in fluoroquinolone-resistant clinical isolates of *Proteus mirabilis*, *J. Antimicrob. Chemother. 58*, 673–677.

65. Ishii, H., Sato, K., Hoshino, K., Sato, M., Yamaguchi, A., Sawai, T., and Osada, Y. (1991) Active efflux of ofloxacin by a highly quinolone-resistant strain of *Proteus vulgaris*, *J. Antimicrob. Chemother. 28*, 827–836.

66. Ishida, H., Fuziwara, H., Kaibori, Y., Horiuchi, T., Sato, K., and Osada, Y. (1995) Cloning of multidrug resistance gene *pqrA* from *Proteus vulgaris*, *Antimicrob. Agents Chemother. 39*, 453–457.

67. Navia, M. M., Ruiz, J., Ribera, A., de Anta, M. T., and Vila, J. (1999) Analysis of the mechanisms of quinolone resistance in clinical isolates of *Citrobacter freundii*, *J. Antimicrob. Chemother. 44*, 743–748.

68. Tavío, M., Vila, J., Ruiz, J., Amicosante, G., Franceschini, N., Martín-Sánchez, A. M., and de Anta, M. T. (2000) In vitro selected fluoroquinolone-resistant mutants of *Citrobacter freundii:* analysis of the quinolone resistance acquisition, *J. Antimicrob. Chemother. 45*, 521–524.

69. Sánchez-Céspedes, J., and Vila, J. (2007) Partial characterisation of the *acrAB* locus in two *Citrobacter freundii* clinical isolates, *Int. J. Antimicrob. Agents 30*, 259–263.

70. Quinn, J. P. (1998) Clinical problems posed by multiresistant nonfermenting gram-negative pathogens, *Clin. Infect. Dis. 27*(Suppl. 1), S117–S124.

71. Ferrara, A. M. (2006) Potentially multidrug-resistant non-fermentative gram-negative pathogens causing nosocomial pneumonia, *Int. J. Antimicrob. Agents 27*, 183–195.

72. McGowan, J. E., Jr., (2006) Resistance in nonfermenting gram-negative bacteria: multidrug resistance to the maximum, *Am. J. Infect. Control 34*, S29–S37; discussion S64–S73.

73. Alonso, A., Rojo, F., and Martínez, J. L. (1999) Environmental and clinical isolates of *Pseudomonas aeruginosa* show pathogenic and biodegradative properties irrespective of their origin, *Environ. Microbiol. 1*, 421–430.

74. Morales, G., Wiehlmann, L., Gudowius, P., van Delden, C., Tummler, B., Martínez, J. L., and Rojo, F. (2004) Structure of *Pseudomonas aeruginosa* populations analyzed by single nucleotide polymorphism and pulsed-field gel electrophoresis genotyping, *J. Bacteriol. 186*, 4228–4237.

75. Martínez, J. L. (2008) Antibiotics and antibiotic resistance genes in natural environments, *Science 321*, 365–367.

76. Martínez, J. L., Fajardo, A., Garmendia, L., Hernández, A., Linares, J. F., Martínez-Solano, L., and Sánchez, M. B. (2009) A global view of antibiotic resistance, *FEMS Microbiol. Rev. 33*, 44–65.

77. Martínez, J. L., Sánchez, M. B., Martínez-Solano, L., Hernández, A., Garmendia, L., Fajardo, A., and Alvarez-Ortega, C. (2009) Functional role of bacterial multidrug efflux pumps in microbial natural ecosystems, *FEMS Microbiol. Rev. 33*, 430–449.

78. Li, X. Z., Nikaido, H., and Poole, K. (1995) Role of *mexA-mexB-oprM* in antibiotic efflux in *Pseudomonas aeruginosa*, *Antimicrob. Agents Chemother. 39*, 1948–1953.

79. Poole, K., Gotoh, N., Tsujimoto, H., Zhao, Q., Wada, A., Yamasaki, T., Neshat, S., Yamagishi, J., Li, X. Z., and Nishino, T. (1996) Overexpression of the *mexC–mexD–oprJ* efflux operon in *nfxB*-type multidrug-resistant strains of *Pseudomonas aeruginosa*, *Mol. Microbiol. 21*, 713–724.

80. Köhler, T., Michea-Hamzehpour, M., Henze, U., Gotoh, N., Curty, L. K., and Pechere, J. C. (1997) Characterization of MexE–MexF–OprN, a positively regulated multidrug efflux system of *Pseudomonas aeruginosa*, *Mol. Microbiol. 23*, 345–354.

81. Aires, J. R., Köhler, T., Nikaido, H., and Plésiat, P. (1999) Involvement of an active efflux system in the natural resistance of *Pseudomonas aeruginosa* to aminoglycosides, *Antimicrob. Agents Chemother. 43*, 2624–2628.

82. Mine, T., Morita, Y., Kataoka, A., Mizushima, T., and Tsuchiya, T. (1999) Expression in *Escherichia coli* of a new multidrug efflux pump, MexXY, from *Pseudomonas aeruginosa*, *Antimicrob. Agents Chemother. 43*, 415–417.

83. Chuanchuen, R., Narasaki, C. T., and Schweizer, H. P. (2002) The MexJK efflux pump of *Pseudomonas aeruginosa* requires OprM for antibiotic efflux but not for efflux of triclosan, *J. Bacteriol. 184*, 5036–5044.

84. Li, Y., Mima, T., Komori, Y., Morita, Y., Kuroda, T., Mizushima, T., and Tsuchiya, T. (2003) A new member of the tripartite multidrug efflux pumps, MexVW–OprM, in *Pseudomonas aeruginosa*, *J. Antimicrob. Chemother. 52*, 572–575.

85. Mima, T., Sekiya, H., Mizushima, T., Kuroda, T., and Tsuchiya, T. (2005) Gene cloning and properties of the RND-type multidrug efflux pumps MexPQ-OpmE and MexMN–OprM from *Pseudomonas aeruginosa*, *Microbiol. Immunol. 49*, 999–1002.

86. Aendekerk, S., Diggle, S. P., Song, Z., Hoiby, N., Cornelis, P., Williams, P., and Camara, M. (2005) The MexGHI–OpmD multidrug efflux pump controls growth, antibiotic susceptibility and virulence in *Pseudomonas aeruginosa* via 4-quinolone-dependent cell-to-cell communication, *Microbiology 151*, 1113–1125.

87. Morita, Y., Kimura, N., Mima, T., Mizushima, T., and Tsuchiya, T. (2001) Roles of MexXY– and MexAB–multidrug efflux pumps in intrinsic multidrug resistance of *Pseudomonas aeruginosa* PAO1, *J. Gen. Appl. Microbiol. 47*, 27–32.

88. Grkovic, S., Brown, M. H., and Skurray, R. A. (2002) Regulation of bacterial drug export systems, *Microbiol. Mol. Biol. Rev. 66*, 671–701, table of contents.

89. Oh, H., Stenhoff, J., Jalal, S., and Wretlind, B. (2003) Role of efflux pumps and mutations in genes for topoisomerases II and IV in fluoroquinolone-resistant *Pseudomonas aeruginosa* strains, *Microb. Drug Resist. 9*, 323–328.

90. Köhler, T., Michea-Hamzehpour, M., Plésiat, P., Kahr, A. L., and Pechere, J. C. (1997) Differential selection of multidrug efflux systems by quinolones in *Pseudomonas aeruginosa*, *Antimicrob. Agents Chemother. 41*, 2540–2543.

91. Join-Lambert, O. F., Michea-Hamzehpour, M., Köhler, T., Chau, F., Faurisson, F., Dautrey, S., Vissuzaine, C., Carbon, C., and Pechere, J. (2001) Differential selection of multidrug efflux mutants by trovafloxacin and ciprofloxacin in an experimental model of *Pseudomonas aeruginosa* acute pneumonia in rats, *Antimicrob. Agents Chemother. 45*, 571–576.

92. Macia, M. D., Borrell, N., Segura, M., Gomez, C., Perez, J. L., and Oliver, A. (2006) Efficacy and potential for resistance selection of antipseudomonal treatments in a mouse model of lung infection by hypermutable *Pseudomonas aeruginosa*, *Antimicrob. Agents Chemother. 50*, 975–983.

93. Cosson, P., Zulianello, L., Join-Lambert, O., Faurisson, F., Gebbie, L., Benghezal, M., Van Delden, C., Curty, L. K., and Köhler, T. (2002) *Pseudomonas aeruginosa* virulence analyzed in a *Dictyostelium discoideum* host system, *J. Bacteriol. 184*, 3027–3033.

94. Sanchez, P., Linares, J. F., Ruiz-Diez, B., Campanario, E., Navas, A., Baquero, F., and Martínez, J. L. (2002) Fitness of in vitro selected *Pseudomonas aeruginosa* nalB and nfxB multidrug resistant mutants, *J. Antimicrob. Chemother. 50*, 657–664.

95. Hirakata, Y., Srikumar, R., Poole, K., Gotoh, N., Suematsu, T., Kohno, S., Kamihira, S., Hancock, R. E., and Speert, D. P. (2002) Multidrug efflux systems play an important role in the invasiveness of *Pseudomonas aeruginosa, J. Exp. Med. 196*, 109–118.

96. Evans, K., Passador, L., Srikumar, R., Tsang, E., Nezezon, J., and Poole, K. (1998) Influence of the MexAB–OprM multidrug efflux system on quorum sensing in *Pseudomonas aeruginosa, J. Bacteriol. 180*, 5443–5447.

97. Köhler, T., van Delden, C., Curty, L. K., Hamzehpour, M. M., and Pechere, J. C. (2001) Overexpression of the MexEF–OprN multidrug efflux system affects cell-to-cell signaling in *Pseudomonas aeruginosa, J. Bacteriol. 183*, 5213–5222.

98. Jeannot, K., Elsen, S., Köhler, T., Attree, I., Van Delden, C., and Plésiat, P. (2008) Resistance and virulence of *Pseudomonas aeruginosa* clinical strains overproducing the MexCD–OprJ efflux pump, *Antimicrob. Agents Chemother. 52*, 2455–2462.

99. Hocquet, D., Berthelot, P., Roussel-Delvallez, M., Favre, R., Jeannot, K., Bajolet, O., Marty, N., Grattard, F., Mariani-Kurkdjian, P., Bingen, E., et al. (2007) *Pseudomonas aeruginosa* may accumulate drug resistance mechanisms without losing its ability to cause bloodstream infections, *Antimicrob. Agents Chemother. 51*, 3531–3536.

100. Ventre, I., Goodman, A. L., Vallet-Gely, I., Vasseur, P., Soscia, C., Molin, S., Bleves, S., Lazdunski, A., Lory, S., and Filloux, A. (2006) Multiple sensors control reciprocal expression of *Pseudomonas aeruginosa* regulatory RNA and virulence genes, *Proc. Natl. Acad. Sci. USA 103*, 171–176.

101. Martínez-Solano, L., Macia, M. D., Fajardo, A., Oliver, A., and Martínez, J. L. (2008) Chronic *Pseudomonas aeruginosa* infection in chronic obstructive pulmonary disease, *Clin. Infect. Dis. 47*, 1526–1533.

102. Jain, M., Bar-Meir, M., McColley, S., Cullina, J., Potter, E., Powers, C., Prickett, M., Seshadri, R., Jovanovic, B., Petrocheilou, A., et al. (2008) Evolution of *Pseudomonas aeruginosa* type III secretion in cystic fibrosis: a paradigm of chronic infection, *Transl. Res. 152*, 257–264.

103. Burke, V., Robinson, J. O., Richardson, C. J., and Bundell, C. S. (1991) Longitudinal studies of virulence factors of *Pseudomonas aeruginosa* in cystic fibrosis, *Pathology 23*, 145–148.

104. Linares, J. F., Lopez, J. A., Camafeita, E., Albar, J. P., Rojo, F., and Martínez, J. L. (2005) Overexpression of the multidrug efflux pumps MexCD–OprJ and MexEF–OprN is associated with a reduction of type III secretion in *Pseudomonas aeruginosa, J. Bacteriol. 187*, 1384–1391.

105. Vettoretti, L., Plésiat, P., Muller, C., El Garch, F., Phan, G., Attree, I., Ducruix, A., and Llanes, C. (2009) Efflux unbalance in *Pseudomonas aeruginosa* isolates from cystic fibrosis patients, *Antimicrob. Agents Chemother. 53*, 1987–1997.

106. Hocquet, D., Muller, A., Blanc, K., Plésiat, P., Talon, D., Monnet, D. L., and Bertrand, X. (2008) Relationship between antibiotic use and incidence of MexXY–OprM overproducers among clinical isolates of *Pseudomonas aeruginosa, Antimicrob. Agents Chemother. 52*, 1173–1175.

107. Henrichfreise, B., Wiegand, I., Pfister, W., and Wiedemann, B. (2007) Resistance mechanisms of multiresistant *Pseudomonas aeruginosa* strains from Germany and correlation with hypermutation, *Antimicrob. Agents Chemother. 51*, 4062–4070.

108. Llanes, C., Neuwirth, C., El Garch, F., Hocquet, D., and Plésiat, P. (2006) Genetic analysis of a multiresistant strain of *Pseudomonas aeruginosa* producing PER-1 beta-lactamase, *Clin. Microbiol. Infect. 12*, 270–278.

109. Rolston, K. V., Kontoyiannis, D. P., Yadegarynia, D., and Raad, II. (2005) Nonfermentative gram-negative bacilli in cancer patients: increasing frequency of infection and antimicrobial susceptibility of clinical isolates to fluoroquinolones, *Diagn. Microbiol. Infect. Dis. 51*, 215–218.

110. Perz, J. F., Craig, A. S., Stratton, C. W., Bodner, S. J., Phillips, W. E., Jr., and Schaffner, W. (2005) *Pseudomonas putida* septicemia in a special care nursery due to contaminated flush solutions prepared in a hospital pharmacy, *J. Clin. Microbiol. 43*, 5316–5318.

111. Teran, W., Felipe, A., Segura, A., Rojas, A., Ramos, J. L., and Gallegos, M. T. (2003) Antibiotic-dependent induction of *Pseudomonas putida* DOT–T1E TtgABC efflux pump is mediated by the drug binding repressor TtgR, *Antimicrob. Agents Chemother. 47*, 3067–3072.

112. Noble, R. C., and Overman, S. B. (1994) *Pseudomonas stutzeri* infection: a review of hospital isolates and a review of the literature, *Diagn. Microbiol. Infect. Dis. 19*, 51–56.

113. Jude, F., Arpin, C., Brachet-Castang, C., Capdepuy, M., Caumette, P., and Quentin, C. (2004) TbtABM, a multidrug efflux pump associated with tributyltin resistance in *Pseudomonas stutzeri, FEMS Microbiol. Lett. 232*, 7–14.

114. Ribera, A., Ruiz, J., Jiminez de Anta, M. T., and Vila, J. (2002) Effect of an efflux pump inhibitor on the MIC of nalidixic acid for *Acinetobacter baumannii* and *Stenotrophomonas maltophilia* clinical isolates, *J. Antimicrob. Chemother. 49*, 697–698.

115. Vila, J., Ribera, A., Marco, F., Ruiz, J., Mensa, J., Chaves, J., Hernandez, G., and Jimenez de Anta, M. T. (2002) Activity of clinafloxacin, compared with six other quinolones, against *Acinetobacter baumannii* clinical isolates, *J. Antimicrob. Chemother. 49*, 471–477.

116. Fournier, P. E., Vallenet, D., Barbé, V., Audic, S., Ogata, H., Poirel, L., Richet, H., Robert, C., Mangenot, S., Abergel, C., et al. (2006) Comparative genomics of multidrug resistance in *Acinetobacter baumannii, PLoS Genet. 2*, e7.

117. Damier-Piolle, L., Magnet, S., Bremont, S., Lambert, T., and Courvalin, P. (2008) AdeIJK, a resistance–nodulation–cell division pump effluxing multiple antibiotics in *Acinetobacter baumannii, Antimicrob. Agents Chemother. 52*, 557–562.

118. Barbé, V., Vallenet, D., Fonknechten, N., Kreimeyer, A., Oztas, S., Labarre, L., Cruveiller, S., Robert, C., Duprat, S., Wincker, P., et al. (2004) Unique features revealed by the genome sequence of *Acinetobacter* sp. ADP1, a versatile and naturally transformation competent bacterium, *Nucleic Acids Res. 32*, 5766–5779.

119. Chu, Y. W., Chau, S. L., and Houang, E. T. (2006) Presence of active efflux systems AdeABC, AdeDE and AdeXYZ in different *Acinetobacter* genomic DNA groups, *J. Med. Microbiol. 55*, 477–478.

120. Magnet, S., Courvalin, P., and Lambert, T. (2001) Resistance-nodulation-cell division-type efflux pump involved in aminoglycoside resistance in *Acinetobacter baumannii* strain BM4454, *Antimicrob. Agents Chemother. 45*, 3375–3380.

121. Marchand, I., Damier-Piolle, L., Courvalin, P., and Lambert, T. (2004) Expression of the RND-type efflux pump AdeABC in *Acinetobacter baumannii* is regulated by the AdeRS two-component system, *Antimicrob. Agents Chemother. 48*, 3298–3304.

122. Nemec, A., Maixnerova, M., van der Reijden, T. J., van den Broek, P. J., and Dijkshoorn, L. (2007) Relationship between the AdeABC efflux system gene content, netilmicin susceptibility and multidrug resistance in a genotypically diverse collection of *Acinetobacter baumannii* strains, *J. Antimicrob. Chemother.* 60, 483–489.

123. Chau, S. L., Chu, Y. W., and Houang, E. T. (2004) Novel resistance–nodulation–cell division efflux system AdeDE in *Acinetobacter* genomic DNA group 3, *Antimicrob. Agents Chemother.* 48, 4054–4055.

124. Lin, L., Ling, B. D., and Li, X. Z. (2009) Distribution of the multidrug efflux pump genes, *adeABC*, *adeDE* and *adeIJK*, and class 1 integron genes in multiple-antimicrobial-resistant clinical isolates of *Acinetobacter baumannii–Acinetobacter calcoaceticus* complex, *Int. J. Antimicrob. Agents 33*, 27–32.

125. Su, X. Z., Chen, J., Mizushima, T., Kuroda, T., and Tsuchiya, T. (2005) AbeM, an H^+-coupled *Acinetobacter baumannii* multidrug efflux pump belonging to the MATE family of transporters, *Antimicrob. Agents Chemother.* 49, 4362–4364.

126. Crossman, L. C., Gould, V. C., Dow, J. M., Vernikos, G. S., Okazaki, A., Sebaihia, M., Saunders, D., Arrowsmith, C., Carver, T., Peters, N., et al. (2008) The complete genome, comparative and functional analysis of *Stenotrophomonas maltophilia* reveals an organism heavily shielded by drug resistance determinants, *Genome Biol. 9*, R74.

127. Ryan, R. P., Monchy, S., Cardinale, M., Taghavi, S., Crossman, L., Avison, M. B., Berg, G., van der Lelie, D., and Dow, J. M. (2009) The versatility and adaptation of bacteria from the genus *Stenotrophomonas*, *Nat. Rev. Microbiol. 7*, 514–525.

128. Alonso, A., and Martínez, J. L. (1997) Multiple antibiotic resistance in *Stenotrophomonas maltophilia*, *Antimicrob. Agents Chemother. 41*, 1140–1142.

129. Alonso, A., and Martínez, J. L. (2000) Cloning and characterization of SmeDEF, a novel multidrug efflux pump from *Stenotrophomonas maltophilia*, *Antimicrob. Agents Chemother. 44*, 3079–3086.

130. Alonso, A., and Martínez, J. L. (2001) Expression of multidrug efflux pump SmeDEF by clinical isolates of *Stenotrophomonas maltophilia*, *Antimicrob. Agents Chemother. 45*, 1879–1881.

131. Alonso, A., Morales, G., Escalante, R., Campanario, E., Sastre, L., and Martínez, J. L. (2004) Overexpression of the multidrug efflux pump SmeDEF impairs *Stenotrophomonas maltophilia* physiology, *J. Antimicrob. Chemother. 53*, 432–434.

132. Zhang, L., Li, X. Z., and Poole, K. (2001) SmeDEF multidrug efflux pump contributes to intrinsic multidrug resistance in *Stenotrophomonas maltophilia*, *Antimicrob. Agents Chemother. 45*, 3497–3503.

133. Zhang, L., Li, X. Z., and Poole, K. (2000) Multiple antibiotic resistance in *Stenotrophomonas maltophilia:* involvement of a multidrug efflux system, *Antimicrob. Agents Chemother. 44*, 287–293.

134. Li, X. Z., Zhang, L., and Poole, K. (2002) SmeC, an outer membrane multidrug efflux protein of *Stenotrophomonas maltophilia*, *Antimicrob. Agents Chemother. 46*, 333–343.

135. Sanchez, P., Alonso, A., and Martínez, J. L. (2002) Cloning and characterization of SmeT, a repressor of the *Stenotrophomonas maltophilia* multidrug efflux pump Sme-DEF, *Antimicrob. Agents Chemother. 46*, 3386–3393.

136. Hernández, A., Mate, M. J., Sanchez-Diaz, P. C., Romero, A., Rojo, F., and Martínez, J. L. (2009) Structural and functional analysis of SmeT, the repressor of the *Stenotrophomonas maltophilia* multidrug efflux pump SmeDEF, *J. Biol. Chem. 284*, 14428–14438.

137. Sanchez, P., Alonso, A., and Martínez, J. L. (2004) Regulatory regions of smeDEF in *Stenotrophomonas maltophilia* strains expressing different amounts of the multidrug efflux pump SmeDEF, *Antimicrob. Agents Chemother. 48*, 2274–2276.

138. Ribera, A., Domenech-Sanchez, A., Ruiz, J., Benedi, V. J., Jimenez de Anta, M. T., and Vila, J. (2002) Mutations in *gyrA* and *parC* QRDRs are not relevant for quinolone resistance in epidemiological unrelated *Stenotrophomonas maltophilia* clinical isolates, *Microb. Drug Resist. 8*, 245–251.

139. Valdezate, S., Vindel, A., Saez-Nieto, J. A., Baquero, F., and Canton, R. (2005) Preservation of topoisomerase genetic sequences during in vivo and in vitro development of high-level resistance to ciprofloxacin in isogenic *Stenotrophomonas maltophilia* strains, *J. Antimicrob. Chemother. 56*, 220–223.

140. Sánchez, M. B., Hernández, A., Rodríguez-Martínez, J. M., Martínez-Martínez, L., and Martínez, J. L. (2008) Predictive analysis of transmissible quinolone resistance indicates *Stenotrophomonas maltophilia* as a potential source of a novel family of Qnr determinants, *BMC Microbiol. 8*, 148.

141. Shimizu, K., Kikuchi, K., Sasaki, T., Takahashi, N., Ohtsuka, M., Ono, Y., and Hiramatsu, K. (2008) Smqnr, a new chromosome-carried quinolone resistance gene in *Stenotrophomonas maltophilia*, *Antimicrob. Agents Chemother. 52*, 3823–3825.

142. Jacoby, G., Cattoir, V., Hooper, D., Martínez-Martínez, L., Nordmann, P., Pascual, A., Poirel, L., and Wang, M. (2008) *qnr* Gene nomenclature, *Antimicrob. Agents Chemother. 52*, 2297–2299.

143. Martínez-Martínez, L., Pascual, A., and Jacoby, G. A. (1998) Quinolone resistance from a transferable plasmid, *Lancet 351*, 797–799.

144. Sanchez, P., Moreno, E., and Martínez, J. L. (2005) The biocide triclosan selects *Stenotrophomonas maltophilia* mutants that overproduce the SmeDEF multidrug efflux pump, *Antimicrob. Agents Chemother. 49*, 781–782.

145. Valvano, M. A., Keith, K. E., and Cardona, S. T. (2005) Survival and persistence of opportunistic *Burkholderia* species in host cells, *Curr. Opin. Microbiol. 8*, 99–105.

146. Holden, M. T., Titball, R. W., Peacock, S. J., Cerdeno-Tarraga, A. M., Atkins, T., Crossman, L. C., Pitt, T., Churcher, C., Mungall, K., Bentley, S. D., et al. (2004) Genomic plasticity of the causative agent of melioidosis, *Burkholderia pseudomallei*, *Proc. Natl. Acad. Sci. USA 101*, 14240–14245.

147. Kumar, A., Mayo, M., Trunck, L. A., Cheng, A. C., Currie, B. J., and Schweizer, H. P. (2008) Expression of resistance–nodulation–cell-division efflux pumps in commonly used *Burkholderia pseudomallei* strains and clinical isolates from northern Australia, *Trans. R. Soc. Trop. Med. Hyg. 102*(Suppl. 1), S145–S151.

148. Chan, Y. Y., Tan, T. M., Ong, Y. M., and Chua, K. L. (2004) BpeAB-OprB, a multidrug efflux pump in *Burkholderia pseudomallei*, *Antimicrob. Agents Chemother. 48*, 1128–1135.

149. Kumar, A., Chua, K. L., and Schweizer, H. P. (2006) Method for regulated expression of single-copy efflux pump genes in a surrogate *Pseudomonas aeruginosa* strain:

identification of the BpeEF–OprC chloramphenicol and trimethoprim efflux pump of *Burkholderia pseudomallei* 1026b, *Antimicrob. Agents Chemother. 50*, 3460–3463.

150. Chan, Y. Y., Ong, Y. M., and Chua, K. L. (2007) Synergistic interaction between phenothiazines and antimicrobial agents against *Burkholderia pseudomallei*, *Antimicrob. Agents Chemother. 51*, 623–630.

151. Guglierame, P., Pasca, M. R., De Rossi, E., Buroni, S., Arrigo, P., Manina, G., and Riccardi, G. (2006) Efflux pump genes of the resistance–nodulation–division family in *Burkholderia cenocepacia* genome, *BMC Microbiol. 6*, 66.

152. Wigfield, S. M., Rigg, G. P., Kavari, M., Webb, A. K., Matthews, R. C., and Burnie, J. P. (2002) Identification of an immunodominant drug efflux pump in *Burkholderia cepacia*, *J. Antimicrob. Chemother. 49*, 619–624.

153. Piddock, L. J. (2006) Multidrug-resistance efflux pumps—not just for resistance, *Nat. Rev. Microbiol. 4*, 629–636.

154. Andriole, V. T. (2005) The quinolones: past, present, and future, *Clin. Infect. Dis. 41* (Suppl. 2), S113–S119.

155. Yoshida, H., Bogaki, M., Nakamura, S., Ubukata, K., and Konno, M. (1990) Nucleotide sequence and characterization of the *Staphylococcus aureus norA* gene, which confers resistance to quinolones, *J. Bacteriol. 172*, 6942–6949.

156. Gill, M. J., Brenwald, N. P., and Wise, R. (1999) Identification of an efflux pump gene, *pmrA*, associated with fluoroquinolone resistance in *Streptococcus pneumoniae*, *Antimicrob. Agents Chemother. 43*, 187–189.

157. Neyfakh, A. A. (1992) The multidrug efflux transporter of *Bacillus subtilis* is a structural and functional homolog of the *Staphylococcus* NorA protein, *Antimicrob. Agents Chemother. 36*, 484–485.

158. Neyfakh, A. A., Borsch, C. M., and Kaatz, G. W. (1993) Fluoroquinolone resistance protein NorA of *Staphylococcus aureus* is a multidrug efflux transporter, *Antimicrob. Agents Chemother. 37*, 128–129.

159. Ng, E. Y., Trucksis, M., and Hooper, D. C. (1994) Quinolone resistance mediated by *norA*: physiologic characterization and relationship to *flqB*, a quinolone resistance locus on the *Staphylococcus aureus* chromosome, *Antimicrob. Agents Chemother. 38*, 1345–1355.

160. Kaatz, G. W., Thyagarajan, R. V., and Seo, S. M. (2005) Effect of promoter region mutations and *mgrA* overexpression on transcription of *norA*, which encodes a *Staphylococcus aureus* multidrug efflux transporter, *Antimicrob. Agents Chemother. 49*, 161–169.

161. Fournier, B., Aras, R., and Hooper, D. C. (2000) Expression of the multidrug resistance transporter NorA from *Staphylococcus aureus* is modified by a two-component regulatory system, *J. Bacteriol. 182*, 664–671.

162. Fournier, B., and Hooper, D. C. (2000) A new two-component regulatory system involved in adhesion, autolysis, and extracellular proteolytic activity of *Staphylococcus aureus*, *J. Bacteriol. 182*, 3955–3964.

163. Truong-Bolduc, Q. C., Zhang, X., and Hooper, D. C. (2003) Characterization of NorR protein, a multifunctional regulator of *norA* expression in *Staphylococcus aureus*, *J. Bacteriol. 185*, 3127–3138.

164. Luong, T. T., Newell, S. W., and Lee, C. Y. (2003) Mgr, a novel global regulator in *Staphylococcus aureus, J. Bacteriol. 185*, 3703–3710.

165. Truong-Bolduc, Q. C., and Hooper, D. C. (2007) The transcriptional regulators NorG and MgrA modulate resistance to both quinolones and beta-lactams in *Staphylococcus aureus, J. Bacteriol. 189*, 2996–3005.

166. Noguchi, N., Tamura, M., Narui, K., Wakasugi, K., and Sasatsu, M. (2002) Frequency and genetic characterization of multidrug-resistant mutants of *Staphylococcus aureus* after selection with individual antiseptics and fluoroquinolones, *Biol. Pharm. Bull. 25*, 1129–1132.

167. Kaatz, G. W., Moudgal, V. V., and Seo, S. M. (2002) Identification and characterization of a novel efflux-related multidrug resistance phenotype in *Staphylococcus aureus, J. Antimicrob. Chemother. 50*, 833–838.

168. Truong-Bolduc, Q. C., Dunman, P. M., Strahilevitz, J., Projan, S. J., and Hooper, D. C. (2005) MgrA is a multiple regulator of two new efflux pumps in *Staphylococcus aureus, J. Bacteriol. 187*, 2395–2405.

169. Truong-Bolduc, Q. C., Strahilevitz, J., and Hooper, D. C. (2006) NorC, a new efflux pump regulated by MgrA of *Staphylococcus aureus, Antimicrob. Agents Chemother. 50*, 1104–1107.

170. Yamada, Y., Hideka, K., Shiota, S., Kuroda, T., and Tsuchiya, T. (2006) Gene cloning and characterization of SdrM, a chromosomally-encoded multidrug efflux pump, from *Staphylococcus aureus, Biol. Pharm. Bull. 29*, 554–556.

171. Yamada, Y., Shiota, S., Mizushima, T., Kuroda, T., and Tsuchiya, T. (2006) Functional gene cloning and characterization of MdeA, a multidrug efflux pump from *Staphylococcus aureus, Biol. Pharm. Bull. 29*, 801–804.

172. Huang, J., O'Toole, P. W., Shen, W., Amrine-Madsen, H., Jiang, X., Lobo, N., Palmer, L. M., Voelker, L., Fan, F., Gwynn, M. N., and McDevitt, D. (2004) Novel chromosomally encoded multidrug efflux transporter MdeA in *Staphylococcus aureus, Antimicrob. Agents Chemother. 48*, 909–917.

173. McAleese, F., Petersen, P., Ruzin, A., Dunman, P. M., Murphy, E., Projan, S. J., and Bradford, P. A. (2005) A novel MATE family efflux pump contributes to the reduced susceptibility of laboratory-derived *Staphylococcus aureus* mutants to tigecycline, *Antimicrob. Agents Chemother. 49*, 1865–1871.

174. Kaatz, G. W., McAleese, F., and Seo, S. M. (2005) Multidrug resistance in *Staphylococcus aureus* due to overexpression of a novel multidrug and toxin extrusion (MATE) transport protein, *Antimicrob. Agents Chemother. 49*, 1857–1864.

175. Kaatz, G. W., DeMarco, C. E., and Seo, S. M. (2006) MepR, a repressor of the *Staphylococcus aureus* MATE family multidrug efflux pump MepA, is a substrate-responsive regulatory protein, *Antimicrob. Agents Chemother. 50*, 1276–1281.

176. DeMarco, C. E., Cushing, L. A., Frempong-Manso, E., Seo, S. M., Jaravaza, T. A., and Kaatz, G. W. (2007) Efflux-related resistance to norfloxacin, dyes, and biocides in bloodstream isolates of *Staphylococcus aureus, Antimicrob. Agents Chemother. 51*, 3235–3239.

177. Piddock, L. J., Johnson, M. M., Simjee, S., and Pumbwe, L. (2002) Expression of efflux pump gene *pmrA* in fluoroquinolone-resistant and -susceptible clinical isolates of *Streptococcus pneumoniae, Antimicrob. Agents Chemother. 46*, 808–812.

178. Martinez-Garriga, B., Vinuesa, T., Hernandez-Borrell, J., and Vinas, M. (2007) The contribution of efflux pumps to quinolone resistance in *Streptococcus pneumoniae* clinical isolates, *Int. J. Med. Microbiol. 297*, 187–195.

179. Robertson, G. T., Doyle, T. B., and Lynch, A. S. (2005) Use of an efflux-deficient *Streptococcus pneumoniae* strain panel to identify ABC-class multidrug transporters involved in intrinsic resistance to antimicrobial agents, *Antimicrob. Agents Chemother. 49*, 4781–4783.

180. Brenwald, N. P., Appelbaum, P., Davies, T., and Gill, M. J. (2003) Evidence for efflux pumps, other than PmrA, associated with fluoroquinolone resistance in *Streptococcus pneumoniae*, *Clin. Microbiol. Infect. 9*, 140–143.

181. Pestova, E., Millichap, J. J., Siddiqui, F., Noskin, G. A., and Peterson, L. R. (2002) Non-PmrA-mediated multidrug resistance in *Streptococcus pneumoniae*, *J. Antimicrob. Chemother. 49*, 553–556.

182. Marrer, E., Satoh, A. T., Johnson, M. M., Piddock, L. J., and Page, M. G. (2006) Global transcriptome analysis of the responses of a fluoroquinolone-resistant *Streptococcus pneumoniae* mutant and its parent to ciprofloxacin, *Antimicrob. Agents Chemother. 50*, 269–278.

183. Garvey, M. I., and Piddock, L. J. (2008) The efflux pump inhibitor reserpine selects multidrug-resistant *Streptococcus pneumoniae* strains that overexpress the ABC transporters PatA and PatB, *Antimicrob. Agents Chemother. 52*, 1677–1685.

184. Marrer, E., Schad, K., Satoh, A. T., Page, M. G., Johnson, M. M., and Piddock, L. J. (2006) Involvement of the putative ATP-dependent efflux proteins PatA and PatB in fluoroquinolone resistance of a multidrug-resistant mutant of *Streptococcus pneumoniae*, *Antimicrob. Agents Chemother. 50*, 685–693.

185. Avrain, L., Garvey, M., Mesaros, N., Glupczynski, Y., Mingeot-Leclercq, M. P., Piddock, L. J., Tulkens, P. M., Vanhoof, R., and Van Bambeke, F. (2007) Selection of quinolone resistance in *Streptococcus pneumoniae* exposed in vitro to subinhibitory drug concentrations, *J. Antimicrob. Chemother. 60*, 965–972.

186. Jones, H. E., Brenwald, N. P., Owen, K. A., and Gill, M. J. (2003) A multidrug efflux phenotype mutant of *Streptococcus pyogenes*, *J. Antimicrob. Chemother. 51*, 707–710.

187. Leclercq, R. (2009) Epidemiological and resistance issues in multidrug-resistant staphylococci and enterococci, *Clin. Microbiol. Infect. 15*, 224–231.

188. Witte, W., Cuny, C., Klare, I., Nubel, U., Strommenger, B., and Werner, G. (2008) Emergence and spread of antibiotic-resistant gram-positive bacterial pathogens, *Int. J. Med. Microbiol. 298*, 365–377.

189. Top, J., Willems, R., and Bonten, M. (2008) Emergence of CC17 *Enterococcus faecium*: from commensal to hospital-adapted pathogen, *FEMS Immunol. Med. Microbiol. 52*, 297–308.

190. Arsene, S., and Leclercq, R. (2007) Role of a *qnr-like* gene in the intrinsic resistance of *Enterococcus faecalis* to fluoroquinolones, *Antimicrob. Agents Chemother. 51*, 3254–3258.

191. Jonas, B. M., Murray, B. E., and Weinstock, G. M. (2001) Characterization of *emeA*, a NorA homolog and multidrug resistance efflux pump, in *Enterococcus faecalis*, *Antimicrob. Agents Chemother. 45*, 3574–3579.

192. Lee, E. W., Chen, J., Huda, M. N., Kuroda, T., Mizushima, T., and Tsuchiya, T. (2003) Functional cloning and expression of *emeA*, and characterization of EmeA, a multidrug efflux pump from *Enterococcus faecalis*, *Biol. Pharm. Bull. 26*, 266–270.

193. Lee, E. W., Huda, M. N., Kuroda, T., Mizushima, T., and Tsuchiya, T. (2003) EfrAB, an ABC multidrug efflux pump in *Enterococcus faecalis*, *Antimicrob. Agents Chemother. 47*, 3733–3738.

194. Oyamada, Y., Ito, H., Fujimoto, K., Asada, R., Niga, T., Okamoto, R., Inoue, M., and Yamagishi, J. (2006) Combination of known and unknown mechanisms confers high-level resistance to fluoroquinolones in *Enterococcus faecium*, *J. Med. Microbiol. 55*, 729–736.

195. Nishioka, T., Ogawa, W., Kuroda, T., Katsu, T., and Tsuchiya, T. (2009) Gene cloning and characterization of EfmA, a multidrug efflux pump, from *Enterococcus faecium*, *Biol. Pharm. Bull. 32*, 483–488.

196. Ahmed, M., Borsch, C. M., Taylor, S. S., Vazquez-Laslop, N., and Neyfakh, A. A. (1994) A protein that activates expression of a multidrug efflux transporter upon binding the transporter substrates, *J. Biol. Chem. 269*, 28506–28513.

197. Ahmed, M., Lyass, L., Markham, P. N., Taylor, S. S., Vazquez-Laslop, N., and Neyfakh, A. A. (1995) Two highly similar multidrug transporters of *Bacillus subtilis* whose expression is differentially regulated, *J. Bacteriol. 177*, 3904–3910.

198. Holmes, D. J., Caso, J. L., and Thompson, C. J. (1993) Autogenous transcriptional activation of a thiostrepton-induced gene in *Streptomyces lividans*, *EMBO J. 12*, 3183–3191.

199. Summers, A. O. (1992) Untwist and shout: a heavy metal-responsive transcriptional regulator, *J. Bacteriol. 174*, 3097–3101.

200. Baranova, N. N., Danchin, A., and Neyfakh, A. A. (1999) Mta, a global MerR-type regulator of the *Bacillus subtilis* multidrug-efflux transporters, *Mol. Microbiol. 31*, 1549–1559.

201. Ohki, R., and Murata, M. (1997) *bmr3*, a third multidrug transporter gene of *Bacillus subtilis*, *J. Bacteriol. 179*, 1423–1427.

202. Ohki, R., and Tateno, K. (2004) Increased stability of *bmr3* mRNA results in a multidrug-resistant phenotype in *Bacillus subtilis*, *J. Bacteriol. 186*, 7450–7455.

203. Godreuil, S., Galimand, M., Gerbaud, G., Jacquet, C., and Courvalin, P. (2003) Efflux pump Lde is associated with fluoroquinolone resistance in *Listeria monocytogenes*, *Antimicrob. Agents Chemother. 47*, 704–708.

204. Rautelin, H., and Hanninen, M. L. (2000) Campylobacters: the most common bacterial enteropathogens in the Nordic countries, *Ann. Med. 32*, 440–445.

205. Hanninen, M. L., and Hannula, M. (2007) Spontaneous mutation frequency and emergence of ciprofloxacin resistance in *Campylobacter jejuni* and *Campylobacter coli*, *J. Antimicrob. Chemother. 60*, 1251–1257.

206. Luo, N., Sahin, O., Lin, J., Michel, L. O., and Zhang, Q. (2003) In vivo selection of *Campylobacter* isolates with high levels of fluoroquinolone resistance associated with *gyrA* mutations and the function of the CmeABC efflux pump, *Antimicrob. Agents Chemother. 47*, 390–394.

207. Ge, B., McDermott, P. F., White, D. G., and Meng, J. (2005) Role of efflux pumps and topoisomerase mutations in fluoroquinolone resistance in *Campylobacter jejuni* and *Campylobacter coli*, *Antimicrob. Agents Chemother.* 49, 3347–3354.

208. Lin, J., Michel, L. O., and Zhang, Q. (2002) CmeABC functions as a multidrug efflux system in *Campylobacter jejuni*, *Antimicrob. Agents Chemother.* 46, 2124–2131.

209. Lin, J., Akiba, M., Sahin, O., and Zhang, Q. (2005) CmeR functions as a transcriptional repressor for the multidrug efflux pump CmeABC in *Campylobacter jejuni*, *Antimicrob. Agents Chemother.* 49, 1067–1075.

210. Rafii, F., Park, M., and Wynne, R. (2005) Evidence for active drug efflux in fluoroquinolone resistance in *Clostridium hathewayi*, *Chemotherapy 51*, 256–262.

211. Perez-Vazquez, M., Roman, F., Garcia-Cobos, S., and Campos, J. (2007) Fluoroquinolone resistance in *Haemophilus influenzae* is associated with hypermutability, *Antimicrob. Agents Chemother. 51*, 1566–1569.

212. Xu, X. J., Su, X. Z., Morita, Y., Kuroda, T., Mizushima, T., and Tsuchiya, T. (2003) Molecular cloning and characterization of the HmrM multidrug efflux pump from *Haemophilus influenzae* Rd, *Microbiol. Immunol. 47*, 937–943.

213. Singh, A. K., Haldar, R., Mandal, D., and Kundu, M. (2006) Analysis of the topology of *Vibrio cholerae* NorM and identification of amino acid residues involved in norfloxacin resistance, *Antimicrob. Agents Chemother. 50*, 3717–3723.

214. Giraud, E., Blanc, G., Bouju-Albert, A., Weill, F. X., and Donnay-Moreno, C. (2004) Mechanisms of quinolone resistance and clonal relationship among *Aeromonas salmonicida* strains isolated from reared fish with furunculosis, *J. Med. Microbiol. 53*, 895–901.

215. Jacobs, L., and Chenia, H. Y. (2007) Characterization of integrons and tetracycline resistance determinants in *Aeromonas* spp. isolated from South African aquaculture systems, *Int. J. Food Microbiol. 114*, 295–306.

216. Sánchez-Céspedes, J., Blasco, M. D., Marti, S., Alba, V., Alcalde, E., Esteve, C., and Vila, J. (2008) Plasmid-mediated QnrS2 determinant from a clinical *Aeromonas veronii* isolate, *Antimicrob. Agents Chemother. 52*, 2990–2991.

217. Kuroda, T., and Tsuchiya, T. (2009) Multidrug efflux transporters in the MATE family, *Biochim. Biophys. Acta 1794*, 763–768.

218. Colmer, J. A., Fralick, J. A., and Hamood, A. N. (1998) Isolation and characterization of a putative multidrug resistance pump from *Vibrio cholerae*, *Mol. Microbiol. 27*, 63–72.

219. Woolley, R. C., Vediyappan, G., Anderson, M., Lackey, M., Ramasubramanian, B., Jiangping, B., Borisova, T., Colmer, J. A., Hamood, A. N., McVay, C. S., and Fralick, J. A. (2005) Characterization of the *Vibrio cholerae* vceCAB multiple-drug resistance efflux operon in *Escherichia coli*, *J. Bacteriol. 187*, 5500–5503.

220. Huda, N., Lee, E. W., Chen, J., Morita, Y., Kuroda, T., Mizushima, T., and Tsuchiya, T. (2003) Molecular cloning and characterization of an ABC multidrug efflux pump, VcaM, in Non-O1 *Vibrio cholerae*, *Antimicrob. Agents Chemother. 47*, 2413–2417.

221. Morita, Y., Kodama, K., Shiota, S., Mine, T., Kataoka, A., Mizushima, T., and Tsuchiya, T. (1998) NorM, a putative multidrug efflux protein, of *Vibrio parahaemolyticus* and its homolog in *Escherichia coli*, *Antimicrob. Agents Chemother. 42*, 1778–1782.

222. Brown, M. H., Paulsen, I. T., and Skurray, R. A. (1999) The multidrug efflux protein NorM is a prototype of a new family of transporters, *Mol. Microbiol. 31*, 394–395.

223. Morita, Y., Kataoka, A., Shiota, S., Mizushima, T., and Tsuchiya, T. (2000) NorM of *Vibrio parahaemolyticus* is an Na(+)-driven multidrug efflux pump, *J. Bacteriol. 182*, 6694–6697.

224. Wexler, H. M. (2007) *Bacteroides*: the good, the bad, and the nitty-gritty, *Clin. Microbiol. Rev. 20*, 593–621.

225. Ueda, O., Wexler, H. M., Hirai, K., Shibata, Y., Yoshimura, F., and Fujimura, S. (2005) Sixteen homologs of the *mex*-type multidrug resistance efflux pump in *Bacteroides fragilis*, *Antimicrob. Agents Chemother. 49*, 2807–2815.

226. Pumbwe, L., Ueda, O., Yoshimura, F., Chang, A., Smith, R. L., and Wexler, H. M. (2006) *Bacteroides fragilis* BmeABC efflux systems additively confer intrinsic antimicrobial resistance, *J. Antimicrob. Chemother. 58*, 37–46.

227. Pumbwe, L., Chang, A., Smith, R. L., and Wexler, H. M. (2006) Clinical significance of overexpression of multiple RND-family efflux pumps in *Bacteroides fragilis* isolates, *J. Antimicrob. Chemother. 58*, 543–548.

228. Miyamae, S., Nikaido, H., Tanaka, Y., and Yoshimura, F. (1998) Active efflux of norfloxacin by *Bacteroides fragilis*, *Antimicrob. Agents Chemother. 42*, 2119–2121.

229. Miyamae, S., Ueda, O., Yoshimura, F., Hwang, J., Tanaka, Y., and Nikaido, H. (2001) A MATE family multidrug efflux transporter pumps out fluoroquinolones in *Bacteroides thetaiotaomicron*, *Antimicrob. Agents Chemother. 45*, 3341–3346.

230. Ikeda, T., and Yoshimura, F. (2002) A resistance-nodulation-cell division family xenobiotic efflux pump in an obligate anaerobe, *Porphyromonas gingivalis*, *Antimicrob. Agents Chemother. 46*, 3257–3260.

231. Dridi, L., Tankovic, J., and Petit, J. C. (2004) CdeA of *Clostridium difficile*, a new multidrug efflux transporter of the MATE family, *Microb. Drug Resist. 10*, 191–196.

232. Rafii, F., and Park, M. (2008) Detection and characterization of an ABC transporter in *Clostridium hathewayi*, *Arch. Microbiol. 190*, 417–426.

233. Liu, J., Takiff, H. E., and Nikaido, H. (1996) Active efflux of fluoroquinolones in *Mycobacterium smegmatis* mediated by LfrA, a multidrug efflux pump, *J. Bacteriol. 178*, 3791–3795.

234. Takiff, H. E., Cimino, M., Musso, M. C., Weisbrod, T., Martinez, R., Delgado, M. B., Salazar, L., Bloom, B. R., and Jacobs, W. R., Jr., (1996) Efflux pump of the proton antiporter family confers low-level fluoroquinolone resistance in *Mycobacterium smegmatis*, *Proc. Natl. Acad. Sci. USA 93*, 362–366.

235. Li, X. Z., Zhang, L., and Nikaido, H. (2004) Efflux pump-mediated intrinsic drug resistance in *Mycobacterium smegmatis*, *Antimicrob. Agents Chemother. 48*, 2415–2423.

236. Sander, P., De Rossi, E., Boddinghaus, B., Cantoni, R., Branzoni, M., Bottger, E. C., Takiff, H., Rodriquez, R., Lopez, G., and Riccardi, G. (2000) Contribution of the multidrug efflux pump LfrA to innate mycobacterial drug resistance, *FEMS Microbiol. Lett. 193*, 19–23.

237. Buroni, S., Manina, G., Guglierame, P., Pasca, M. R., Riccardi, G., and De Rossi, E. (2006) LfrR is a repressor that regulates expression of the efflux pump LfrA in *Mycobacterium smegmatis, Antimicrob. Agents Chemother. 50*, 4044–4052.

238. De Rossi, E., Arrigo, P., Bellinzoni, M., Silva, P. A., Martin, C., Ainsa, J. A., Guglierame, P., and Riccardi, G. (2002) The multidrug transporters belonging to major facilitator superfamily in *Mycobacterium tuberculosis, Mol. Med. 8*, 714–724.

239. Bhatt, K., Banerjee, S. K., and Chakraborti, P. K. (2000) Evidence that phosphate specific transporter is amplified in a fluoroquinolone resistant *Mycobacterium smegmatis, Eur. J. Biochem. 267*, 4028–4032.

240. Banerjee, S. K., Bhatt, K., Misra, P., and Chakraborti, P. K. (2000) Involvement of a natural transport system in the process of efflux-mediated drug resistance in *Mycobacterium smegmatis, Mol. Gen. Genet. 262*, 949–956.

241. Pasca, M. R., Guglierame, P., Arcesi, F., Bellinzoni, M., De Rossi, E., and Riccardi, G. (2004) Rv2686c–Rv2687c–Rv2688c, an ABC fluoroquinolone efflux pump in *Mycobacterium tuberculosis, Antimicrob. Agents Chemother. 48*, 3175–3178.

242. Courvalin, P. (1990) Plasmid-mediated 4-quinolone resistance: a real or apparent absence? *Antimicrob. Agents Chemother. 34*, 681–684.

243. Gomez-Gomez, J. M., Blazquez, J., Espinosa De Los Monteros, L. E., Baquero, M. R., Baquero, F., and Martínez, J. L. (1997) In vitro plasmid-encoded resistance to quinolones, *FEMS Microbiol. Lett. 154*, 271–276.

244. Martínez, J. L., Alonso, A., Gomez-Gomez, J. M., and Baquero, F. (1998) Quinolone resistance by mutations in chromosomal gyrase genes: Just the tip of the iceberg? *J. Antimicrob. Chemother. 42*, 683–688.

245. Robicsek, A., Strahilevitz, J., Jacoby, G. A., Macielag, M., Abbanat, D., Park, C. H., Bush, K., and Hooper, D. C. (2006) Fluoroquinolone-modifying enzyme: a new adaptation of a common aminoglycoside acetyltransferase, *Nat. Med. 12*, 83–88.

246. Yamane, K., Wachino, J., Suzuki, S., Kimura, K., Shibata, N., Kato, H., Shibayama, K., Konda, T., and Arakawa, Y. (2007) New plasmid-mediated fluoroquinolone efflux pump, QepA, found in an *Escherichia coli* clinical isolate, *Antimicrob. Agents Chemother. 51*, 3354–3360.

247. Kim, H. B., Park, C. H., Kim, C. J., Kim, E. C., Jacoby, G. A., and Hooper, D. C. (2009) Prevalence of plasmid-mediated quinolone resistance determinants over a 9-year period, *Antimicrob. Agents Chemother. 53*, 639–645.

248. Yamane, K., Wachino, J., Suzuki, S., and Arakawa, Y. (2008) Plasmid-mediated *qepA* gene among *Escherichia coli* clinical isolates from Japan, *Antimicrob. Agents Chemother. 52*, 1564–1566.

249. Liu, J. H., Deng, Y. T., Zeng, Z. L., Gao, J. H., Chen, L., Arakawa, Y., and Chen, Z. L. (2008) Coprevalence of plasmid-mediated quinolone resistance determinants QepA, Qnr, and AAC(6′)-Ib-cr among 16S rRNA methylase RmtB-producing *Escherichia coli* isolates from pigs, *Antimicrob. Agents Chemother. 52*, 2992–2993.

250. Ma, J., Zeng, Z., Chen, Z., Xu, X., Wang, X., Deng, Y., Lu, D., Huang, L., Zhang, Y., Liu, J., and Wang, M. (2009) High prevalence of plasmid-mediated quinolone resistance determinants *qnr, aac(6′)-Ib-cr*, and *qepA* among ceftiofur-resistant Enterobacteriaceae

isolates from companion and food-producing animals, *Antimicrob. Agents Chemother.* *53*, 519–524.

251. Martínez, J. L. (2009) The role of natural environments in the evolution of resistance traits in pathogenic bacteria, *Proc. Biol. Sci. 276*, 2521–2530.

252. Perichon, B., Courvalin, P., and Galimand, M. (2007) Transferable resistance to aminoglycosides by methylation of G1405 in 16S rRNA and to hydrophilic fluoroquinolones by QepA-mediated efflux in *Escherichia coli, Antimicrob. Agents Chemother. 51*, 2464–2469.

253. Cattoir, V., Poirel, L., and Nordmann, P. (2008) Plasmid-mediated quinolone resistance pump QepA2 in an *Escherichia coli* isolate from France, *Antimicrob. Agents Chemother. 52*, 3801–3804.

254. Hansen, L. H., Johannesen, E., Burmolle, M., Sorensen, A. H., and Sorensen, S. J. (2004) Plasmid-encoded multidrug efflux pump conferring resistance to olaquindox in *Escherichia coli, Antimicrob. Agents Chemother. 48*, 3332–3337.

255. Hansen, L. H., Jensen, L. B., Sorensen, H. I., and Sorensen, S. J. (2007) Substrate specificity of the OqxAB multidrug resistance pump in *Escherichia coli* and selected enteric bacteria, *J. Antimicrob. Chemother. 60*, 145–147.

256. Hansen, L. H., Sorensen, S. J., Jorgensen, H. S., and Jensen, L. B. (2005) The prevalence of the OqxAB multidrug efflux pump amongst olaquindox-resistant *Escherichia coli* in pigs, *Microb. Drug Resist. 11*, 378–382.

257. Kim, H. B., Wang, M., Park, C, H., Kim, E. C., Jacoby, G. A., and Hooper, D. C. (2009) *oqxAB* encoding a multidrug efflux pump in human clinical isolates of *Enterobacteriaceae, Antimicrob. Agents Chemother. 53*, 3582–3584.

258. Vila, J., and Martínez, J. L. (2008) Clinical impact of the over-expression of efflux pump in nonfermentative gram-negative bacilli, development of efflux pump inhibitors, *Curr. Drug Targets 9*, 797–807.

259. Zhang, L., and Mah, T. F. (2008) Involvement of a novel efflux system in biofilm-specific resistance to antibiotics, *J. Bacteriol. 190*, 4447–4452.

260. Ruzin, A., Keeney, D., and Bradford, P. A. (2007) AdeABC multidrug efflux pump is associated with decreased susceptibility to tigecycline in *Acinetobacter calcoaceticus–Acinetobacter baumannii* complex, *J. Antimicrob. Chemother. 59*, 1001–1004.

261. Viktorov, D. V., Zakharova, I. B., Podshivalova, M. V., Kalinkina, E. V., Merinova, O. A., Ageeva, N. P., Antonov, V. A., Merinova, L. K., and Alekseev, V. V. (2008) High-level resistance to fluoroquinolones and cephalosporins in *Burkholderia pseudomallei* and closely related species, *Trans. R. Soc. Trop. Med. Hyg. 102*(Suppl. 1), S103–S110.

262. Poole, K. (2004) Efflux-mediated multiresistance in gram-negative bacteria, *Clin. Microbiol. Infect. 10*, 12–26.

263. Middlemiss, J. K., and Poole, K. (2004) Differential impact of MexB mutations on substrate selectivity of the MexAB–OprM multidrug efflux pump of *Pseudomonas aeruginosa, J. Bacteriol. 186*, 1258–1269.

264. He, G. X., Kuroda, T., Mima, T., Morita, Y., Mizushima, T., and Tsuchiya, T. (2004) An H(+)-coupled multidrug efflux pump, PmpM, a member of the MATE family of transporters, from *Pseudomonas aeruginosa, J. Bacteriol. 186*, 262–265.

265. Hearn, E. M., Dennis, J. J., Gray, M. R., and Foght, J. M. (2003) Identification and characterization of the *emhABC* efflux system for polycyclic aromatic hydrocarbons in *Pseudomonas fluorescens* cLP6a, *J. Bacteriol. 185*, 6233–6240.

266. Hearn, E. M., Gray, M. R., and Foght, J. M. (2006) Mutations in the central cavity and periplasmic domain affect efflux activity of the resistance–nodulation–division pump EmhB from *Pseudomonas fluorescens* cLP6a, *J. Bacteriol. 188*, 115–123.

267. Duque, E., Segura, A., Mosqueda, G., and Ramos, J. L. (2001) Global and cognate regulators control the expression of the organic solvent efflux pumps TtgABC and TtgDEF of *Pseudomonas putida*, *Mol. Microbiol. 39*, 1100–1106.

268. Huda, M. N., Morita, Y., Kuroda, T., Mizushima, T., and Tsuchiya, T. (2001) Na^+-driven multidrug efflux pump VcmA from *Vibrio cholerae* non-O1, a non-halophilic bacterium, *FEMS Microbiol. Lett. 203*, 235–239.

269. Begum, A., Rahman, M. M., Ogawa, W., Mizushima, T., Kuroda, T., and Tsuchiya, T. (2005) Gene cloning and characterization of four MATE family multidrug efflux pumps from *Vibrio cholerae* non-O1, *Microbiol. Immunol. 49*, 949–957.

XENOBIOTIC EFFLUX IN BACTERIA AND FUNGI: A GENOMICS UPDATE

By RAVI D. BARABOTE, *Department of Plant Sciences, University of California, Davis, California,* JOSE THEKKINIATH, RICHARD E. STRAUSS, *Department of Biological Sciences, Texas Tech University, Lubbock, Texas,* GOVINDSAMY VEDIYAPPAN, JOE A. FRALICK, *Department of Microbiology and Immunology, Texas Tech University Health Sciences Center, Lubbock, Texas,* and MICHAEL J. SAN FRANCISCO, *Department of Biological Sciences, Texas Tech University, and Department of Microbiology and Immunology, Texas Tech University Health Sciences Center, Lubbock, Texas*

CONTENTS

Advances in Enzymology and Related Areas of Molecular Biology, Volume 77
Edited by Eric J. Toone   Copyright © 2011 John Wiley & Sons, Inc.

I. Introduction

"...not enough to kill the streptococci but enough to educate them to resist penicillin" (Alexander Fleming, Nobel Prize lecture, Dec. 11, 1945). These prophetic words underscore the arms race in which we find ourselves today. Large populations and mutable genomes give microbes a profound capacity to respond to changing environmental conditions. The misuse of antibiotics in human health and agriculture has contributed to continuing microbial drug resistance. Thus, 65 years later, in 2010, we continue to battle microorganisms and strive to design novel and useful antimicrobial agents (1).

II. Bacterial Efflux Pumps

A. TYPES AND SUBSTRATES

The efflux of antibiotics was discovered in 1980 by Steward Levy and co-workers, who were studying the mechanism of tetracycline resistance in *Escherichia coli* (2). Since that time it has been demonstrated that a single efflux pump can provide resistance to multiple antibiotics (MDR efflux pumps). It has also been found that MDR-like transporters are highly abundant and ubiquitous in nature and represent, on average, more than 10% of the total number of transporters per organism (3).

Although MDR efflux pumps play an important role in the inherent resistance of bacteria to antibiotics, these pumps appear to be evolutionarily ancient transporters that have a wide variety of physiological functions, beyond antibiotic resistance, which contributes to adaptation to a wide variety of environments (4–6).

Phylogenetically, bacterial antibiotic efflux pumps belong to one of five families (Figure 1): (1) SMR or small multidrug resistance subfamily of the DNT (drug/metabolite transporters) superfamily (3, 7); (2) MATE (multi-antimicrobial toxic compound extrusion) subfamily of the MOP

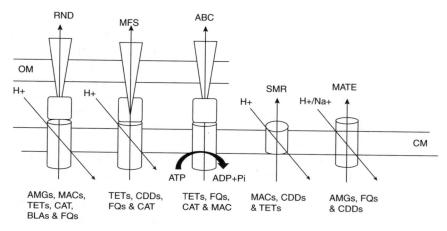

Figure 1.   Bacterial antibiotic drug transporters: the five classes of MDR efflux pumps. Not all MFS and ABC MDR pumps are tripartite in structure. MFS, major facilitator superfamily; RND, resistance–nodulation–division; MATE, multidrug and toxic compound extrusion; SMR, small multi-drug resistance; ABC, ATP-binding cassette; OM, outer membrane; P, periplasm; CM, cytoplasmic membrane; TETs, tetracyclines; CAT, chloramphenicol; FQs, fluoroquinolones; CDDs, cationic dyes and detergents; AMGs, aminoglycosides; MACs, macrolides; BLAs, β-lactams. [Adapted from (22).]

(multidrug/oligosaccharidyl-lipid/polysaccharide flipases) superfamily (8, 9); (3) MFS (major facilitator superfamily) (10); (4) RND (resistance–nodulation–division) superfamily (11); and (5) ABC (ATP-binding cassette) superfamily (12). MDR transporters can also be classified into two main groups based on the mode of energy coupling for transport/efflux: (1) primary active transporters that belong to the ABC superfamily and utilize ATP hydrolysis to expel the drug from the cell, and (2) secondary transporters that utilize the proton-motive force or ion gradient for drug expulsion. SMR, MATE, MFS, and RND pumps are secondary transporters or antiporters. Other classifications have been proposed (13, 14), and several reviews have been dedicated to the classification and descriptions of MDR transporter families (15–22).

1.   ABC Pumps

ABC (ATP-binding cassette) transporters are found in all living organisms and are classified as primary active transporters that belong to the

ABC superfamily and utilize the free energy of ATP hydrolysis to expel the drug from the cell. Historically, the first MDR transporters characterized were members of the ABC superfamily of eukaryotic origin such as P-glycoprotein (23–25). Since that time, ABC pumps are also found in pathogenic fungi and parasitic protozoa, where they impart resistance to antimicrobial drugs (26). The first bacterial ABC MDR pump (LmrA, *Lactococcus lactis*) was reported in 1996 (27). ABC pumps have been found to be widespread among bacteria (14) and appear to play an important role in drug resistance in some pathogenic bacteria, such as *Enterococcus faecalis* and *Streptococcus pneumoniae* (28) (Table 1). Many of these MDR efflux pumps are homologs of the heterodimeric LmrCD pump of *L. lactis*.

ABC transporters have structural characteristics that set them apart from other efflux pumps. They usually have two similar halves, each containing two parts: a transmembrane domain (TMD) that is usually arranged into six transmembrane-spanning α-helices, and a nulecotide-binding domain (NBD), also known as the ATP-binding cassette domain, containing the Walker A, Walker B, and ABC signature motifs (29). The NBDs are responsible for the binding and hydrolysis of ATP and hence generation of the energy of the translocation of the substrate, while the TMDs form the translocation pathway for the transported substrates to cross the cytoplasmic membrane. In most cases a single protein contains a TMD–NBD–TMD–NBD structure (24). However, in bacterial transporters, a TMD fused to a NBD forms a half-transporter, which homo- or hetero-dimerizes with another half-transporter to form a functional full-size transporter (30): for example, the homodimeric LmrA or heterodimeric LmrCD ABC pump of *L. lactis*. There are exceptions, however; DrrA and DrrB found in *Streptomyces peucetius* contain a single NBD or TMD domain, respectively, and are each thought to function as a tetramer (31).

The generally accepted mechanism by which ABC transporters function is often explained in terms of a two-cylinder engine mechanism (30, 32). In this model, drugs enter a high-affinity site in the TMDs of the ABC transporter from the cytosol side of the cytoplasmic membrane and then, upon the binding/hydrolysis of ATP (at the NBDs to provide the power stroke) and through conformational cycling, the substrate is occluded at the high-affinity site and progresses to a low-affinity release site on the extracellular side of the membrane (Figure 2). As depicted in Figure 1, in some gram-negative bacteria, ABC MDR efflux pumps have periplasmic accessory or membrane fusion proteins (MFPs). These MFPs interact with

TABLE 1
Examples of ABC MDR Efflux Pumps and Their Substrates

Efflux System	Substrates[a]	Gene Location	Organism	Refs.
Msr(A)	ML, SG-B	Plasmid	*Staphylococcus* spp.	(36)
Msr(C)	ML, SG-B	Chromosome	*Enterococcus faecium*	(37, 38)
Msr(D)	ML, KL	Chromosome	*Streptococcus pneumoniae*	(39)
Vga(A/B)	SG-A	Plasmid	*Staphylococcus aureus*	(40)
Lsa	LS, SG	Chromosome	*Enterococcus faecalis*	(39–41)
EfrAB	CP, NOR, TC	Chromosome	*E. faecalis*	(42)
Lsa(B)	CD	Plasmid	*Staphylococcus sciuri*	(43)
MacAB-TolC	ML, CD, OFL	Chromosome	*Eschericia coli*	(44)
LmrA	BL, TC	Chromosome	*Lactococcus lactis*	(45)
LmrA	AG, CP, OFL	Chromosome	*L. lactis*	(45)
VcaM	CP, NOR, TC	Chromosome	*Vibrio cholerae*	(46)
Rv2686c–Rv2687c–Rv2688c	FQ	Chromosome	*Mycobacterium tuberculosis*	(47)
MD1, MD2	CP	Chromosome	*Mycoplasma hominis*	(48)

[a]AC, acriflavine; AG, aminoglycoside; AH, aromatic hydrocarbons; AZ, azithromycin; BER, berberine; BL, β-lactams; BS, bile salts; CCCP, carbonyl cyanide *m*-chlorophenylhydrazone; CH, cholate; CLH, chlorhexidine; CL, cerulenin; CD, clindamycin; CM, chloramphenicol; CP, ciprofloxacin; CPC, cetylpyridinium chloride; CV, crystal violet; DAR, daunorubicin; DOC, deoxycholate; DXR, doxorubicin; DAPI, 6-diamidino-2-phenyl indole dihydrochloride; EB, ethidium bromide; ER, ethyromycin; FA, fatty acids; FQ, fluoroquinolones; FU, fusidic acid; FL, florfenicol; GL, glycylcyclines; GM, gentamicin; KL, ketolides; KM, kanamycin; LC, lipophilic cations; LS, linocosamides; ML, macrolides; MOX, moxifloxacin; MN, minocycline; MV, methyl viologen; NAL, nalidixic acid; NOR, norfloxacin; NV, novobiocin; OFL, ofloxacin; OZ, oxazolidinones; PCP, pentachlorophenol; PI, pentamidine isothionate; PMA, phenylmercuric acetate; PM, puromycin; PY, pyronine; QA, quaternary amine compounds; RD, rhodamine; RF, rifampicin; RM, roxythromycin; SAL, salicylate; SDS, sodium dodecl sulfate; SG-A, type A streptogramins; SG-B, type B streptogramins; SM, sulfonamides; SP, spiramycin; SPR, sparfloxacin; STM, streptomycin; SO, safranin O; TC, tetracycline; TL, thiolactomycin; TG, tigecycline; TS, triclosan; TP, trimethoprim; TX, Triton X-100.

outer membrane channel or efflux proteins (OEPs) to form a tripartite pump that bridges both membranes of the gram-negative envelope to mediate the extrusion of substrates from the cell, such as the MacAB–TolC ABC MDR pump. MacA is a periplasmic protein of the MFP family, TolC is an outer membrane channel protein, and MacB is a half-type ABC transporter with four putative TMD segments and an N-terminal NBD (34, 35). Table 1 lists examples of ABC MDR pumps found in bacteria and their substrates.

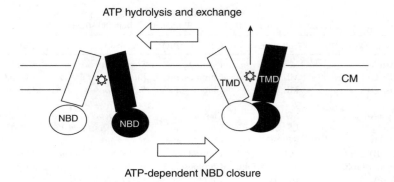

Figure 2. Hypothetical mechanism of ABC transporters: the closure and dimerization of cytosolic NBDs, which provides the "power stroke" of the two-cylinder engine mechanism for ABC transporters (30, 32) that pulls the TMDs from an inward-to-outward facing conformation. The ABC transporter is heterodimeric (black and white). CM, cytoplasmic membrane; TMD, transmembrane domain; NBD, nucleotide-binding domain. [Adapted from (33).]

2.   SMR Pumps

SMR (small multidrug resistance)-mediated multiple drug resistance is widespread among bacteria (16). These bacterial MDR efflux pumps are among the smallest known pumps and are made up of proteins that are typically 100 to 115 amino acid residues in length. SMR pumps have a four-transmembrane α-helical topology (7) and a highly conserved residue, Glu14 (49, 50), that has been shown to be essential and involved directly in drug and proton binding (51, 52). It has been assumed that these pumps function as oligomeric complexes, perhaps as dimers, and that the Glu14 of both protomers in a dimer form a shared binding pocket (51, 53). The genes encoding SMR pumps are found in a variety of plasmids from clinical isolates of *S. aureus* and other staphylocci (54–56) as well in the chromosomes of many bacteria (57–59) (Table 2). The substrates of SMR pumps typically share similar physical properties but may differ in size and shape and are almost exclusively monovalent hydrophobic cations.

The SMR pump EmrE of *E. coli* is one of the better characterized of the SMR pumps (49, 52, 60). EmrE transports a diverse array of aromatic, positively charged substrates in exchange for protons (61). A model for its translocation cycle has been suggested in which the binding of the drug to its binding site (Glu14) deprotonates the Glu14 residues in the functional

TABLE 2
Examples of SMR Efflux Pumps and Their Substrates

Efflux System	Substrates[a]	Gene Location	Organisms	Ref.
Mmr	CP, NOR, AC, EB	Chromosome	*Mycobacterium smegmatis*	(62)
Smr/QacC	EB, CV, MV, QA	Plasmid	*Staphylococcus aureus*	(55)
EmrE	EB, AC, MV	Chromosome	*Escherichia coli*	(63)
EbrAB	EB, AC, PY, SO	Chromosome	*Bacillus subtilis*	(58)
YkkCD	EB, CV, PY, MV, TC, SP, STM, PM, CPC	Chromosome	*B. subtilis*	(64)
Tbsmr	AC, EB, MV	Chromosome	*Mycobacterium tuberculosis*	(59)
Pasmr	AC, EB, MV	Chromosome	*Pseudomonas aeruginosa*	(65)
QacE	EB, QA, SM	Plasmid	Gram-negative bacteria	(66)

[a]See Table 1 footnote for substrate abbreviations.

antiparallel dimer and causes the transporter to undergo a conformational change in which the binding sites close behind the substrate and open in front of the substrate to expose it to the outer face of the cytoplasmic membrane (Figure 3). The release of the drug is thought to be catalyzed by protons that protonate the two Glu14's, thus coupling drug export to H^+ import. However, the precise mechanism by which the proton-induced conformational changes bring about translocation of substrate across the membrane remains to be deciphered. Table 2 includes some examples of SMR pumps and their substrates.

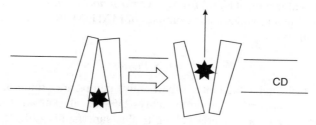

Figure 3. Hypothetical mechanism of SMR transporters: the opening and closing of the antiparallel SMR dimmer. The substrate (stars) binds and deprotanates Glu14 residues in the binding pocket of the SMR dimer, causing the transporter to undergo conformational changes, which, in turn, cause the transporter to close behind the substrate and open on the other side of the cytoplasmic membrane (CM). The release of the substrate is coupled to the protonation of the two Glu14 of the dimer, thus coupling drug export to H^+ import.

3. MATE Pumps

MATE (multidrug and toxic compound extrusion) MDR pumps are a relatively new family of secondary efflux pumps (67). Early members, NorM and its *E. coli* homolog YdhG, were originally classified as MFS pumps because they possessed 12 putative transmembrane domains (68). However, it was later discovered that there was little sequence homology of these pumps with other MDR efflux pumps, and they were reclassified as members of a new family of MDR pumps (MATE) (8). Since that time over 1000 members of this family, including representatives from all three kingdoms (Eukarya, Archaea, and Eubacteria) have been identified (8, 19) and placed into three large subfamilies: (1) bacterial (2) eukaryotic, and (3) bacterial and archaebacterial transporters.   MATE proteins range in size from 400 to 700 amino acid residues with 12 putative transmembrane α-helices. No apparent consensus sequence is found in all MATE sequences but they do share approximately 40% sequence similarity (67).

Bacterial MATE MDR pumps are energized by either H^+- or Na^+-coupled antiporters (68–70). These secondary MDR pumps can remove cationic drugs such as ethidium bromide, tetraphenylphosphonium, acriflavine, norfloxacin, and berberine, and at least one, MepA of *S. aureus*, when overexpressed, can confer resistance to tigecycline, a new glycylcycline antibiotic that is effective against methicillin resistant *S. aureus* cells (MRSA) (71). Because of their recent discovery, MATE MDR efflux pumps are the least well characterized and relatively little is known about their structure–function relationship. Most studies have focused on describing their presence and the antibiotic resistance they provide. However, with the recent crystallization of NorM from *Neisseria gonorrhoeae* (72, 73), this may change. Table 3 provides some examples of MATE MDR efflux pumps and their substrates.

4. MFS Pumps

The MFS (major facilitator superfamily) pumps are proton-dependent secondary transporters. Approximately 25% of all known membrane transport proteins in bacteria belong to this superfamily (20, 85), which contains over 50 distinct families and more than 1500 members (15, 17, 86, 87). Structurally, most MFS transporters are 400 to 600 amino acid residues in length and contain 12 or 14 TMDs (17), although there are at least two exceptions: one family has only six TMD and another 24 (15). There is also good evidence for an internal tandem gene duplication, indicating

TABLE 3
Examples of MATE Transporters, Substrates and Gene Locations

Efflux System	Substrates[a]	Gene Location	Organisms	Refs.
MepA	GL, EB	Chromosome	*Staphylococcus aureus*	(71, 74)
NorE (YdhE)	NOR, AC, CP, KM, STP	Chromosome	*Escherichia coli*	(68, 75)
NorM	NOR, CP, AC, EB, BER	Chromosome	*Neisseria gonorrhoeae, Neisseria Meningitidis*	(73)
AbeM	NOR, CP, KM, ER, CM, AC, DAR, DXR, TS, GM	Chromosome	*Acinetobacter baumannii*	(70)
BexA	NOR, AC, EB, MV	Chromosome	*Bacteroides thetaiotaomicron*	(76)
HmrM	NOR	Chromosome	*Haemophilis influenzae*	(77)
PmpM	NOR	Chromosome	*Pseudomonas aeruginosa*	(78)
NorM	NOR, CP, EB	Chromosome	*Vibrio parahaemolyticus*	(68)
VmrA	AC, EB, DAPI	Chromosome	*V. parahaemolyticus*	(79)
CdeA	NOR, AC, EB	Chromosome	*Clostridium difficile*	(80)
VcmA	NOR, CP, EB, KM, STM, DAR, DXR	Chromosome	*Vibrio cholerae*	(81)
VcrM	CM, AC, EB	Chromosome	*V. cholerae*	(82)
VcmB, D, H and VcmN	CP, NOR, KM, EB, OFL	Chromosome	*V. cholerae*	(83)
MdtK	NOR, DXR, AC	Chromosome	*Salmonella typhimurium*	(84)

[a]See Table 1 footnote for substrate abbreviations.

a common ancestral gene. The best characterized MDR MFS pump protein is EmrD, an efflux pump that exports amphipathic compounds, such as carbonylcyanide-*m*-chlorophenylhydrazone (CCCP), across the *E. coli* cytoplasmic membrane. Its crystal structure has been solved to 3.5-Å resolution (88). It is 394 amino acids in length and has 12 TMDs organized as a pair of six-helix domains that surrounds a hydrophobic pore. Two long loops extend into the inner leaflet of the cytoplasmic membrane that are thought to determine substrate specificity, called a *substrate specificity filter*, and which may facilitate transport (88). It has been postulated that the mechanism by which MFS transporters move substrates across a membrane is via a rocker switch mechanism alternating-access model coupled with an H^+ antiport (Figure 4). This model was originally based on the crystal structures of GlpT (89), LacY (90), and more recently, EmrD (88). Table 4 presents some examples of MFS MDR efflux pumps and their substrates.

TABLE 4
Examples of MFS MDR Transporters, Substrates, and Their Gene Locations

Efflux System	Substrates[a]	Gene Location	Organisms	Refs.
Cml, CmlA, CmlB	CM	Mostly plasmid; some chromosome	*Pseudomonas aeruginosa, Enterobacter aerogenes, Klebsiella pneumonia, Salmonella enterica serovar typhimurium*	(95)
Cml, Cmlv	CM	Plasmid	*Streptomyces* spp., *Corynebacterium* spp.	(95)
Mef(A)	ML	Chromosome	*Streptococcus* spp., *Corynebacterium* spp., *Acinetobacter* spp., *Enterococcus* spp., *Staphylococcus* spp., *Neisseria* spp., etc.	(38, 96)
PmrA	CP, NOR	Chromosome	*Streptococcus pneumoniae*	(97, 98)
MdfA	CM, ER	Chromosome	*Escherichia coli*	(99, 100)
EmrAB–TolC	LC, CCCP, NAL, TL	Chromosome	*E. coli*	(93)
EmrKY–TolC	ML	Chromosome	*E. coli*	(101)
Flo, FloR	CM	Plasmid	*E. coli, K. pneumoniae,*	(95)
pp-Flo	FL	Chromosome	*Vibrio cholerae, S. enterica* serovar typhimurium	
Mef(A)	ML	Chromosome	*Streptococcus* spp.	(38, 96)
EmeA	CP, NOR	Chromosome	*Enterococcus faecalis*	(102, 103)
Lde	CP, NOR	Chromosome	*Listeria monocytogenes*	(104)
Bmr, Bmr3, and Blt	FQ	Chromosome	*Bacillus subtilis*	(97, 105)
NorA	NOR, CP	Chromosome	*Staphylococcus aureus, Bacteroides fragilis*	(97, 106, 107)
NorB	NOR, CP, MOX, SPR	Chromosome	*S. aureus*	(108)

246

MdeA	ML, LS, SG-A	Chromosome	*Staphylococcus aureus, S. hemolyticus, Bacillus cereus, B. subtilis*	(109)
QacA	EB, QA, CLH, PI	Plasmid	*S. aureus*	(110)
LmrB	LS	Chromosome	*B. subtilis, Corynebacterium glutamicum*	(111–113)
Cme	ER	Chromosome	*Clostridium difficile*	(114)
VceAB–VceC	DOC, CCCP, NAL, SAL, CM, ER, PMA, PCP	Chromosome	*V. cholerae*	(91, 92)
Tet (A, B, C, D, E, G, H, J, & Y, Z	TC	Plasmid	Gram-negative bacteria	(115–117)
Tet (K, L)	TC	Plasmid/chromosome	Gram-positive bacteria	(115, 116)
Tet38	TC	Chromosome	*S. aureus*	(108)
Tet(V)	TC	Chromosome	*Mycobacterium tuberculosis, Mycobacterium fortuitum*	(118–120)
Rv1258/Tap	OFL	Chromosome	*M. tuberculosis*	(121)
Rv1634 FQ	—	Chromosome	*M. tuberculosis*	(119)
P55/Rv1410	TC	Chromosome	*M. tuberculosis, Mycobacterium bovis*	(122)
EfpA, LfrA	FQ, CP, NOR	Chromosome	*Mycobacterium smegmatis*	(62)

aSee Table 1 footnote for substrate abbreviations.

247

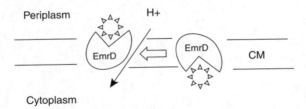

Figure 4. Hypothetical mechanism for substrate transport by EmrD: The drug can enter the hydrophobic internal cavity of EmrD either from the cytoplasm or from the inner leaflet of the cytoplasmic membrane (CM). The drug is then transported through a rocker-switch alternating-access process coupled through a proton antiport. [Adapted from (88).]

Some multidrug-resistant MFS systems have a tripartite structure (as do some ABC and RND pumps) such as the VceABC pump of *Vibrio cholerae* (91, 92) and the EmrAB–TolC pump of *E. coli* (93, 94). These pumps are responsible for the removal of substrates across both membranes of the gram-negative bacterial envelope. These systems are comprised of an MFS transporter containing 14 TMD (EmrB, VceB), a periplasmic MFP (EmrA, VceA), and an OEP (TolC, VceC). Interestingly, unlike the tripartite AcrAB–TolC pump, where AcrB and TolC have been shown to be trimers, electron microscopy studies of reconstructed EmrAB suggest that they exist as dimers (94). More structural studies are warranted to decipher the interactions and architecture of MDR MFS tripartite efflux pumps.

5. RND Pumps

RND superfamily transporters are found in Eukarya, Archaea, and Eubacteria and have been placed into eight phylogenetically distinct families that correlate with their substrate specificity (123). RND transporters are larger than MFS transporters and range in size from 700 to over 1300 amino acid residues in length (124). Like MFS transporters, RND transporters are predicted to adopt a 12-TMD structure, and the sequences of the first and second halves of the RND transporter are similar, suggesting that they have also arisen through gene duplication (124). However, unlike MFS transporters, RND transporters possess large periplasmic domains (34, 125, 126).

The crystal structure of the RND transporter, AcrB, has been solved (125, 127–129). It is a homotrimer, the monomer of which is 1049 amino acids in length and contains 12 transmembrane α-helices and two expansive periplasmic loops that determine substrate specificity (126, 130). Topologically, the core of AcrB is formed by a bundle of the 12 transmembrane α-helices, two of which (TM4 and TM10) extend approximately 70 Å into the periplasm, forming a distal TolC docking domain and a porter/pore domain, the latter being closest to the plane of the outer leaflet of the cytoplasmic membrane. In the center of the trimer, the TolC docking domain produces a funnel-shaped structure with a narrow diameter that leads to a central pore that is located in the porter domain. It is this domain with which AcrB docks to TolC. The central pore leads to a central cavity approximately 35 Å in width. Three vestibules located near the cytoplasmic membrane have been implicated as entrances by which substrates may gain access to the central cavity (129).

The initial structural studies were conducted on crystals with threefold symmetry. However, recently, reports describing an asymmetric AcrB trimer have been published (131, 132). The asymmetric structure reveals three different monomer conformations, presumably representing three consecutive states in the transport cycle and suggest a model for drug transport based on conformational cycling of the monomers by the RND pump (131–134). The three different monomer conformations are designated as loose (L), tight (T), and open (O) (131, 132, 135, 136). In this model, conformational changes from loose to tight to open and then back to loose enable the substrate access to a tunnel through which substrates are translocated to the outside via the TolC channel. The mechanism by which this is accomplished is based on occlusion migrating from the entrance toward the central tunnel similar to that of a peristaltic pump. The energy for the conformational cycling is envisioned to be provided by electronmotive force (emf) to this transporter [see the article by Pos (135) for an excellent review of this mechanism].

6.   Tripartite Pumps

RND, MFS, and ABC transporters can form tripartite pumps in gram-negative bacteria. The components of these tripartite MDR pumps, as described previously, are periplasmic MFPs, outer membrane OEPs, and the respective transporter protein (RND/MFS/ABC). All three components are essential for their function. The composite structure spans the cytoplasmic membrane, the periplasmic space, and the outer membrane allowing for the removal of the substrate from the cytosol/cytoplasmic membrane to the outside of the cell envelope (Figure 5). This provides a huge advantage

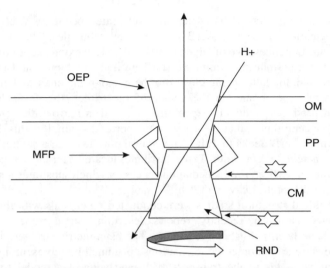

Figure 5.   Tripartite RND efflux pump: a tripartite MDR efflux pump consisting of an RND transporter, a periplasmic (PP) membrane fusion protein (MFP), and an outer membrane (OM) channel or efflux protein (OEP). The circular arrow depicts the rotational conformational changes of the substrate-binding sites in the monomers of the transporter, leading to the peristaltic mechanism of transport.

for the gram-negative bacterial cell because once exported, the drug must negotiate the outer membrane barrier to reenter the cell. Thus, as was so insightfully pointed out by Nikaido, these MDR pumps work synergistically with the outer membrane barrier (137, 138). That both the outer membrane barrier and MDR efflux pumps play an important role in the intrinsic resistance to various hydrophobic inhibitors was shown by the additive effect of deep rough (affects outer membrane permeability to hydrophobic agents) and MDR efflux (TolC) mutants (139).

The most extensively studied tripartite pumps are the MexA–MexB–OprM pump of *Pseudomonas aeruginosa* and the AcrA–AcrB–TolC pump of *E. coli*, both of which are considered to play a major role in antibiotic resistance for their respective bacteria. The RND transporters MexB–AcrB determine the substrate specificity of their tripartite pump, which is quite large compared to other MDR pumps and combined include bile salts, organic solvents, dyes, and compounds that are anionic, cationic, zwitterionic, and a broad range of different antibiotics (133) (Table 5).

TABLE 5

Examples of RND Family of MDR Efflux Pumps and Their Substrates

Efflux System	Substrates[a]	Gene Location	Organisms	Refs.
AcrAB–TolC	ML, ER, EB, LS, DOC, GL, BS, CV, CP, NV, KL, AC, AZ, BL, CH, OZ, FQ, CM	Chromosome	*Escherichia coli* and other gram-negative bacteria	(100, 138, 155, 159, 161–166)
AcrEF–TolC	FA, FQ, FU, NAL, NV, RF, TC, SDS, TX, OZ	Chromosome	*E. coli, Salmonella* spp.	(166, 167)
MexAB–OprM	AC, AH, BL, CL, CM, CV, EB, FQ, ER, NV, SM, SDS, TG, RD, TL, TS	Chromosome	*Pseudomonas aeruginosa*	(65, 99, 168)
MexCD–OprJ	ER, SDS, TL, TP, FQ, RM, CM	Chromosome	*P. aeruginosa*	(99, 169)
MexEF–OprN	TP, TS, FQ	Chromosome	*P. aeruginosa*	(99, 169)
MexJK–OprM	AH, CM, ER, FQ, TP, TS	Chromosome	*P. aeruginosa*	(99, 170)
MexXY–OprN	AG, ER, FQ, TC	Chromosome	*P. aeruginosa*	(99, 171)
MexHI–OpmD	EB, AC, NOR, RD	Chromosome	*P. aeruginosa*	(172)
MexVW–OprM	FQ, CM, TC	Chromosome	*P. aeruginosa*	(173)
SmeABC and SmeDEF	FQ, TC, ER, BL	Chromosome	*Stenotrophomonas maltophilia*	(100, 174)
CmeABC and CmeDEF	FQ	Chromosome	*Campylobacter jejuni*	(175)
SdeAB	FQ	Chromosome	*Serratia marcescens*	(176)
SdeXY	NOR	Chromosome	*S. marcescens*	(177)
MtrCDEML	FQ	Chromosome	*Neisseria gonorrhoeae*	(178–180)
CeoAB–OpcM	FQ	Chromosome	*Burkholderia cepacia (cenocepacia)*	(181)
ActAB–TolC	BL, SDS, AC, MN	Chromosome	*Haemophilis influenzae*	(99, 182)
AdeABC	FQ, CP, TC, CM, ER	Chromosome	*Acinetobacter baumannii*	(99, 183)

[a]See Table 1 footnote for substrate abbreviations.

i. MFPs. Membrane fusion proteins are associated with their cytoplasmic membrane as either a lipoprotein or via a TMD near the N-terminal, with the preponderance of the protein residing in the periplasm (181). Partial crystal structures of membrane fusion proteins MexA (140), AcrA (141), and MacA (142) are available (i.e., missing their extreme N- and C-terminal regions). The structures of AcrA and MexA share a significant degree of sequence and structure similarity. Both are elongated, sickle-shaped molecules composed of three domains: a β-barrel domain, a centrally located lipoyl domain, and a coiled-coil α-helical hairpin at the other end of the molecule. Chemical cross-linkage and mutagenesis studies have shown that the α-helical coiled-coil hairpin of AcrA–MexA docks with the coiled coils of the OEP (TolC–OprM) (143, 144). The β-barrel domain is the probable site of interaction with the transporter protein (AcrB–MexB). The stoichiometry and oligomeric state of the assembled MFPs are unknown [see the article by Zgurskaya et al. (145) for an excellent review on this topic].

ii. OEPs. The architecture of the OEPs whose crystal structures have been solved (*E. coli* TolC, *P. aeruginosa* OprM, and *Vibrio cholerae* VceC) are remarkably similar, even though their amino acid sequence identity or similarity is quite low (146–148). In each case the homotrimers of these proteins make up a long cannon-shaped structure consisting of a 40 Å long β-barrel, which passes through the outer membrane and a 100 Å long α-helical barrel which projects into the periplasm, which is closed at its periplasmic end (146). Based on this structure and the crystal structures of AcrB (129) and AcrA (141) and with the evidence that TolC could be cross-linked independently to either AcrA or AcrB (149–152), models have been proposed which attempt to explain the assembly and function of MDR pumps (143, 152–158). In such models, the periplasmic ends of a trimeric AcrB and a trimeric TolC are envisioned to dock in such a manner as to form a continuous channel that crosses the periplasm and spans the outer membrane. The periplasmic contact between AcrB and TolC has been suggested to involve the TolC entrance coils and the apex (TolC docking domain) of AcrB (129, 154). In these models this connection is bridged and stabilized by the MFP, which is anchored to the cytoplasmic membrane and may play a role in the recruitment of TolC to the AcrB antiporter (159). During assembly of the MDR pump, the periplasmic end of the OEP must open in order for the pump to function. This transition to the open state has been likened to an "irislike" realignment of the entrance helices (146, 160).

This opening of TolC is thought to occur through conformational changes in TolC via its interaction(s) with either AcrB or AcrA or both (146). However, the details by which tripartite pumps are assembled is only beginning to be deciphered and is currently an active area of investigation [see the literature (138, 155) for excellent reviews on this topic].

There are several different mechanisms by which an organism can become resistant to antimicrobial drugs. However, resistance mediated by multidrug-resistance (MDR) efflux pumps appears to be a dominant paradigm among microbial human pathogens. Mobile genetic elements such as plasmids and transposons carrying genes encoding MDR pumps are thought to play an important role in the lateral acquisition of drug resistance by bacteria. The emergence and increasing numbers of drug-resistant pathogenic bacteria pose a great threat to human health. Therefore, it is imperative to study the origin, evolution, and organismal distribution of these xenobiotic transporters, especially in order to develop effective strategies to combat human diseases.

B. PHYLOGENY AND EVOLUTION OF BACTERIAL EFFLUX PUMPS

As noted above, one-tenth of the transporters encoded in a bacterial genome are involved in MDR efflux. The transporter classification system (14) currently recognizes approximately 650 transporter families that include about a dozen large superfamilies (http://www.tcdb.org/). A majority of these families are associated with solute uptake, and only a few constitute exporters. Furthermore, only a half-dozen exporter families (that include four of the superfamilies) contain members that extrude xenobiotic compounds. Despite the fewer proportions of these xenobiotic-extruding transporters in organisms and their relatively high importance, these classes of exporters remain less well understood than were the uptake systems (13).

The capability to export xenobiotic drugs appears to have evolved across independent lineages of transporters. As described above, thus far, functionally characterized bacterial drug exporters and their sequence homologs identified from genome analyses belong to one of five phylogenetically distinct and ubiquitously found transporter families: the MFS superfamily, the RND superfamily, the DMT superfamily, the MATE family, and the ABC superfamily (13). It should be noted that not all families within each of these superfamilies are involved in drug efflux, and that many carry out solute uptake or export. It is likely that the drug efflux pumps function to export cellular metabolites and other molecules, and perhaps simple modifications in their protein sequence can confer additional capabilities to export either specific or multiple xenobiotics.

The MFS superfamily comprises of approximately 65 families of transporters, of which only six contain characterized drug efflux pumps (Table 6). Similarly, only eight families within the ABC, two families within the RND, and only one family in the DMT superfamily (i.e., the SMR family) function in drug efflux in bacteria (Table 6).

III. Genomics of Bacterial Efflux Pumps

Since the first complete genome sequence of *Haemophilus influenzae* that was completed by the Institute for Genomic Research (TIGR), hundreds of bacterial genomes have now been sequenced. Data from these genome sequences are constantly revealing many new uncharacterized transport proteins. In comparison to the few multidrug efflux transporters that have been characterized functionally, the genome-sequencing efforts have identified numerous putative xenobiotic efflux pumps. Previous analyses of complete genomes indicate that approximately 10% of the transporters encoded in bacterial genomes are involved in multidrug efflux (3).

Comparative genomic analyses offer a comprehensive overview of the distribution of transporters across a wide group of organisms. Such analyses offer distinct advantages that can help answer many biological questions. For example:

1. Are xenobiotic efflux pumps encoded in all bacterial genomes or do they occur only in certain species?
2. Is the distribution of the efflux types uniform or species specific?
3. Did specific efflux transporters arise specifically in response to the use of drugs?

In this section we focus on answering some of these questions.

A. METHODOLOGY

Sequences of all the proteins predicted in the complete genome sequences (which include chromosomes and plasmids) of 854 bacteria were downloaded onto our local computer from the National Center for Biotechnology Information (NCBI) through their ftp site (ftp.ncbi.nih.gov). A total of 3,092,197 proteins sequences were obtained. The Basic Local

TABLE 6
Phylogeny of the Bacterial Drug Efflux Pump Families[a]

TC Family (TC#)	Examples
(1) Major facilitator superfamily (2.A.1)	
Drug:H$^+$ antiporter-1 (DHA1) family (2.A.1.2)	CmlA of *Pseudomonas aeruginosa*
Drug:H$^+$ antiporter-2 (DHA2) family (2.A.1.3)	QacA of *Staphylococcus aureus*
Sugar efflux transporter (SET) family (2.A.1.20)	SetA of *Escherichia coli*
Drug:H$^+$ antiporter-3 (DHA3) family (2.A.1.21)	Cmr of *Corynebacterium glutamicum*
Putative aromatic compound/drug exporter (ACDE) family (2.A.1.32)	YitG of *Bacillus subtilis*
Fosmidomycin resistance (Fsr) family (2.A.1.35)	Fsr of *E. coli*
(2) Resistance–nodulation–cell division (RND) superfamily (2.A.6)	
(Largely gram-negative bacterial) hydrophobe/amphiphile efflux-1 (HAE1) family (2.A.6.2)	AcrAB of *E. coli*
(Gram-positive bacterial putative) hydrophobe/amphiphile efflux-2 (HAE2) family (2.A.6.5)	MmpL7 of *Mycobacterium tuberculosis*
(3) Drug/metabolite transporter (DMT) superfamily (2.A.7)	
4 TMD small multidrug resistance (SMR) family (2.A.7.1)	Smr of *S. aureus*
(4) Multi-antimicrobial extrusion (MATE) family (2.A.66.1)	NorM of *Vibrio parahaemolyticus*
(5) ATP-binding cassette (ABC) superfamily (3.A.1)	
Drug exporter-1 (DrugE1) family (3.A.1.105)	DrrAB of *Streptomyces peucetius*
Lipid exporter (LipidE) family (3.A.1.106)	Sav1866 of *S. aureus*
Drug exporter-2 (DrugE2) family (3.A.1.117)	LmrA of *Lactococcus lactis*
Drug/siderophore exporter-3 (DrugE3) family (3.A.1.119)	TetAB of *Corynebacterium striatum*
(Putative) drug resistance ATPase-1 (Drug RA1) family (3.A.1.120)	SrmB of *Streptomyces ambofaciens*
(Putative) drug resistance ATPase-2 (Drug RA2) family (3.A.1.121)	MsrA of *Staphylococcus epidermidis*
Multidrug/hemolysin exporter (MHE) family (3.A.1.130)	CylA/B of *Streptococcus agalactia*
Drug exporter-4 (DrugE4) family (3.A.1.135)	YdaG/YdbA of *L. lactis*

[a]More information on each of the drug transporter families is available at the TCDB web site (http://www.tcdb.org).

Alignment Search Tool (BLAST) was also downloaded from the NCBI ftp site. All transporter proteins present in the Transporter Classification Database [TCDB; (184)] were also downloaded onto our local machine. A total number of 5238 functionally described transport proteins were obtained from TCDB. Of these, 236 transport proteins belong in the drug efflux transporter families described in Table 6. All 3,092,197 predicted proteins were searched against the 5238 known transport proteins using the BLAST tool on our local machine. The expected value (E-value) cutoff of 10^{-6} was used for the BLAST search. This cutoff has been found from our previous studies to yield true hits and minimized false positives (185, 186). Proteins that showed one of the 236 MDR proteins in TCDB as their topmost BLAST hit were identified and inspected carefully for any false positives. A total of 30,564 putative MDR efflux transport proteins from all the five major MDR efflux transporter classes were identified from the 8454 bacterial genomes. The organismal and taxonomical distributions of these transporter proteins are described below. It must be mentioned that a few distant homologs of the MDR pumps that fall below our search threshold (i.e., false negatives) may have been missed. However, from our previous experiences, we anticipate that the numbers of such false negatives would be extremely low.

B. DISTRIBUTION OF DRUG EFFLUX PUMPS IN BACTERIAL GENOMES

The genome sizes of the bacteria analyzed ranged between 0.25 and 10 Mb, with the exception of a myxobacterium belonging to the deltaproteobacteria subdivision, *Sorangium cellulosum* So ce 56, which has a 13-Mb genome that is 71% G + C-rich and encodes 9381 predicted proteins (187). Our computational analyses identified a total of 64 MDR proteins in its genome. The smallest genome (0.25 Mb) analyzed was that of *Candidatus sulcia muelleri* GWSS, a gut symbiont of the blue-green sharpshooter and several other leafhopper species (188). The genome of this organism is only 22% G + C-rich and is predicted to encode just 227 proteins. We could not identify any xenobiotic efflux proteins in the genome of this organism, representing the only organism lacking any recognizable xenobiotic transporters. At least one MDR efflux transport protein was identified in each of the remaining 854 bacterial genomes.

Of the collection of 30,564 drug transport proteins identified in the 854 genomes, approximately 33% (10,013 proteins) belong to the ABC superfamily, 31% (9349 proteins) are MFS type, 27% (8235 proteins) RND type, 6% (1810 proteins) MATE type, and 4% (1157 proteins) are SMR type. A majority of MFS- and MATE-type pumps are generally composed of a single membrane translocator, while RND-type transporters commonly consist of two membrane-associated proteins. Although several SMR pumps are known to contain a single protein, there are many examples of SMR pumps composed of more than one protein. The drug-transporting ABC pumps may be composed of a single protein with both the membrane-spanning and ATP-hydrolyzing domains fused, or may consist of two separate proteins containing the two individual domains. Therefore, based on the adjusted calculations, we estimate that the relative abundance of the different drug efflux pumps would be MFS (41%) > ABC (29%) > RND (18%) > MATE (8%) > SMR (4%). This shows that the MFS- and ABC-type pumps are abundantly present in bacteria, while the MATE and SMR families have restricted representation. All five types were found to occur in a wide variety of bacteria and are not restricted to pathogenic bacteria, as noted above.

C. CORRELATION WITH GENOME SIZE

In general, the total number of proteins dedicated to xenobiotic efflux in the bacterial genomes correlated ($R^2 = 0.72$, $p < 0.05$) with their genome size (Figure 6). The slope of the best-fit line was 10.36 (Figure 6), indicating that approximately 10 drug transport proteins are encoded per megabase of bacteria genome. This is in good agreement with previous observations that 10% of all transport proteins in prokaryotes are involved in multidrug efflux (3). In general, a megabase of genome in bacteria encodes roughly 1000 proteins, and approximately 10 to15% of all proteins predicted in bacterial genomes are transport proteins (189). Therefore, one would expect to find about 100 to 150 transport proteins encoded per megabase of genome, of these 10 to 15 proteins would be involved in xenobiotic efflux. In our analyses, the organism encoding the highest number of drug efflux proteins was *Burkholderia multivorans* ATCC 17616, which has a genome size of 7 Mb (190); it encodes 158 xenobiotic efflux proteins. This number is more than double the number of MDR efflux proteins expected for its genome size. *B. multivorans* is associated with infections in cystic fibrosis patients and is an important opportunistic pathogen that colonizes the lungs. It is associated with a decrease in long-term survival of patients. A minority of patients with *B. multivorans* infection develop cepacia syndrome, which is frequently fatal (191).

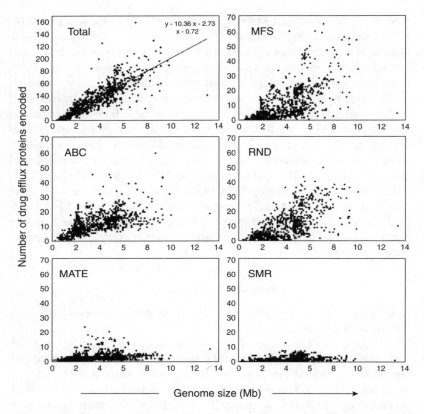

Figure 6. Correlation between the total number of transporters encoded in a genome of 854 bacteria and their genome sizes in megabases. Linear regression of the correlation between the total number of transporters and the genome size was calculated. The R^2 value obtained was 0.72. The equation of the best-fit line is $y = 10.36x - 2.73$. Note that the y-axis scale in the top left panel and the remaining plots is different.

Figure 6 also shows the correlation between genome size and the numbers of each of the five types of xenobiotic efflux pumps. The data show a relative abundance of MFS and ABC proteins followed by the RND homologs. The MATE and SMR types occur in substantially lower proportions in the genomes. There appears to be an expansion of MFS-type transporters with genome size, and some of the larger genomes contain abundant MFS transporters. This is likely because most MFS pumps consist of a single protein, and a single gene duplication can give rise to a functional new pump. Such an expansion is less pronounced with ABC- and RND-type pumps, probably because of the multicomponent nature of these systems. Evolution of a new functional pump would involve coordinated duplication of multiple genes, which perhaps occurs less frequently. *B. multivorans* ATCC 17616 encodes the highest number of MFS and RND proteins (66 and 48 proteins, respectively) among all bacteria compared in our study, this is reflective of the large complement of drug transporters in this opportunistic pathogen. However, *Streptomyces griseus* subsp. *griseus* NBRC 13350, which is slightly larger (8.5 Mb), encodes the largest number of ABC transport proteins (58 homologs). It is followed by seven other actinobacteria, two of which contain smaller genomes. *Corynebacterium glutamicum* ATCC 13032 with a genome size of 3.3 Mb, and *Beutenbergia cavernae* DSM 12333 with a genome size of 4.7 Mb, both encode 44 ABC proteins each. This indicates a relative expansion and likely importance of ABC family drug transporters in actinomycetes, soil-dwelling microbes that are known to produce a wide range of antibiotics. Both SMR and MATE transporters do not appear to increase in numbers linearly with genome size. Most organisms appear to encode two to four MATE-type and one or two SMR-type drug efflux transporters. *Bacillus licheniformis* ATCC 14580 encodes an expanded repertoire of 12 SMR family proteins. This organism is a soil-dwelling endospore-forming microbe that is used extensively in the industrial production of important enzymes such as proteases, penicllinases, and amylases as well as smaller compounds such as the antibiotic bacitracin and various organic metabolites. Its 4.2-Mb genome encodes 125 drug efflux transporter proteins (three times the expected number) representing all five families. The organism with the largest number of MATE family proteins is *Eubacterium eligens* ATCC 27750, a firmicute and member of the normal human gut microflora. This organism (with a 2.8-Mb genome) encodes a total of 46 drug efflux proteins that include the 23 ABC, 21 MATE, and 2 RND family proteins but no MFS or SMR pumps. Thus, expansion of the MATE transporters may compensate for the lack of MFS facilitators in this organism.

D. ORGANISMS LACKING ONE OR MORE TRANSPORTER TYPES

Although just one organism was found to lack any recognizable drug efflux pumps as mentioned above, less than half (47%) of the bacteria analyzed contain all five types of drug transport proteins (Table 7). This

TABLE 7

Enumeration of Organisms Against Different Profiles Showing the Presence of Each of the
Five Transporter Family Homologs[a]

Number of Drug Transporter Types	Organisms		Transporter Type				
	Number	Percent	ABC	MFS	RND	MATE	SMR
Five	398	47	Yes	Yes	Yes	Yes	Yes
Four	170	20	Yes	Yes	Yes	Yes	—
	82	10	Yes	Yes	Yes	—	Yes
	3	0.4	Yes	—	Yes	Yes	Yes
	3	0.4	Yes	Yes	—	Yes	Yes
Three	83	10	Yes	Yes	Yes	—	—
	21	2	Yes	—	Yes	Yes	—
	11	1	Yes	Yes	—	Yes	—
	2	0.2	Yes	Yes	—	—	Yes
Two	26	3	Yes	Yes	—	—	—
	7	1	—	Yes	Yes	—	—
	4	0.5	Yes	—	Yes	—	—
	3	0.4	Yes	—	—	Yes	—
	1	0.1	—	—	Yes	Yes	—
One	38	4.4	Yes	—	—	—	—
	1	0.1	—	Yes	—	—	—
None	1	0.1	—	—	—	—	—

[a]The table is sorted in the descending order of the number of organisms under each category listed in the first column.

group of bacteria contains representatives from most of the major taxonomic subdivisions of eubacteria. Fifty-three percent of bacteria (455 genomes) were found to lack one or more of the five transporter types (Table 7). Approximately, 5% of bacteria (39 genomes) carry just one type of drug efflux transporter; many of these are intracellular pathogens and obligate symbionts. The genome sizes of these organisms ranged from 0.42 to 1.9 Mb; all except one [*Baumannia cicadellinicola* str. Hc (*Homalodisca coagulata*)] lacked the secondary carriers (i.e., MFS, RND, SMR, and MATE porters) and contained 1 to 10 proteins of just the ABC-type transporters. These organisms include 23 mycoplasmas and 13 chlamydiae that are known to have undergone genome reduction. A reduced genome (0.42 Mb) gammaproteobacterium, *Buchnera aphidicola* str. Cc (*Cinara cedri*), also contains just 2 ABC-type drug transport proteins; however, other *B. aphildicola* strains additionally contain MFS-type transporters. A bacteroidetes with 1.9-Mb genome, *Candidatus amoebophilus asiaticus* 5a2, encodes 10 ABC-type drug efflux proteins, but none of the other types. This organism is an obligate endosymbiont specific to its protozoan host, *Acanthamoeba* sp. TUMSJ-321, isolated from lake sediment in Malaysia (192). *B. cicadellinicola* str. Hc, a gammaproteobacterium, has a 0.69-Mb genome (193) and encodes just three MFS-type drug efflux proteins; it lacks any recognizable drug transporters of the other four types. This newly discovered organism is an obligate endosymbiont of the leafhopper insect *Homalodisca coagulata* (Say), also known as the glassy-winged sharpshooter, which feeds on the xylem of plants.

Forty-one (5%) organisms encode only two out of the five efflux types, many of which also are intracellular pathogens and endosymbionts. Of these, 26 contain just MFS- and ABC-type drug efflux transporters, seven encode MFS- and RND-type pumps, four encode ABC- and RND-type pumps, three contain ABC- and MATE-type transporters, and only one organism encodes the RND and MATE family of transporters but not the other types. The organisms containing both and MFS- and ABC-type transporters comprise predominantly of firmicutes (17 species, mostly *Streptococcus pyogenes* strains), gammaproteobacteria (four *B. aphidicola* strains), Alphaproteobacteria (three), and Actinobacteria (two *Tropheryma whipplei* strains). The remaining organisms that do not contain either MFS- or ABC-type efflux pumps are alphaproteobacteria (seven), spirochetes (two), bacteroidetes (two), gammproteobacteria (two), a firmicute, and an unclassified bacterium; most encode fewer than 10 drug transporter proteins. However, two organisms encoded as many as 19 and 30 efflux proteins each and are discussed briefly below. *Halorhodospira halophila* SL1, a gammaproteobacterium, has a genome of 2.7-Mb genome and encodes eight ABC-type drug transport proteins and 11 RND-type efflux transporter proteins. *H. halophila* SL1, formerly *Ectothiorhodospira halophila*, was isolated from salt lake mud and is one of the most halophilic eubacteria known (194). The other organism, *Eubacterium rectale* ATCC 33656, with a 3.4-Mb genome, encodes an equal number (15 each) of ABC- and MATE-type drug transport proteins. *E. rectale* ATCC 33656, a firmicute, was isolated from human feces. *Eubacterium* spp. are thought to play a beneficial role in maintaining the normal ecology of the large intestine, in part by producing chemicals such as butyric acid which act to inhibit the growth of other bacteria. These organisms are occasionally isolated from wounds and abscesses and may be an opportunistic pathogen. This genus has also been isolated from sewage and soil.

One hundred and seventeen (14%) of the bacteria that contain three types of drug efflux transporters are predominated by bacteria (83 genomes) that contain MFS, RND, and ABC proteins (83 organisms), while 21 bacteria lack the MFS and SMR homologs, 11 lack the RND and SMR pumps, and two organisms lack RND and MATE transport proteins. The organisms that lack the MFS proteins comprise of spirochaetes (eight), firmicutes (four), gammaproteobacteria (four), bacteroidetes/chlorobi (two), thermotogae (two), and fusobacteria (one). Finally, 30% (258) of bacteria lack just one type of transporter, of which a majority lack SMR-type pumps (170 genomes), followed by MATE-type pumps (82 genomes). The MFS and RND pumps are absent in three genomes each, while there were no organisms that lacked just the ABC-type drug efflux transporters.

E. TAXONOMICAL DISTRIBUTION OF DRUG EFFLUX PUMPS

The bacteria surveyed in our present study can be classified into 19 different taxonomical groups. Gammaproteobacteria, firmicutes, and alphaproteobacteria are well represented, with 221, 189, and 111 bacterial genomes, respectively. While planctomycetes, fusobacteria, and acidobacteria are poorly represented, having just one or two completely sequenced members, another 13 taxonomic subdivisions contain five to 69 bacterial species with complete genomes. The relatively well-represented groups offer an opportunity to assess any taxonomic bias with respect to the occurrence of drug efflux pumps. The average number of drug efflux transport proteins per megabase of genome within each group of bacteria is shown in Figure 7. These data show that both firmicutes and gammaproteobacteria contain the highest density of drug transport proteins per megabase of genome, while the genomes of chlamydiae/verrucomicrobia contain a fourfold lower density of these proteins (Figure 7). Spirochaetes and cyanobacteria encode half as many drug transporters as firmicutes per unit length of the genome. Interestingly, the relative density of each of the five transporter families varies remarkably across the bacterial groups

→

Figure 7. Average number of transporters encoded per megabase of genome in the different taxonomical groups of bacteria. The 19 taxonomical groups (and the number of bacterial genomes sequenced in each group), numbered 1 to 19 (bottom to top) on the y-axis, are firmicutes (189), gammaproteobacteria (221), betaproteobacteria (67), actinobacteria (69), bacteroidetes/chlorobi (27), epsilonproteobacteria (25), alphaproteobacteria (111), fusobacteria (1), thermotogae (10), other bacteria (16), deltaproteobacteria (27), acidobacteria (2), aquificae (5), deinococcus-thermus (5), chloroflexi (10), spirochaetes (18), cyanobacteria (36), planctomycetes (1), and chlamydiae/verrucomicrobia (14), respectively.

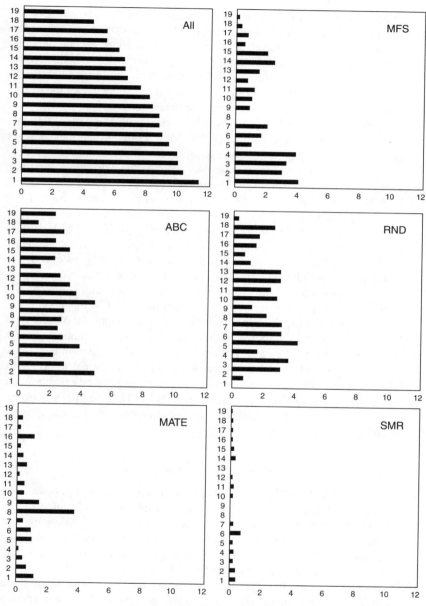

Average number of drug efflux proteins
encoded per Mb of genome

(Figure 7). Firmicutes and thermatogae encode more ABC proteins per megabase of genome, whereas betaproteobacteria, epsilonproteobacteria, spirochaetes, chlamydiae/verrucomicrobia, and aquificae encode only half as many of these transport proteins. Although spirochaetes and chlamydiae/verrucomicrobia encode fewer ABC transporters, the largest number of drug transporters in these bacteria is the ABC type, reflective of the overall lower abundance of drug transporters in these bacteria.

Actinobacteria and betaproteobacteria along, with firmicutes and gammaproteobacteria, encode three to five MFS proteins per megabase of genome, higher than most other groups. The highest number of drug transporters in the deinococcus-thermus group is the MFS type, and these bacteria encode two to three MFS transporters per megabase of genome. Bacteroidetes/chlorobi as well as all proteobacteria encode more RND transporters per unit of genome, while firmicutes have a much lower representation of RND pumps in their genomes. MATE and SMR pumps generally occur in substantially lower proportions in bacterial genomes. Interestingly, the single member of the fusobacteria analyzed encodes as many as four MATE family proteins per megabase of genome in comparison to the zero to two homologs in other groups of bacteria. *Fusobacterium nucleatum* subsp. *nucleatum* ATCC 25586 has a 2.17-Mb genome and encodes 19 drug efflux transport proteins, of which eight are MATE type, six are ABC proteins, and five are RND-family proteins. It appears to lack MFS- or SMR-type drug transporters completely. *F. nucleatum* belongs to the normal microflora of the human oral and gastrointestinal tracts (195). This bacterium is capable of forming coaggregates with other pathogenic and nonpathogenic bacteria in the mouth. Its MATE-type efflux transporters may play a role in resistance to antimicrobials produced by other inhabitants of the oral microflora.

F. DIFFERENCES AMONG STRAINS

Approximately, half of the bacterial species analyzed (431 genomes) are represented by a single sequenced strain, while 111 different bacterial species have multiple strains (ranging between two and 27 strains) whose genomes have been sequenced completely, accounting for the remaining 423 genomes. Thirty-seven bacterial species with four or more completely sequenced strains are discussed below (Table 8).

Escherichia coli has the most strains sequenced (27 strains); these include both pathogenic and nonpathogenic strains. Six other species contains 10 or more sequenced strains: *Salmonella enterica* (15 strains), *Staphylococcus aureus* (14 strains), *Streptococcus pyogenes* (13 strains), *Prochlorococcus marinus* (12 strains), *Streptococcus pneumoniae* (11 strains), and *Clostridium botulinum* (10 strains). Surprisingly, the strain-level variation in the numbers of the different drug efflux pumps appears to be very low in *S. aureus* and *S. pyogenes*, both pathogenic in humans.

Six sequenced strains of *Acinetobacter baumannii* show large variations in the numbers of drug transporters (Table 8). Most notable are the differences in the numbers of MFS transporters, followed by RND, SMR, and ABC proteins in descending order. *A. baumannii* is an aerobic gram-negative bacillus that is an opportunistic pathogen in humans. Infections by this organism are becoming increasingly problematic, due to the high number of resistance genes found in clinical isolates. Some strains are now resistant to all known antibiotics. Most of these resistance genes appear to have been transferred horizontally from other organisms. Many of them cluster into a single genomic island in strain AYE, as compared to strain SDF.

A similar high level of variation is observed in six strains of a nonpathogenic bacterium, *Rhodopseudomonas palustris*, which is commonly found in soil and water environments (Table 8). However, the most noted differences were observed in the numbers of ABC- and RND-type pumps, and to a lesser extent in the number of MFS drug transporters. Intermediate-level differences are found among strains of other pathogenic species, such as *Bacillus anthracis*, *Bacillus cereus*, *Coxiella burnetii*, *Escherichia coli*, *Pseudomonas aeruginosa*, and *Yersinia pestis*, as well as nonpathogenic species such as *Pseudomonas putida*, *Shewanella* sp., and *Synechococcus* sp. (Table 8). In the six pathogens cited above, the differences are mostly in the numbers of MFS and ABC transporters, except in the case of *C. burnetii* and *E. coli*, whose strains mostly differ in the complements of MFS and RND families of proteins. A few additional pathogens with interesting profiles are as follows: *S. enterica* strains differ in the number of SMR proteins besides MFS and RND; *C. botulinum* strains show differences in the number of MATE homologs in addition to ABC and MFS proteins; strains of *Francisella tularensis* and *Burkholderia cenocepacia* vary mostly in the number of MFS-type drug efflux pumps predicted, and *S. pneumoniae* strains mostly show differences in the number of ABC proteins (Table 8). Strains of all other species show relatively fewer differences in the number of drug efflux transporters encoded in their genome (Table 8).

TABLE 8

Strain-Level Variation in the Number of Drug Efflux Proteins Encoded[a]

Genus and Species	Number of Strains	MFS	ABC	RND	MATE	SMR	Total MDR Proteins
Acinetobacter baumannii	6	16 (6–21)	10 (8–12)	14 (10–17)	2 (2–3)	4 (2–7)	46 (30–54)
Bacillus anthracis	5	43 (42–45)	22 (20–26)	5 (5–5)	5 (4–6)	6 (6–6)	81 (77–88)
B. cereus	9	46 (42–51)	23 (21–26)	5 (4–5)	4 (4–5)	6 (5–7)	84 (78–91)
Buchnera aphidicola	6	1 (0–1)	3 (2–3)	0 (0–1)	0 (0–1)	0 (0–0)	4 (2–4)
Burkholderia cenocepacia	4	48 (45–50)	13 (12–14)	35 (34–37)	3 (3–3)	3 (2–3)	102 (98–104)
B. mallei	4	26 (24–27)	9 (7–11)	16 (15–17)	2 (2–3)	2 (2–2)	55 (52–60)
B. pseudomallei	4	28 (28–28)	11 (11–11)	22 (22–22)	3 (3–3)	2 (2–2)	66 (66–66)
Campylobacter jejuni	5	5 (4–6)	4 (4–4)	5 (4–5)	2 (1–2)	4 (4–4)	19 (17–21)
Chlamydia trachomatis	5	0 (0–0)	2 (2–2)	0 (0–0)	0 (0–0)	0 (0–0)	2 (2–2)
Chlamydophila pneumoniae	4	0 (0–0)	2 (2–2)	0 (0–0)	0 (0–0)	0 (0–0)	2 (2–2)
Clostridium botulinum	10	4 (3–8)	19 (15–23)	1 (1–2)	11 (10–14)	0 (0–0)	35 (31–39)
Coxiella burnetii	5	12 (8–14)	6 (5–8)	6 (4–8)	1 (0–1)	0 (0–1)	25 (19–31)
Cyanothece sp.	4	3 (2–5)	11 (8–15)	10 (9–11)	1 (0–2)	0 (0–0)	25 (22–29)
Escherichia coli	27	21 (18–25)	15 (14–16)	8 (7–13)	3 (2–4)	4 (3–6)	51 (46–58)
Francisella tularensis	7	11 (9–15)	7 (5–8)	6 (5–6)	0 (0–0)	0 (0–0)	23 (20–29)
Haemophilus influenzae	4	3 (2–3)	7 (6–8)	3 (3–3)	1 (1–1)	0 (0–0)	14 (12–15)
Helicobacter pylori	7	2 (2–2)	2 (2–3)	3 (3–3)	2 (2–2)	0 (0–0)	9 (9–10)

Legionella pneumophila	4	15 (14–15)	16 (14–16)	15 (13–17)	0 (0–0)	1 (1–1)	46 (43–49)
Mycobacterium tuberculosis	5	13 (12–13)	12 (12–13)	15 (14–16)	0 (0–0)	1 (1–1)	41 (39–43)
Neisseria meningitidis	4	3 (3–3)	5 (5–6)	5 (5–5)	1 (1–1)	1 (0–1)	15 (14–16)
Prochlorococcus marinus	12	1 (1–1)	8 (6–9)	3 (2–6)	0 (0–0)	0 (0–0)	12 (10–16)
Pseudomonas aeruginosa	4	23 (20–24)	15 (13–19)	34 (33–35)	3 (3–3)	6 (6–6)	81 (76–86)
P. putida	4	19 (16–22)	16 (13–18)	27 (22–30)	2 (2–2)	3 (2–4)	67 (57–73)
Rhodobacter sphaeroides	4	6 (5–6)	12 (9–14)	8 (8–9)	2 (1–2)	2 (2–3)	30 (28–31)
Rhodopseudomonas palustris	6	10 (7–12)	17 (11–27)	29 (23–34)	2 (2–3)	3 (2–3)	60 (50–77)
Salmonella enterica	15	21 (18–22)	15 (14–16)	9 (7–11)	2 (1–3)	4 (2–7)	51 (46–55)
Shewanella baltica	4	12 (11–12)	14 (14–15)	24 (23–25)	7 (6–7)	1 (1–1)	58 (57–60)
Shewanella sp.	4	12 (9–17)	13 (13–14)	31 (28–36)	6 (6–6)	1 (1–1)	63 (60–73)
Staphylococcus aureus	14	23 (22–24)	4 (4–6)	3 (3–3)	1 (1–1)	0 (0–0)	31 (30–33)
Streptococcus pneumoniae	11	1 (1–2)	14 (12–16)	1 (1–2)	3 (3–3)	0 (0–0)	20 (17–21)
S. pyogenes	13	5 (4–6)	10 (9–12)	0 (0–0)	0 (0–1)	0 (0–0)	16 (14–18)
S. suis	5	1 (1–2)	21 (19–23)	1 (1–1)	4 (3–6)	0 (0–0)	27 (24–32)
Synechococcus sp.	9	2 (1–4)	9 (7–16)	5 (3–8)	0 (0–0)	1 (0–1)	17 (12–25)
Xanthomonas campestris	4	9 (8–9)	9 (8–9)	23 (21–27)	2 (2–2)	2 (1–2)	44 (42–46)
Xylella fastidiosa	4	1 (0–3)	5 (4–5)	8 (7–9)	1 (1–1)	0 (0–0)	14 (13–18)
Yersinia pestis	7	15 (13–17)	11 (7–16)	11 (10–12)	2 (2–2)	4 (3–4)	44 (37–51)
Y. pseudotuberculosis	4	17 (17–18)	13 (13–13)	13 (12–13)	2 (2–2)	4 (4–4)	49 (48–49)

aTotal number of completely sequenced strains for each organism is provided in column 2. The table lists the average number of transport proteins of each type in each species as well as provides in parentheses the minimum and maximum numbers of these transporters across the strains in a given species.

The organization differences between prokaryotic (bacterial) and eukaryotic cells (animals, plants, fungi) in part impose requirements for transport proteins to move molecules into and out of cells. Gram-positive bacteria are enclosed by a single cytoplasmic membrane, whereas gram-negative bacteria possess two membranes: an inner cytoplasmic membrane and a lipopolysaccharide-containing outer membrane. Movement of molecules into and out of gram-positive bacteria generally requires single-component transporters. Transport of molecules in gram-negative bacteria requires passage through the periplasm enclosed between the two membranes. Thus, many transporters in gram-negative bacteria are multicomponent, as described in this chapter. The cytoplasm of fungi is enclosed by a single plasma membrane, and therefore most transporters are products of a single gene (described below). However, unlike gram-positive bacteria, fungi also possess transport proteins in membranes of intracellular organelles. While the majority of fungal efflux pumps reside in the cytoplasmic membrane, some pumps are located in the membranes of vacuoles (196, 197).

IV. Fungi

The fungi represent a large, diverse group of eukaryotic microorganisms that range in size from macroscopic multicellular mushrooms to microscopic unicellular forms. Fungi inhabit most environments on the planet, but their primary reservoir is the soil. Although most may appear invisible, many are essential for carbon and nutrient recycling in nature, others are important in the food, pharmaceutical, and beverage industries, and many are serious pathogens of plants and animals. Fungi are also known to synthesize a vast array of compounds, many of which are toxins. Fungi may form threadlike tubular walled structures, hyphae that branch and anastomose into complex matlike structures known as mycelia. Hyphae may be segmented into individual cell-like units connected by pores that permit movement of organelles, nuclei, and cytoplasm (198). Some fungi may exist as unicellular forms with cell walls and a single nucleus per cell. These yeast forms occur in many fungal groups. Dimorphic fungi possess both yeast and mycelial stages in their life histories, and each growth form may provide novel functional capabilities to the organism (199–202). Thus plant- and animal-pathogenic fungi may utilize one developmental form during interaction with the host and the other during growth outside the host. Some fungi have no known mycelial stage and form saclike cells which may be multinucleate (yeasts). Some of these fungi have motile flagellated forms as part of their life cycle, and these often play a role in the initial interaction with the host (203). Almost all groups of fungi have members that are important pathogens of plants, animals, and humans. Over the past two decades we have observed a sharp increase in the occurrence of invasive fungal infections of humans, many of which are associated with morbidity and mortality (204). More recently, fungi have been implicated as the causative agents in the white-nose bat syndrome (*Gleomyces* sp.) and in the global extinction crisis of amphibians (*Batrachochytrium dendrobatidis*) (205, 206).

A.   ANTIFUNGAL AGENTS AND THEIR USE

Because of their eukaryotic nature, fungi are inherently difficult to treat without causing damage to the host. Furthermore, their relatively slow growth rate (compared to bacteria) adds to the loss in efficacy of certain antifungal agents. These features therefore restrict the array of antifungal drugs that can be used therapeutically (207). The limited number of useful drugs to treat fungal infections is under the additional burden because of rapidly developing antifungal drug resistance (207, 208). Thus, the development of new drugs that impair uniquely fungal biological processes with limited side effects and ease of delivery is of vital importance. Seven major classes of drugs are currently used to treat fungal infections therapeutically: the triazoles, polyenes, echinocandins, allylamines, nucleoside analogs, morpholines, and griseofulvin-type. Among drugs currently used to treat fungal infections are those that target an enzyme in a unique fungal sterol biosynthesis pathway absent in plants and animals. The azoles (fluconozole, ketokonozole, itraconozole) are inhibitors of fungal $P450_{14DM}$ (lanesterol 14α-demethylase), which causes the accumulation of C-14 methyl sterols in fungi and impairs normal membrane function (209–214). The polyenes (amphotericin B and nystatin) intercalate into membranes containing the unique fungal sterol ergosterol, causing ion leakage and cell death (215). Echinocandins (e.g., caspofungin and micafungin) inhibit 1,3-β-D-glucan synthase, required for fungal cell wall synthesis. Loss of wall integrity can result in cell lysis and death. The azoles, polyenes, and echinocandins are the only three of the seven antifungal drug classes that can be used to treat systemic infections (216). The allylamines (naftifine and terbinafine) inhibit squalene oxidase, a key step in ergosterol biosynthesis (required for fungal plasma membranes). Morpholine (amorolfine) impairs ergosterol biosynthesis by inhibiting cytochrome P450 enzymes and nucleoside analogs (flucytosine), and griseofulvin-type substances (grifulvinV, fulvicin U/F) interfere with DNA and RNA synthesis and mitotic spindle formation, respectively (215). In nature, fungi encounter a large variety of antifungal substances that are made by a broad spectrum of organisms (217, 218). These compounds include peptides, fatty acids, proteins, alkaloids, quinones, and statins. Survival therefore necessitates employment of effective antitoxin mechanisms. The most common processes used by fungi to become resistant to antifungal agents are destruction of the agent, change in the target enzyme or pathway by mutation, and active efflux to maintain low intracellular concentrations (213, 219).

B. FUNGAL EFFLUX TRANSPORT AND GENOMICS

Data from genome sequences available at the Broad Institute and the J. Craig Venter Institute (Transporter Protein Analysis Database) were used in this chapter (http://www.broadinstitute.org/; http://www.membranetransport.org/). A significant proportion of fungal genomes are devoted to transport proteins. Transport proteins are important for nutrient uptake, intracellular ion concentration maintenance, secretion of proteins, secretion of secondary metabolites, and efflux of toxins and xenobiotics. In the yeasts, *Saccharomyces* sp., and in filamentous fungi, *Aspergillus* sp., *Neurospora* sp. and *Cryptococcus* sp., the number of transport proteins per megabase of genome is between 13 and 30 (220). By comparison, the closely related fungal group, the oomycetes, have fewer than 5, and the model eukaryotes *Arabidopsis thaliana* have 9.7, *Caenorhabditis elegans* 6.7, *Drosophila melanogaster* 5, *Entamoeba histolytica* 8.5, and *Mus musculus* 0.4 (220). This underscores the importance of transport proteins to the survival of fungi in different environments. For example, in the soil, fungi have to contend with bacterial and fungal antibiotics, plant root exudates, chemicals from pollution, protozoans, and insects. Pathogenic fungi have to survive in potentially toxic host plant or animal environments which necessitate effective efflux processes. Additionally, many fungi produce toxins and secondary metabolites that must be secreted into their hosts or the environment (221). Two major classes of transport proteins are found in fungi: the ATP-binding cassette superfamily (ABC) and the major facilitator superfamily (MFS) (222).

These two superfamilies of transporters make up between 12 and 22% and 76 and 85%, respectively, of the total number of transporters in many fungi. The ABC transporters belong to a large superfamily of membrane proteins that are found in other eukaryotes and bacteria. The nucleotide-binding domain (ATP-binding cassette) is the most highly conserved region among members of this superfamily. The energy that drives the movement of molecules across the membrane through these transporters is derived from ATP. Members of the superfamily are important in the import of sugars, amino acids, peptides, ions and efflux of proteins, secondary metabolites, and xenobiotics. All fungi examined thus far possess ABC superfamily transporters. Table 9 shows the proportion of fungal and related oomycetes and *Phytophthora infestans* efflux transporters out of the total number of transporters of each superfamily, ABC and MFS, from a distribution of pathogenic and nonpathogenic fungi. Greater than 50% and up to 75% of the ABC transporters in these organisms are devoted to efflux purposes. Although the number of *P. infestans* ABC transporters is large, they actually represent 0.67 per megabase genome and the number of ABC efflux transporters per megabase genome is 0.5. On the other hand, the fungi *Aspergillus fumigatus, A. nidulans, C. neoformans*, and *S. cerevisiae* encode approximately 1 ABC efflux transporter per megabase of genome. However, *Neurospora crassa* encodes just 0.4 ABC efflux transporters per megabase of genome. These observations suggest that a sizable portion of the transport protein-encoding genome in some fungi is committed to maintenance of an intracellular environment low in potentially harmful

TABLE 9
Distribution of Efflux Transporters in Some Fungi

Organism	Total ABC	Efflux ABC	$\%^a$	Total MFS	Efflux MFS	$\%^b$	P/N^c	Genome (Mb)
Saccharomyces cerevisiae	24	13	54	85	18	21	N	13
Aspergillus fumigatus	45	35	75	275	96	35	P	33
Cryptococcus neoformans	29	19	65	192	54	28	P	19
Neurospora crassa	31	17	55	141	55	39	N	40
Aspergillus nidulans	45	35	75	356	99	28	N	31
Aspergillus oryzae	72	60	83	507	180	35	P	32
Phytophthora infestans	160	116	72	102	2	<2	P	237

^{a}*Source*: Data acquired from Transporter Protein Analysis Database (http://www.membranetransport.org/).
^{b}Percentage ABC efflux transporters out of total ABC transporters.
^{c}Percentage MFS efflux transporters out of total MFS transporters.
^{d}P/N, pathogen/nonpathogen.

metabolites and xenobiotics or for secretion of toxins. Most of these data are based on bioinformatic evidence, and more experimental evidence is required to support these observations. Interestingly, there appears to be no correlation between the percentage of efflux transporters and whether a fungus is a pathogen or a nonpathogen (Table 9). This suggests that efflux pumps play a variety of roles in addition to those required to survive in a toxic host environment. It is conceivable that resistance capabilities may have evolved from proteins required for other cellular and ecological processes. A similar concept had previously been discussed regarding fungi, where efflux pump gene expression is required during mitosis (223) and where important physiological substrates such as steroids may be transported by efflux pumps (224, 225). Similar suggestions have also been made regarding bacteria, where efflux pumps play roles in endurance in their ecological niches, such as attachment, invasion, colonization, and persistence (226).

1. Fungal ABC Transporters

ABC transporters may be organized into different configurations; transmembrane domains (TMDs) followed by nucleotide-binding domains (NBDs), (TMD–NBD)$_2$, reverse (NBD–TMD)$_2$ or NBD–TMD. Each nucleotide-binding domain contains characteristic sequences [Gx4GK (ST)] Walker A box and [(RK)X3GX3L(hydrophobic)] Walker B box, separated by 90 to 120 amino acids (227). In fungi, full-size transporters typically comprise 1200 amino acids, between 12 and 20 TMDs, and two NBDs (TMD–NBD)$_2$ or (NBD–TMD)$_2$, whereas half-size transporters have between 5 and 10 TMDs and one NBD (NBD–TMD) (228). While the TMDs probably function in substrate translocation across the membrane, the driving energy derived from ATP is harvested by the NBDs. The ABC superfamily of transporters comprises seven families: ABCA, ABCB, ABCC, ABCD, ABCE, ABCF, and ABCG. Of these, families ABCB, ABCC, and ABCG are implicated in active efflux and are also known as the multidrug resistance (MDR), multidrug resistance–associated protein (MRP), and the pleitropic drug resistance (PDR) families (229–231). Families ABCE and ABCF that do not have TMDs are not discussed further (232, 233).

Saccharomyces cerevisiae is the best and most extensively studied model fungus. Consequently, we have much biochemical, physiological, and molecular biological evidence to support the bioinformatic information on the organism. The *S. cerevisiae* genome encodes 24 ABC superfamily transporters, and these represent approximately 34% of the ATP-dependent transporters (220). Representatives of the ABCG family (PDR5) are in a (NBD–TMD)$_2$ configuration and are involved in multidrug resistance. In this fungus, the family is also known as cluster I and includes Pdr5p, Snq2p, and YOl075C proteins (228, 234). *S. cerevisiae* ABCC family representatives (MRP; cluster II.1) Ycg1p, Btp1p, Ybt1p/Bat1p, and Yor1 have a (TMD–NBD)$_2$ configuration and are full-size transporters. The *S. cerevisiae* ABCB family representative (cluster II.2) Ste6p is a full-size transporter. Cluster II.3 representatives of this family, Atm1p, Mdl1p, and Mdl2p, are half-size transporters.

i. Saccharomyces Species. Substrates of efflux proteins Pdr5p, Snq2p, and Yor1 were studied using single, double, and triple mutants with 349 substrates (235). This study demonstrated that these pumps share overlapping, though not identical substrate preferences. Triple mutants showed full sensitivity to itraconozole, miconozole, nystatin, antimycin, and tetradecylammonium bromide. Pdr5p was observed to provide resistance to cycloheximide, benomyl, and phenaprmyl, Snq2p to resazurin and quinoline oxides, and Yor1p to propanil, ferbam, oligomycin, and thiram. In an elegant study using point mutations with Pdr5p, it was demonstrated that substrate specificity is in part determined by protein folding and that pump inhibitor sites are functionally separated from substrate interacting sites (235). Pdr5p has also been shown to mediate resistance to certain mycotoxins and is involved in the transport of gulcocorticoids (236, 237). ABCC family proteins of cluster II.1—Ycg1p, Btp1p, and Ybt1p/Bat1p— are capable of transporting bile acids and glutathione conjugates. ABCB (cluster II.2) family protein Ste6p is required for transport of mating pheromone (factor α). Cluster II.3 proteins are all localized to the mitochondrial membranes and play a role in export of mitochondrial peptides. While most of the *S. cerevisiae* ABC efflux transporters are localized to the plasma membrane (e.g., Pdr5p, Snq2, Yor1p, Pdr10p, Pdr15p, Ste6p), others may be localized to vacuolar membranes (e.g., Btp1p, Ybt1p, Ycf1p) (230).

ii. Aspergillus Species. The genus *Aspergillus* contains members that have considerable impact on our lives and environment. *Aspergillus fumigatus* is an opportunistic pathogen of animals, while *A. flavus* infects grain crops and is responsible for the production of aflatoxin. *A. nidulans* is generally nonpathogenic and is a soilborne fungus. *A. fumigatus* persists in the environment as airborne spores, and consequently, the human respiratory tract is constantly exposed to the fungus. Infections of immunocompromized individuals by *A. fumigatus* are very high, and the mortality rate is as high as 50%. Recent observations of increased *A. fumigatus* resistance to triazoles have been attributed partly to enhanced drug efflux pump activity (238, 239). This fungus encodes 45 ABC transporters, of which 75% are involved in efflux. Of these, 12 pumps belong to the ABCG family, 13 to the ABCB family, and 10 to the ABCC family. BLAST pairwise alignment shows that 12 efflux transporters in *A. fumigatus*, two transporters from *A. nidulans*, and 10 transporters from *A. flavus* are related to the prototypical mammalian P-glycoprotein (ABCB family). *A. flavus* has 23 transporters that belong to the ABCG family based on pairwise alignment, while *A. nidulans* has 14 representatives. ABC proteins AfuMdr1 (*A. fumigatus*) and *A. flavus* AflMdr1 are similar to the *Schizosaccharomyces pombe* leptomycin B resistance protein and also to the human Mdr1. *A. fumigatus*, AfuMdr2, is similar to two MDR-like genes of *S. cerevisiae* and confer resistance to the echinocandin B analog, cilofungin (240).

The genome of the nonpathogen *A. nidulans* encodes 45 ABC transporters, of which 75% are involved in efflux. In *A. nidulans*, ABC superfamily efflux transporters bearing a resemblance to the *S. cerevisiae* Pdr5p and Snq2p transporters have been described experimentally (241). The proteins AtrA and AtrB also share homology to the dimorphic fungus *Candida albicans*, Cdr1, and mammalian P-glycoprotein-type transporters (see below). AtrB expression in *S. cerevisiae* was capable of countering the drug hypersensitivity of a *S. cerevisiae* Pdr5p mutant. Interestingly, transcription of these *A. nidulans* transporters was found to be enhanced following exposure of the fungus to azoles or plant defense chemicals. *A. nidulans* AtrA, B, C, and D genes were expressed differentially when the fungus was grown in the presence of the structurally unrelated compounds camptothecin, imazalil, itraconazole, hygromycin, and 4-nitroquinoline oxide (4-NQO) (242). In the presence of 4-NQO, AtrA expression was increased approximately 14-fold, but AtrB was increased more than 4500-fold. AtrC expression was enhanced more than 62-fold in the presence of hygromycin, and AtrD expression was enhanced more than 250-fold. Itraconozole in the growth medium enhanced AtrB and AtrD expression 39- and 23-fold, respectively, while exposure to camptothecin resulted in decreased expression of AtrA and AtrC. AtrA was also repressed in the presence of imazalil and itraconozole. These observations suggest that resistance to certain compounds may be provided preferentially by certain transporters. Alternatively, induction and repression of transport protein-encoding genes may respond differently to varying stimuli.

iii. Candida Species. *Candida albicans* is a dimorphic fungus capable of causing opportunistic systemic infections in immunocompromised individuals and superficial mucosal infections in healthy individuals (243). Many systemic infections often result in mortality. In fact, 50% of nosocomially acquired fungal infections are caused by this fungus (244, 245). Dimorphism in this fungus is characterized by a yeast-like budding form and a hyphal form that can develop into a mycelium. The hyphal–mycelial stage is often recognized with the onset of pathogenesis. In *C. albicans*, out of 17 predicted ABC proteins, 13 proteins are identified as resembling the ABCG family (Pdr5p-like), two proteins bear significant identity to the ABCB family (α-factor export), and two proteins show significant identity to the ABCC family (oligomycin resistance). In *C. albicans*, Cdr1p (ABCG) provides resistance to azoles and in certain clinical isolates has been shown to be important in resistance to fluconazole, ketoconazole, and itraconazole (245, 246). Indeed, overexpression of the transporter is believed to play a significant role in azole resistance in clinical isolates of the fungus (247). Using photoaffinity labels in competition studies, Cdr1p has been shown to possess separate substrate-binding sites for nystatin and myconazole (248). Site-directed mutagenesis of a conserved cysteine residue in the Walker A motif of Cdr1p suggests that the nucleotide-binding domains respond asymmetrically to substitutions in this amino acid (249–251). Structure and function studies with this protein suggest that specific amino acid residues in the NBD are either indispensable or are important determinants in substrate affinity interactions (251). Furthermore, conformational changes in Cdr1p are unaffected by specific amino acid changes in Cdr1p but are impaired in others (251). In an elegant experiment, Cdr1p was functionally reconstituted into sealed membrane vesicles and shown to carry out drug efflux and also translocate phospholipids (252). Energy-dependent efflux using rhodamine, a fluorescent dye, has been measured in *Candida* species (253).

While the ABC efflux transporters of *C. albicans* provide resistance to azoles, they have not been implicated in resistance to the cell wall inhibiting echinocandins (254). Interestingly, in *C. dublinienisis*, however, the role of Cdr1p in fluconazole resistance has been questioned (255). This suggests that utilization of certain efflux pumps by different species of a fungal genus may vary and may provide clues to alternate mechanisms or pumps employed for drug resistance.

Compared to the animal pathogenic fungi, biochemical and molecular biological experimental data in the plant pathogens are lacking. However, many fungal plant pathogen genomes have been sequenced and provide useful information. Plant pathogen efflux transporters have recently been reviewed (229).

Two broad categories of these pumps have been designated: those involved in secretion of virulence factors and toxins (mycotoxins, e.g., aflatoxin and gliotoxin, and host-specific toxins, e.g., vitorin, botrydil, and cercosporin) and those involved in the removal of plant-derived antifungal agents. The efflux transporters of the first class are not described further in this chapter. Fungal plant pathogens have been shown to gain resistance to fungicides when exposed to low levels of the antifungal agents. Strobilurin is a *quinine outside inhibiting* (QoI) *fungicide* and is particularly useful against two different classes of fungi, the ascomycetes and the basidiomycetes, as well a related group, the oomycetes. The target of QoI fungicides is the mitochondrion, and fungi have evolved different mechanisms to provide resistance to these compounds (256).

In *Pyrenophora tritici-repentis*, exposure of the fungus to sublethal concentrations of QoI fungicides showed increased efflux-based resistance to strobilurin and azole fungicides (257). Analysis of the *P. tritici-repentis* genome shows that it encodes 15 ABC transporters that belong to the ABCG subfamily, 47 ABCB subfamily proteins, and 38 ABCC proteins ($E < 10^{-12}$). *Puccinia graminis-tritici*, a pathogen of wheat, barley, and oats (258), encodes approximately 10 ABCG subfamily proteins, 15 ABCB proteins, and 15 ABCC subfamily transporters. The plant pathogen *Magnaporthe grisea* also shows enhanced ABC efflux transporter gene expression following low-level exposure to antifungal agents (259). This ABC efflux transporter has been shown to be an important pathogenicity factor during the infection of rice (259). Gene expression of ABC1 was enhanced following exposure to toxins and a rice antifungal phytoalexin, sakuranetin. Interestingly, a mutation in the gene did not result in hypersensitivity to antifungal agents tested (inhibitors of protein synthesis, sterol biosynthesis, protein secretion, and a phytoalexin). The mutant was, however, unable to cause disease in rice plant bioassays. These results suggest that in vitro susceptibility testing may not always provide an accurate perspective of the role of a protein or proteins during infection and/or that the MgAbc1 protein may play additional roles during the infection process (e.g., secretion of toxins or efflux of compounds not tested in vitro). The ABC3 transporter in *M. grisea* has been shown through mutational analysis to provide fungal resistance to peroxide and other cytotoxic agents of plant origin during the early stages of infection (260). Analysis of the genome of *M. grisea* indicates that it encodes approximately 15 proteins of the ABCG subfamily, 66 proteins of the ABCB subfamily, and 52 of the ABCC subfamily ($E < 10^{-12}$). The genus *Fusarium* causes disease in over 200 species of plants, many of which are of agricultural importance. Among these are legumes, cereals, and wheat. Three genera—*F. oxysporum, F. graminearium*, and *F. verticilloides*—encode 50, 36, and 31 proteins that belong to the ABCG subfamily, 56, 57, and 55 of the ABCB family, and 56, 57, and 55 ABCC proteins, respectively, signifying the importance of these efflux transporters to these pathogens. Recent documentation of increased benzimidazole resistance in *F. graminearium* in China (261) may be due to enhanced activity of these transporters. *Verticillium dalliae* and *V. albo-atrum* pathogens of woody plants (e.g., ash, elm, oak, maple) encode 22 and 27 ABCG proteins, 49 and 48 ABCB transporters, and 34 and 35 ABCC transporters, respectively. *Ustilago maydis*, a basidiomycete pathogen on grain crops, encodes 13 ABCG proteins and 44 ABCB transporters ($E < 10^{-12}$). *Mycospharella graminicola*, a pathogen of wheat, has been shown experimentally to have resistance to azoles conferred by ABC transporters (262). *Botrytis cinerea*, a pathogen of grapes, ornamentals, fruits, and vegetables encodes approximately 21 ABCG subfamily proteins, 47 ABCB proteins, and 47 ABCC transporters ($E < 10^{-12}$). Reduced susceptibility of this pathogen to the fungicides fludixonil and fenpiclonil is moderated by the ABC efflux transporter, BcatrB (263). A mutant with a disruption in the BcATRB gene showed higher accumulation of fludixonil and reduced accumulation in strains in which the gene was overexpressed. The use of azoles to control plant pathogenic fungi has been considered from the perspective of enhancing resistance to these compounds in animal pathogenic fungi (233, 264). Although the debate is far from over, it is apparent that use of antimicrobial agents in agriculture may result in them leaching into water systems and soils, thereby exposing other pathogens to sublethal concentrations that may result in increased resistance broadly.

The amphibian pathogen *Batrachochytrium dendrobatidis*, a chytrid fungus, has been implicated in the global decline of frog populations (265). This pathogen infects the keratinized mouth parts of tadpoles and keratinized epithelial cells of adults. Little is known of how the fungus infects its host, survives within the host epithelial cells, or survives in the environment. Analysis of the genome sequence shows the presence of

17 ABCG efflux transporters, 64 ABCB transporters, and 33 ABCB subfamily transporters $(E < 10^{-12})$. No experimental data are available on the role of these efflux transporters in the survival and infective stages of the pathogen.

iv. **Phylogenetic Analysis of Fungal ABC Transporters.** Because evolutionary converge of function implies convergence of DNA or protein sequences, a phylogenetic perspective can be used to assess whether pathogenic fungal species are more similar in sequence than is expected from their phylogenetic relationships. Several methods are available for evaluating converge within a phylogenetic context (266–268). One such method is illustrated here, based on a phylogenetic hypothesis (Figure 8) of the relationships among a set of 27 species of ascomycota fungi based on amino acid sequences of the *Candida* ABC efflux transporter (*Cdr1p*), which has been characterized biochemically. About half of the species (14 of 27, indicated by a "P" following the strain identifier) are pathogenic; the remaining 13 species (indicated by "N") are nonpathogens. To the extent that the pathogens are convergent in protein sequence, they should tend to cluster together on the tree; if their sequences are not convergent, the tree should, instead, reflect their evolutionary relationships. There is no obvious clustering of pathogenic and nonpathogenic species, independent of taxonomic relationships. The basal branch of the tree (estimated by midpoint rooting) divides the species into the Saccharomycetales and eurotiomycetidales (Ascomycetes). Although some of the relationships among genera are unexpected (e.g., the inclusion of *Coccidioides*, *Penicillium*, and *Neosartorya* within *Aspergillus*), such a lack of resolution is common in single-gene analyses.

Figure 8. Phylogenetic relationships among 27 species (35 strains) of ascomycota fungi based on amino acid sequences of the Cdr1p encoding gene. Sequences were obtained from BLAST comparisons with the *Candida* Cdr1p sequence and were aligned using the MEGA v4.0.2 interface to ClustalW (275) with a gap-opening penalty of 15 and a gap extension penalty of 6.6 (276). The mean overall Jukes–Cantor distance was 0.06, indicating consistency with the parsimony algorithm. Shown is the bootstrap concensus tree (277), based on 500 bootstrap replicates of the three most parsimonious trees (length = 6130) with missing and ambiguous states deleted (278). Of the 1267 positions in the final data set, 871 were parsimony informative. Consistency index = 0.57, retention index = 0.73, and composite index = 0.41. Maximum parsimony trees were estimated using the close-neighbor-interchange algorithm (276) with search level 2, in which initial trees were obtained with the random addition of sequences (10 replicates). Phylogenetic analyses were conducted in MEGA4 (279).

Xenobiotic efflux in bacteria and fungi

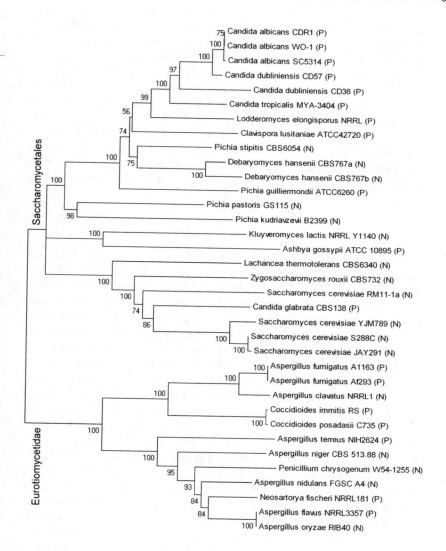

Ideally, the tree should be compared against a comparable phylogenetic tree, for the same species, based on multigene or phylogenomic analyses (269–273). Because such a phylogenetic study has yet to be done, the tree of Figure 8 can be compared more crudely with a classification based on the Linnean hierarchy (derived from the Global Biodiversity Information Facility; www.gbif.net/species/browse/taxon/13140889), using the method of Podani et al. (274) to detect hierarchical levels at which the two dendrograms show maximum agreement. The resulting comparison indicated strong similarities at all levels in the trees, with marginally significant disagreement only in the dispersion of *Aspergillus* species. Thus, the phylogenetic tree based on *Cdr1p* sequences is strongly consistent with the accepted Linnean classification, with no evidence supporting either sequence convergence or pathogenic patterns conserved among species (273).

2. Fungal MFS Transporters

The major facilitator superfamily (MFS) proteins in the fungi are relatively poorly understood compared to the ABC transport proteins. A recent review of fungal MFS proteins and their roles in fungal physiology underscores this point (280). Major facilitator superfamily efflux transporters possess either 12 (drug/proton antiporter, DHA1) or 14 (DHA2) TMDs, with a large cytoplasmic loop between domains 6 and 7. Analysis of the *S. cerevisiae* genome sequence indicates the presence of approximately 85 MFS transporters, of which 21% are involved in efflux (Table 9). Only three of these proteins have been characterized functionally. Among other fungi the percentages of MFS–MDR transporters range between 29 and 39% (Table 9). Curiously, in the fungal-related group, the oomycetes (*P. infestans*), less than 2% of the MFS proteins are involved in efflux. This is particularly interesting because the percentage of ABC efflux transporters in *P. infestans* is not notably different from the other fungi.

One of the first MFS proteins involved in drug resistance in fungi was identified in *S. cerevisiae* through mutant selection and complementation experiments (281). The MFS efflux transporter Dtr1p, known to provide resistance to organic acids and antimalarial drugs, has also been shown to play an essential role in *S. cerevisiae* spore wall maturation by secreting the building block, bisformyl tyrosine, from the cytoplasm (282). This study provides evidence for an additional role of the multidrug transporter in the development of the fungus. In the fission yeast *S. pombe*, the MFS drug efflux transporter Caf5, together with a brefeldin A–resistance protein-encoding gene, Bfr1, was found to be up-regulated in cells where Int6CT was overexpressed (283). Int6CT is a C-terminal fragment of translation initiation factor Int6 that may promote transcriptional activity of Pap1. Pap1 plays a role in oxidative stress resistance and enhancement of HBA2 efflux pump gene expression (284).

In *A. fumigatus*, itraconozole resistance is moderated in part by MFS AfuMDR3 (285). Expression of this gene was enhanced following exposure to the azole. As a dermatophyte, *A. fumigatus* infects keratinized cells and keratin-rich tissue. Sulfite efflux is required to provide a reducing environment for keratin breakdown where cystine is converted to cysteine and *S*-sulfocysteine. Reduced proteins are more accessible to secreted fungal proteases (286). Sulfite secretion is moderated by a MFS tellurite-resistance/dicarboxylate transporter (TDT) family protein. A study on the transportome of *C. albicans* has recently been published (287). Of the 95 putative MFS-encoding genes, 22 and 9 representatives represent the DHA1 and DHA2 subfamilies, respectively. *Candida* strains demonstrate resistance to a variety of antifungal agents. The protein CaMdr1p (BEN^R) provides resistance to benomyl, fluconozole, and methotrexate (288–290). Analysis of a mutation in MDR1 that impairs expression of the gene has demonstrated its importance in the virulence of the fungus (291, 292). Recent studies to elucidate the structural and functional aspects of domains of CaMdr1p utilized both tagging of the protein with green-fluorescent protein (GFP) and alanine-scanning mutagenesis of transmembrane domain 5, believed to contribute to drug/H$^+$ transport (293). These studies demonstrated that this transmembrane domain, bearing the conserve motif G(X$_6$) G(X$_3$) G(X$_3$) GP(X$_2$) G, is essential for drug/H$^+$ transport. Resistance to fluconazole, cycloheximide, 4-nitroqinolone, and phenanthroline has also been shown to be provided by the FLU1, TMP1, and TMP2 genes of this family (294, 295). An interesting observation using a variety of clinical isolates of *C. albicans* suggests that overexpression of FLU1 did not always correlate with drug resistance (296).

Among the plant pathogens, a novel MFS transporter from *Botrytis cinerea*, Bcmfs1, has been shown to provide tolerance toward the natural toxins camptothecin and cercosporin as well as fungicides (297). In *Mycosphaerella graminicola*, the MFS drug efflux transporter MgMFS1 (DHA14) provides resistance to fungicides and naturally occurring toxins, particularly strobilurin and cercosporin (298). Interestingly, in bioassays, virulence of the isolate harboring a disruption in the gene was observed to be similar to the control parent strain. In field studies, expression of the gene was found to be elevated under conditions where a sublethal concentration of the fungicide trifloxistrobin was present (299). These two studies demonstrate that some drug efflux transporters play important roles in survival of the fungus when present outside the host.

C.   REGULATION OF EFFLUX PUMP GENE EXPRESSION

Recent reviews have discussed the regulation of multidrug gene expression in fungi (207, 229). Gulshan and Moye-Rowley (207) provide an excellent overview of the significant regulatory interactions governing pleitropic drug resistance in *S. cerevisiae*. Coleman and Mylonakis provide a much-needed overview of drug efflux transporters in plant pathogens and their genomics. Because of its clinical importance and relative ease of genetic manipulation, studies with *C. albicans* are in abundance. Our understanding of the regulation of ABC transporters in *C. albicans* has come from studies with clinical isolates showing increased drug resistance and through mutational analyses. The transcriptional regulatory protein Tac1p, known to moderate expression of CDR1 and CDR2 genes, has been shown to harbor a single-point mutation that confers increased drug resistance through these pumps (300). Tac1p has also been shown to play a role in the oxidative stress response and lipid metabolism in part through interaction with its own promoter (301). Transcription factor Ndp80p also regulates Cdr1p (302). A negative regulator, Rep1p, first identified in *S. cerevisiae*, has been shown to moderate Mdr1p efflux pump gene expression in *C. albicans* (261). Overexpression of this negative regulator heterologously in *S. cerevisiae* increased susceptibility to fluconazole. Furthermore, a mutation in the gene encoding the protein in *C. albicans* enhanced drug resistance. Another regulatory protein, Mrr1p (multidrug resistance regulator), has been shown to moderate *C. albicans* MDR1 gene expression. Clinical isolates of the fungus with increased resistance to fluconazole showed an ability to coordinate up-regulation of both the transcription factor and the ABC efflux pump (303).

Analysis of the gene encoding the regulatory protein showed that two-point mutations contribute to high-level constitutive expression of the gene, which results in increased drug resistance. In *C. albicans*, MDR1 is under complex regulation involving oxidative stress response, drug exposure, and multiple transcriptional regulators, including Cap1, Mrr1p, Upc2p, and

Mcm1p (304–306). Other studies with *C. albicans* show that uncoupling oxidative phosphorylation in petite mutants of the fungus resulted in reduced sensitivity to flucoazole and voriconazole (but no change in the resistance to ketoconazole, itraconazole, and amphotericin B), and this phenotype could be attributed to overexpression of MDR1 (291). Recent studies with *C. glabrata* show that CgCdr1-, CgCdr2-, and CgSnq2-encoding genes, which moderate azole resistance, may not be coordinately regulated. Gain-of-function (GOF) mutations in the transcription factor encoding gene CgPDR1 increased azole tolerance in vitro through differential expression of CgCdr1-, CgCdr2-, and CgSnq2-encoding genes. Furthermore, strains carrying the GOF mutations also showed enhanced virulence in an in vivo model compared to wild-type strains (307).

While many studies rely on gene disruptions to assess contributions to a specific phenotype or capability, a recent study with *S. cerevisiae* highlights an important issue related to multiple drug resistance in fungi. Deletion of YOR1- and SNQ2-specific regions resulted in increased efflux in Pdr5p efflux substrates. Additionally, increased transcript production and resistance to Yor1p and Snq1p substrates increased in a PDR5 deletion strain (308).

D. IDENTIFICATION OF NOVEL AND USEFUL EFFLUX PUMP INHIBITORS

The use of chemicals that can work synergistically with useful antifungal agents can reduce concentrations of the drugs currently used, and possibly reduce the likelihood of exposure-based drug resistance. A 1.8 million member D-octapeptide combinatorial peptide library was recently used to screen for inhibitors of an ABC (Pdr5p) hyperexpressing strain of *S. cerevisiae* that carried deletions in five other ABC efflux pumps (309). This study identified a noncompetitive inhibitor of ATPase activity and sensitized the strain to fluconazole and increased the permeability of the plasma membrane to rhodamine. A naturally occurring compound tetrandrine was shown to increase rhodamine 123 accumulation in *C. albicans* (310). Cerulenin is an inhibitor of fatty acid synthesis and substrate for ABC and MFS efflux transport in *C. albicans*. Structural analogs of the compound were screened for their ability to increase sensitivity to brefeldin A, and several were found effective against CaMdr1p-mediated resistance (311). A natural product of turmeric, curcumin, known to block ABC transport activity (ABCB1, ABCC1, and ABCG2) in mammalian cancer cells, was shown to be effective in vitro (312). In these studies, rhodamine 6G (R6G) efflux was measured in *S. cerevisiae* cells expressing *C. albicans* Cdr1p and Cdr2p proteins. Treatment with curcumin resulted in decreased extracellular R6G in both expressing cell lines, suggesting that the compound impaired efflux pump function. More detailed studies with curcumin also revealed that this compound enhances the effectiveness of certain antifungal products (ketoconazole, miconazole, and itraconozole) but not others (fluconazole, voriconazole, anisomycin, and cyclohexamide). These observations suggest that cucurmin may be a valuable additive when used with conventional antifungal drugs. Curcumin may also be used in more detailed structure–function studies of efflux pumps to identify amino acid residues essential for drug interaction and transport.

E. BIOFILMS

Microbes are known to form biofilms on surfaces. Biofilms represent a communal aggregate of microorganisms in a matrix with varying amounts of extracellular polysaccharide, protein, and nucleic acid. Microbes in biofilms show loss of motility functions, lowered metabolism, and en-hanced drug and antibiotic resistance (313). The medical importance of biofilms cannot be overemphasized. Biofilms form on in-dwelling medical devices and in wounds. Thus, treatment of microbes in biofilms with drugs and antibiotics is a continuous challenge in medicine. *A. fumigatus* biofilms were studied in vitro and on bronchial epithelial cells for their resistance to a variety of drugs. Azoles were found to be more effective than echino-candins against mature (48 h) biofilms (314). Overall, less antifungal drug susceptibility was observed in biofilms than in planktonic cells. Antifungal drug resistance in *Candida* species has been reviewed recently (315, 316). The role of drug efflux pumps, in biofilms, however, remains somewhat enigmatic (317). Efflux pump mutants (Cdr1p, Cdr2p, Mdr1p) of *C. albicans* were allowed to form biofilms and were studied with the parental nonmutant at three stages (6, 12, and 48 h) for their resistance to fluconazole (318). The results showed that efflux pumps play a role in resistance to this antibiotic in the early but not at the later 12- and 48-h stages. In planktonic cells, however, expression of the three pumps was observed at only 12 and 48 h, suggesting that in *C. albicans*, efflux pump gene expression is regulated differentially and in a phase-specific manner. Gene expression of CDR- and MDR-encoding genes was examined in planktonic and biofilm cells of *C. albicans* and found to be up-regulated in biofilms (319). When single- and double-mutant cells were examined for fluconazole resistance, planktonic cells showed predicted sensitivity, but biofilm cells were resistant. These results suggest a multifactorial-based mechanism of resistance when cells are in a biofilm (319). Research with *C. albicans* on cell density influence on drug resistance suggests that azole tolerance at high cell densities (found in biofilms) cannot be attributed to drug efflux pumps (320). A strain lacking functional drug efflux pump-encoding genes CDR1, CDR2, and MDR1 was susceptible to azoles at low cell densities but was resistant at high cell densities and when present in a biofilm. Cell wall protein production in *C. albicans* has been the subject of a recent review, and linkage among cell wall proteins, biofilm formation, and drug resistance has been postulated (321).

References

1. Fischbach, M. A., and Walsh, C. T. (2009) Antibiotics for emerging pathogens, *Science* *325*, 1089–1093.

2. McMurry, L., Petrucci, R., and Levy, S. (1980) Active efflux of tetracycline encoded by four genetically different tetracycline resistance determinants in *Escherichia coli*, *Proc. Natl. Acad. Sci. USA 77*, 3974–3977.

3. Paulsen, I. T. (2003) Multidrug efflux pumps and resistance: regulation and evolution, *Curr. Opin. Microbiol. 6*, 446–451.

4. Alonso, A., Rojo, F., and Martínez, J. L. (1999) Environmental and clinical isolates of *Pseudomonas aeruginosa* show pathogenic and biodegradative properties irrespective of their origin, *Environ. Microbiol. 1*, 421–430.

5. Grkovic, S., Brown, M. H., and Skurray, R. A. (2002) Regulation of bacterial drug export systems, *Microbiol. Mol. Biol. Rev. 66*, 671–701.

6. Martínez, J. L., Sánchez, M. B., Martinez-Solano, L., Hernández, A., Garmendia, L., Fajardo, A., and Alvarez-Ortega, C. (2009) Functional role of bacterial multidrug efflux pumps in microbial natural ecosystems, *FEMS Microbiol. Rev. 33*, 430–449.

7. Paulsen, I. T., Skurray, R. A., Tam, R., Saier, M. H., Jr., Turner, R. J., Weiner, J. H., Goldberg, E. B., and Grinius, L. L. (1996) The SMR family: a novel family of multidrug efflux proteins involved with the efflux of lipophilic drugs, *Mol. Microbiol. 19*, 1167–1175.

8. Brown, M. H., Paulsen, I. T., and Skurray, R. A. (1999) The multidrug efflux protein NorM is a prototype of a new family of transporters, *Mol. Microbiol. 31*, 394–395.

9. Hvorup, R. N., Winnen, B., Chang, A. B., Jiang, Y., Zhou, X. F., and Saier, M. H., Jr. (2003) The multidrug/oligosaccharidyl-lipid/polysaccharide (MOP) exporter superfamily, *Eur. J. Biochem. 270*, 799–813.

10. Marger, M. D., and Saier, M. H., Jr. (1993) A major superfamily of transmembrane facilitators that catalyse uniport, symport and antiport, *Trends Biochem. Sci. 18*, 13–20.

11. Saier, M. H., Jr., Tam, R., Reizer, A., and Reizer, J. (1994) Two novel families of bacterial membrane proteins concerned with nodulation, cell division and transport, *Mol. Microbiol. 11*, 841–847.

12. Lubelski, J., Konings, W. N., and Driessen, A. J. (2007) Distribution and physiology of ABC-type transporters contributing to multidrug resistance in bacteria, *Microbiol. Mol. Biol. Rev. 71*, 463–476.

13. Saier, M. H., Jr., and Paulsen, I. T. (2001) Phylogeny of multidrug transporters, *Semin. Cell Dev. Biol. 12*, 205–213.

14. Saier, M. H., Jr. (2000) A functional-phylogenetic classification system for transmembrane solute transporters, *Microbiol. Mol. Biol. Rev. 64*, 354–411.

15. Pao, S. S., Paulsen, I. T., and Saier, M. H., Jr. (1998) Major facilitator superfamily, *Microbiol. Mol. Biol. Rev. 62*, 1–34.

16. Paulsen, I. T., Brown, M. H., and Skurray, R. A. (1996) Proton-dependent multidrug efflux systems, *Microbiol. Rev. 60*, 575–608.

17. Saier, M. H., Jr., Beatty, J. T., Goffeau, A., Harley, K. T., Heijne, W. H., Huang, S. C., Jack, D. L., Jahn, P. S., Lew, K., Liu, J., et al. (1999) The major facilitator superfamily, *J. Mol. Microbiol. Biotechnol. 1*, 257–279.

18. Li, X. Z., and Nikaido, H. (2004) Efflux-mediated drug resistance in bacteria, *Drugs 64*, 159–204.

19. Putman, M., van Veen, H. W., and Konings, W. N. (2000) Molecular properties of bacterial multidrug transporters, *Microbiol. Mol. Biol. Rev. 64*, 672–693.

20. Saier, M. H., Jr., Yen, M. R., Noto, K., Tamang, D. G., and Elkan, C. (2009) The Transporter Classification Database: recent advances, *Nucleic Acids Res. 37*, D274–D278.

21. Van Bambeke, F., Glupczynski, Y., Plésiat, P., Pechere, J. C., and Tulkens, P. M. (2003) Antibiotic efflux pumps in prokaryotic cells: occurrence, impact on resistance and strategies for the future of antimicrobial therapy, *J. Antimicrob. Chemother. 51*, 1055–1065.

22. Lynch, A. S. (2006) Efflux systems in bacterial pathogens: an opportunity for therapeutic intervention? An industry view, *Biochem. Pharmacol. 71*, 949–956.

23. Chen, C. J., Chin, J. E., Ueda, K., Clark, D. P., Pastan, I., Gottesman, M. M., and Roninson, I. B. (1986) Internal duplication and homology with bacterial transport proteins in the *mdr1* (P-glycoprotein) gene from multidrug-resistant human cells, *Cell 47*, 381–389.

24. Gottesman, M. M., Pastan, I., and Ambudkar, S. V. (1996) P-glycoprotein and multidrug resistance, *Curr. Opin. Genet. Dev. 6*, 610–617.

25. Cole, S. P., Bhardwaj, G., Gerlach, J. H., Mackie, J. E., Grant, C. E., Almquist, K. C., Stewart, A. J., Kurz, E. U., Duncan, A. M., and Deeley, R. G. (1992) Overexpression of a transporter gene in a multidrug-resistant human lung cancer cell line, *Science 258*, 1650–1654.

26. Legare, D., Cayer, S., Singh, A. K., Richard, D., Papadopoulou, B., and Ouellette, M. (2001) ABC proteins of *Leishmania*, *J. Bioenerg. Biomembr. 33*, 469–474.

27. van Veen, H. W., Venema, K., Bolhuis, H., Oussenko, I., Kok, J., Poolman, B., Driessen, A. J., and Konings, W. N. (1996) Multidrug resistance mediated by a bacterial homolog of the human multidrug transporter MDR1, *Proc. Natl. Acad. Sci. USA 93*, 10668–10672.

28. Robertson, G. T., Doyle, T. B., and Lynch, A. S. (2005) Use of an efflux-deficient *Streptococcus pneumoniae* strain panel to identify ABC-class multidrug transporters

involved in intrinsic resistance to antimicrobial agents, *Antimicrob. Agents Chemother.* *49*, 4781–4783.

29. Piddock, L. J. (2006) Clinically relevant chromosomally encoded multidrug resistance efflux pumps in bacteria, *Clin. Microbiol. Rev. 19*, 382–402.

30. van Veen, H. W., Margolles, A., Muller, M., Higgins, C. F., and Konings, W. N. (2000) The homodimeric ATP-binding cassette transporter LmrA mediates multidrug transport by an alternating two-site (two-cylinder engine) mechanism, *EMBO J. 19*, 2503–2514.

31. Rosenberg, M. F., Callaghan, R., Ford, R. C., and Higgins, C. F. (1997) Structure of the multidrug resistance P-glycoprotein to 2.5 nm resolution determined by electron microscopy and image analysis, *J. Biol. Chem. 272*, 10685–10694.

32. van Veen, H. W., Higgins, C. F., and Konings, W. N. (2001) Molecular basis of multidrug transport by ATP-binding cassette transporters: a proposed two-cylinder engine model, *J. Mol. Microbiol. Biotechnol. 3*, 185–192.

33. Procko, E., O'Mara, M. L., Bennett, W. F., Tieleman, D. P., and Gaudet, R. (2009) The mechanism of ABC transporters: general lessons from structural and functional studies of an antigenic peptide transporter, *FASEB J. 23*, 1287–1302.

34. Murakami, S., and Yamaguchi, A. (2003) Multidrug-exporting secondary transporters, *Curr. Opin. Struct. Biol. 13*, 443–452.

35. Lin, H. T., Bavro, V. N., Barrera, N. P., Frankish, H. M., Velamakanni, S., van Veen, H. W., Robinson, C. V., Borges-Walmsley, M. I., and Walmsley, A. R. (2009) MacB ABC transporter is a dimer whose ATPase activity and macrolide-binding capacity are regulated by the membrane fusion protein MacA, *J. Biol. Chem. 284*, 1145–1154.

36. Schmitz, F. J., Sadurski, R., Kray, A., Boos, M., Geisel, R., Kohrer, K., Verhoef, J., and Fluit, A. C. (2000) Prevalence of macrolide-resistance genes in *Staphylococcus aureus* and *Enterococcus faecium* isolates from 24 European university hospitals, *J. Antimicrob. Chemother. 45*, 891–894.

37. Portillo, A., Ruiz-Larrea, F., Zarazaga, M., Alonso, A., Martínez, J. L., and Torres, C. (2000) Macrolide resistance genes in *Enterococcus* spp., *Antimicrob. Agents Chemother. 44*, 967–971.

38. Pozzi, G., Iannelli, F., Oggioni, M. R., Santagati, M., and Stefani, S. (2004) Genetic elements carrying macrolide efflux genes in streptococci, *Curr. Drug Targets Infect. Disord. 4*, 203–206.

39. Daly, M. M., Doktor, S., Flamm, R., and Shortridge, D. (2004) Characterization and prevalence of MefA, MefE, and the associated *msr(D)* gene in *Streptococcus pneumoniae* clinical isolates, *J. Clin. Microbiol. 42*, 3570–3574.

40. Reynolds, E., Ross, J. I., and Cove, J. H. (2003) Msr(A) and related macrolide/streptogramin resistance determinants: incomplete transporters? *Int. J. Antimicrob. Agents 22*, 228–236.

41. Dina, J., Malbruny, B., and Leclercq, R. (2003) Nonsense mutations in the lsa-like gene in *Enterococcus faecalis* isolates susceptible to lincosamides and streptogramins A, *Antimicrob. Agents Chemother. 47*, 2307–2309.

42. Lee, E. W., Huda, M. N., Kuroda, T., Mizushima, T., and Tsuchiya, T. (2003) EfrAB, an ABC multidrug efflux pump in *Enterococcus faecalis*, *Antimicrob. Agents Chemother. 47*, 3733–3738.

43. Kehrenberg, C., Ojo, K. K., and Schwarz, S. (2004) Nucleotide sequence and organization of the multiresistance plasmid pSCFS1 from *Staphylococcus sciuri*, *J. Antimicrob. Chemother.* *54*, 936–939.

44. Kobayashi, N., Nishino, K., and Yamaguchi, A. (2001) Novel macrolide-specific ABC-type efflux transporter in *Escherichia coli*, *J. Bacteriol.* *183*, 5639–5644.

45. Poelarends, G. J., Mazurkiewicz, P., and Konings, W. N. (2002) Multidrug transporters and antibiotic resistance in *Lactococcus lactis*, *Biochim. Biophys. Acta 1555*, 1–7.

46. Huda, N., Lee, E. W., Chen, J., Morita, Y., Kuroda, T., Mizushima, T., and Tsuchiya, T. (2003) Molecular cloning and characterization of an ABC multidrug efflux pump, VcaM, in Non-O1 *Vibrio cholerae*, *Antimicrob. Agents Chemother.* *47*, 2413–2417.

47. Pasca, M. R., Guglierame, P., Arcesi, F., Bellinzoni, M., De Rossi, E., and Riccardi, G. (2004) Rv2686c-Rv2687c-Rv2688c, an ABC fluoroquinolone efflux pump in *Mycobacterium tuberculosis*, *Antimicrob. Agents Chemother.* *48*, 3175–3178.

48. Raherison, S., Gonzalez, P., Renaudin, H., Charron, A., Bebear, C., and Bebear, C. M. (2005) Increased expression of two multidrug transporter-like genes associated with ethidium bromide and ciprofloxacin resistance in *Mycoplasma hominis*, *Antimicrob. Agents Chemother.* *49*, 421–424.

49. Muth, T. R., and Schuldiner, S. (2000) A membrane-embedded glutamate is required for ligand binding to the multidrug transporter EmrE, *EMBO J. 19*, 234–240.

50. Gutman, N., Steiner-Mordoch, S., and Schuldiner, S. (2003) An amino acid cluster around the essential Glu-14 is part of the substrate- and proton-binding domain of EmrE, a multidrug transporter from *Escherichia coli*, *J. Biol. Chem. 278*, 16082–16087.

51. Koteiche, H. A., Reeves, M. D., and McHaourab, H. S. (2003) Structure of the substrate binding pocket of the multidrug transporter EmrE: site-directed spin labeling of transmembrane segment 1, *Biochemistry 42*, 6099–6105.

52. Schuldiner, S., Granot, D., Mordoch, S. S., Ninio, S., Rotem, D., Soskin, M., Tate, C. G., and Yerushalmi, H. (2001) Small is mighty: EmrE, a multidrug transporter as an experimental paradigm, *News Physiol. Sci. 16*, 130–134.

53. Soskine, M., Steiner-Mordoch, S., and Schuldiner, S. (2002) Crosslinking of membrane-embedded cysteines reveals contact points in the EmrE oligomer, *Proc. Natl. Acad. Sci. USA 99*, 12043–12048.

54. Leelaporn, A., Paulsen, I. T., Tennent, J. M., Littlejohn, T. G., and Skurray, R. A. (1994) Multidrug resistance to antiseptics and disinfectants in coagulase-negative. staphylococci, *J. Med. Microbiol. 40*, 214–220.

55. Grinius, L., Dreguniene, G., Goldberg, E. B., Liao, C. H., and Projan, S. J. (1992) A staphylococcal multidrug resistance gene product is a member of a new protein family, *Plasmid 27*, 119–129.

56. Littlejohn, T. G., Paulsen, I. T., Gillespie, M. T., Tennent, J. M., Midgley, M., Jones, I. G., Purewal, A. S., and Skurray, R. A. (1992) Substrate specificity and energetics of antiseptic and disinfectant resistance in *Staphylococcus aureus*, *FEMS Microbiol. Lett. 74*, 259–265.

57. Lewis, K. (1994) Multidrug resistance pumps in bacteria: variations on a theme, *Trends Biochem. Sci. 19*, 119–123.

58. Masaoka, Y., Ueno, Y., Morita, Y., Kuroda, T., Mizushima, T., and Tsuchiya, T. (2000) A two-component multidrug efflux pump, EbrAB, in *Bacillus subtilis*, *J. Bacteriol. 182*, 2307–2310.

59. Ninio, S., Rotem, D., and Schuldiner, S. (2001) Functional analysis of novel multidrug transporters from human pathogens, *J. Biol. Chem. 276*, 48250–48256.

60. Rotem, D., Sal-man, N., and Schuldiner, S. (2001) In vitro monomer swapping in EmrE, a multidrug transporter from *Escherichia coli*, reveals that the oligomer is the functional unit, *J. Biol. Chem. 276*, 48243–48249.

61. Rotem, D., and Schuldiner, S. (2004) EmrE, a multidrug transporter from *Escherichia coli*, and transports monovalent and divalent substrates with the same stoichiometry, *J. Biol. Chem. 279*, 48787–48793.

62. Li, X. Z., Zhang, L., and Nikaido, H. (2004) Efflux pump-mediated intrinsic drug resistance in *Mycobacterium smegmatis*, *Antimicrob. Agents Chemother. 48*, 2415–2423.

63. Yerushalmi, H., Lebendiker, M., and Schuldiner, S. (1995) EmrE, an *Escherichia coli* 12-kDa multidrug transporter, exchanges toxic cations and H^+ and is soluble in organic solvents, *J. Biol. Chem. 270*, 6856–6863.

64. Jack, D. L., Storms, M. L., Tchieu, J. H., Paulsen, I. T., and Saier, M. H., Jr. (2000) A broad-specificity multidrug efflux pump requiring a pair of homologous SMR-type proteins, *J. Bacteriol. 182*, 2311–2313.

65. Li, X. Z., Poole, K., and Nikaido, H. (2003) Contributions of MexAB-OprM and an EmrE homolog to intrinsic resistance of *Pseudomonas aeruginosa* to aminoglycosides and dyes, *Antimicrob. Agents Chemother. 47*, 27–33.

66. Kucken, D., Feucht, H., and Kaulfers, P. (2000) Association of qacE and qacEDelta1 with multiple resistance to antibiotics and antiseptics in clinical isolates of gram-negative bacteria, *FEMS Microbiol. Lett. 183*, 95–98.

67. Omote, H., Hiasa, M., Matsumoto, T., Otsuka, M., and Moriyama, Y. (2006) The MATE proteins as fundamental transporters of metabolic and xenobiotic organic cations, *Trends Pharmacol. Sci. 27*, 587–593.

68. Morita, Y., Kodama, K., Shiota, S., Mine, T., Kataoka, A., Mizushima, T., and Tsuchiya, T. (1998) NorM, a putative multidrug efflux protein, of *Vibrio parahaemolyticus* and its homolog in *Escherichia coli*, *Antimicrob. Agents Chemother. 42*, 1778–1782.

69. Kuroda, T., and Tsuchiya, T. (2009) Multidrug efflux transporters in the MATE family, *Biochim. Biophys. Acta 1794*, 763–768.

70. Su, X. Z., Chen, J., Mizushima, T., Kuroda, T., and Tsuchiya, T. (2005) AbeM, an H^+-coupled *Acinetobacter baumannii* multidrug efflux pump belonging to the MATE family of transporters, *Antimicrob. Agents Chemother. 49*, 4362–4364.

71. McAleese, F., Petersen, P., Ruzin, A., Dunman, P. M., Murphy, E., Projan, S. J., and Bradford, P. A. (2005) A novel MATE family efflux pump contributes to the reduced susceptibility of laboratory-derived *Staphylococcus aureus* mutants to tigecycline, *Antimicrob. Agents Chemother. 49*, 1865–1871.

72. Long, F., Rouquette-Loughlin, C., Shafer, W. M., and Yu, E. W. (2008) Functional cloning and characterization of the multidrug efflux pumps NorM from *Neisseria gonorrhoeae* and YdhE from *Escherichia coli*, *Antimicrob. Agents Chemother. 52*, 3052–3060.

73. Rouquette-Loughlin, C., Dunham, S. A., Kuhn, M., Balthazar, J. T., and Shafer, W. M. (2003) The NorM efflux pump of *Neisseria gonorrhoeae* and *Neisseria meningitidis* recognizes antimicrobial cationic compounds, *J. Bacteriol. 185*, 1101–1106.

74. Kaatz, G. W., McAleese, F., and Seo, S. M. (2005) Multidrug resistance in *Staphylococcus aureus* due to overexpression of a novel multidrug and toxin extrusion (MATE) transport protein, *Antimicrob. Agents Chemother. 49*, 1857–1864.

75. Yang, S., Clayton, S. R., and Zechiedrich, E. L. (2003) Relative contributions of the AcrAB, MdfA and NorE efflux pumps to quinolone resistance in *Escherichia coli*, *J. Antimicrob. Chemother. 51*, 545–556.

76. Miyamae, S., Ueda, O., Yoshimura, F., Hwang, J., Tanaka, Y., and Nikaido, H. (2001) A MATE family multidrug efflux transporter pumps out fluoroquinolones in *Bacteroides thetaiotaomicron*, *Antimicrob. Agents Chemother. 45*, 3341–3346.

77. Xu, X. J., Su, X. Z., Morita, Y., Kuroda, T., Mizushima, T., and Tsuchiya, T. (2003) Molecular cloning and characterization of the HmrM multidrug efflux pump from *Haemophilus influenzae* Rd, *Microbiol. Immunol. 47*, 937–943.

78. He, G. X., Kuroda, T., Mima, T., Morita, Y., Mizushima, T., and Tsuchiya, T. (2004) An H(+)-coupled multidrug efflux pump, PmpM, a member of the MATE family of transporters, from *Pseudomonas aeruginosa*, *J. Bacteriol. 186*, 262–265.

79. Chen, J., Morita, Y., Huda, M. N., Kuroda, T., Mizushima, T., and Tsuchiya, T. (2002) VmrA, a member of a novel class of Na(+)-coupled multidrug efflux pumps from *Vibrio parahaemolyticus*, *J. Bacteriol. 184*, 572–576.

80. Dridi, L., Tankovic, J., and Petit, J. C. (2004) CdeA of *Clostridium difficile*, a new multidrug efflux transporter of the MATE family, *Microb. Drug Resist. 10*, 191–196.

81. Huda, M. N., Morita, Y., Kuroda, T., Mizushima, T., and Tsuchiya, T. (2001) Na+-driven multidrug efflux pump VcmA from *Vibrio cholerae* non-O1, a non-halophilic bacterium, *FEMS Microbiol. Lett. 203*, 235–239.

82. Huda, M. N., Chen, J., Morita, Y., Kuroda, T., Mizushima, T., and Tsuchiya, T. (2003) Gene cloning and characterization of VcrM, a Na+-coupled multidrug efflux pump, from *Vibrio cholerae* non-O1, *Microbiol. Immunol. 47*, 419–427.

83. Begum, A., Rahman, M. M., Ogawa, W., Mizushima, T., Kuroda, T., and Tsuchiya, T. (2005) Gene cloning and characterization of four MATE family multidrug efflux pumps from *Vibrio cholerae* non-O1, *Microbiol. Immunol. 49*, 949–957.

84. Nishino, K., Latifi, T., and Groisman, E. A. (2006) Virulence and drug resistance roles of multidrug efflux systems of *Salmonella enterica* serovar *typhimurium*, *Mol. Microbiol. 59*, 126–141.

85. Saier, M. H., Jr. (2003) Tracing pathways of transport protein evolution, *Mol. Microbiol. 48*, 1145–1156.

86. Ren, Q., and Paulsen, I. T. (2007) Large-scale comparative genomic analyses of cytoplasmic membrane transport systems in prokaryotes, *J. Mol. Microbiol. Biotechnol. 12*, 165–179.

87. Maloney, P. C. (1990) Microbes and membrane biology, *FEMS Microbiol. Rev. 7*, 91–102.

88. Yin, Y., He, X., Szewczyk, P., Nguyen, T., and Chang, G. (2006) Structure of the multidrug transporter EmrD from *Escherichia coli*, *Science 312*, 741–744.

89. Huang, Y., Lemieux, M. J., Song, J., Auer, M., and Wang, D. N. (2003) Structure and mechanism of the glycerol-3-phosphate transporter from *Escherichia coli*, *Science 301*, 616–620.

90. Abramson, J., Smirnova, I., Kasho, V., Verner, G., Kaback, H. R., and Iwata, S. (2003) Structure and mechanism of the lactose permease of *Escherichia coli*, *Science 301*, 610–615.

91. Woolley, R. C., Vediyappan, G., Anderson, M., Lackey, M., Ramasubramanian, B., Jiangping, B., Borisova, T., Colmer, J. A., Hamood, A. N., McVay, C. S., and Fralick, J. A. (2005) Characterization of the *Vibrio cholerae* vceCAB multiple-drug resistance efflux operon in *Escherichia coli*, *J. Bacteriol. 187*, 5500–5503.

92. Colmer, J. A., Fralick, J. A., and Hamood, A. N. (1998) Isolation and characterization of a putative multidrug resistance pump from *Vibrio cholerae*, *Mol. Microbiol. 27*, 63–72.

93. Lomovskaya, O., and Lewis, K. (1992) Emr, an *Escherichia coli* locus for multidrug resistance, *Proc. Natl. Acad. Sci. USA 89*, 8938–8942.

94. Tanabe, M., Szakonyi, G., Brown, K. A., Henderson, P. J., Nield, J., and Byrne, B. (2009) The multidrug resistance efflux complex, EmrAB from *Escherichia coli* forms a dimer in vitro, *Biochem. Biophys. Res. Commun. 380*, 338–342.

95. Schwarz, S., Kehrenberg, C., Doublet, B., and Cloeckaert, A. (2004) Molecular basis of bacterial resistance to chloramphenicol and florfenicol, *FEMS Microbiol. Rev. 28*, 519–542.

96. Haroche, J., Morvan, A., Davi, M., Allignet, J., Bimet, F., and El Solh, N. (2003) Clonal diversity among streptogramin A-resistant *Staphylococcus aureus* isolates collected in French hospitals, *J. Clin. Microbiol. 41*, 586–591.

97. Poole, K. (2000) Efflux-mediated resistance to fluoroquinolones in gram-positive bacteria and the mycobacteria, *Antimicrob. Agents Chemother. 44*, 2595–2599.

98. Gill, M. J., Brenwald, N. P., and Wise, R. (1999) Identification of an efflux pump gene, *pmrA*, associated with fluoroquinolone resistance in *Streptococcus pneumoniae*, *Antimicrob. Agents Chemother. 43*, 187–189.

99. Poole, K. (2004) Efflux-mediated multiresistance in gram-negative bacteria, *Clin. Microbiol. Infect. 10*, 12–26.

100. Poole, K. (2005) Efflux-mediated antimicrobial resistance, *J. Antimicrob. Chemother. 56*, 20–51.

101. Tanabe, H., Yamasak, K., Furue, M., Yamamoto, K., Katoh, A., Yamamoto, M., Yoshioka, S., Tagami, H., Aiba, H. A., and Utsumi, R. (1997) Growth phase-dependent transcription of *emrKY*, a homolog of multidrug efflux *emrAB* genes of *Escherichia coli*, is induced by tetracycline, *J. Gen. Appl. Microbiol. 43*, 257–263.

102. Jonas, B. M., Murray, B. E., and Weinstock, G. M. (2001) Characterization of emeA, a NorA homolog and multidrug resistance efflux pump, in *Enterococcus faecalis*, *Antimicrob. Agents Chemother. 45*, 3574–3579.

103. Lee, E. W., Chen, J., Huda, M. N., Kuroda, T., Mizushima, T., and Tsuchiya, T. (2003) Functional cloning and expression of emeA, and characterization of EmeA, a multidrug efflux pump from *Enterococcus faecalis*, *Biol. Pharm. Bull. 26*, 266–270.

104. Godreuil, S., Galimand, M., Gerbaud, G., Jacquet, C., and Courvalin, P. (2003) Efflux pump Lde is associated with fluoroquinolone resistance in *Listeria monocytogenes*, *Antimicrob. Agents Chemother. 47*, 704–708.

105. Ohki, R., and Murata, M. (1997) *bmr3*, a third multidrug transporter gene of *Bacillus subtilis*, *J. Bacteriol. 179*, 1423–1427.

106. Borges-Walmsley, M. I., McKeegan, K. S., and Walmsley, A. R. (2003) Structure and function of efflux pumps that confer resistance to drugs, *Biochem. J. 376*, 313–338.

107. Yoshida, H., Bogaki, M., Nakamura, S., Ubukata, K., and Konno, M. (1990) Nucleotide sequence and characterization of the *Staphylococcus aureus norA* gene, which confers resistance to quinolones, *J. Bacteriol. 172*, 6942–6949.

108. Truong-Bolduc, Q. C., Dunman, P. M., Strahilevitz, J., Projan, S. J., and Hooper, D. C. (2005) MgrA is a multiple regulator of two new efflux pumps in *Staphylococcus aureus*, *J. Bacteriol. 187*, 2395–2405.

109. Huang, J., O'Toole, P. W., Shen, W., Amrine-Madsen, H., Jiang, X., Lobo, N., Palmer, L. M., Voelker, L., Fan, F., Gwynn, M. N., and McDevitt, D. (2004) Novel chromosomally encoded multidrug efflux transporter MdeA in *Staphylococcus aureus*, *Antimicrob. Agents Chemother. 48*, 909–917.

110. Mitchell, B. A., Paulsen, I. T., Brown, M. H., and Skurray, R. A. (1999) Bioenergetics of the staphylococcal multidrug export protein QacA: identification of distinct binding sites for monovalent and divalent cations, *J. Biol. Chem. 274*, 3541–3548.

111. Kim, H. J., Kim, Y., Lee, M. S., and Lee, H. S. (2001) Gene lmrB of *Corynebacterium glutamicum* confers efflux-mediated resistance to lincomycin, *Mol. Cells 12*, 112–116.

112. Murata, M., Ohno, S., Kumano, M., Yamane, K., and Ohki, R. (2003) Multidrug resistant phenotype of *Bacillus subtilis* spontaneous mutants isolated in the presence of puromycin and lincomycin, *Can. J. Microbiol. 49*, 71–77.

113. Kumano, M., Fujita, M., Nakamura, K., Murata, M., Ohki, R., and Yamane, K. (2003) Lincomycin resistance mutations in two regions immediately downstream of the -10 region of lmr promoter cause overexpression of a putative multidrug efflux pump in *Bacillus subtilis* mutants, *Antimicrob. Agents Chemother. 47*, 432–435.

114. Lebel, S., Bouttier, S., and Lambert, T. (2004) The cme gene of *Clostridium difficile* confers multidrug resistance in *Enterococcus faecalis*, *FEMS Microbiol. Lett. 238*, 93–100.

115. Roberts, M. C. (1996) Tetracycline resistance determinants: mechanisms of action, regulation of expression, genetic mobility, and distribution, *FEMS Microbiol. Rev. 19*, 1–24.

116. Butaye, P., Cloeckaert, A., and Schwarz, S. (2003) Mobile genes coding for efflux-mediated antimicrobial resistance in gram-positive and gram-negative bacteria, *Int. J. Antimicrob. Agents 22*, 205–210.

117. Agerso, Y., and Guardabassi, L. (2005) Identification of Tet 39, a novel class of tetracycline resistance determinant in *Acinetobacter* spp. of environmental and clinical origin, *J. Antimicrob. Chemother. 55*, 566–569.

118. Ainsa, J. A., Blokpoel, M. C., Otal, I., Young, D. B., De Smet, K. A., and Martin, C. (1998) Molecular cloning and characterization of Tap, a putative multidrug efflux pump present in *Mycobacterium fortuitum* and *Mycobacterium tuberculosis*, *J. Bacteriol. 180*, 5836–5843.

119. De Rossi, E., Arrigo, P., Bellinzoni, M., Silva, P. A., Martin, C., Ainsa, J. A., Guglierame, P., and Riccardi, G. (2002) The multidrug transporters belonging to major facilitator superfamily in *Mycobacterium tuberculosis*, *Mol. Med. 8*, 714–724.

120. Viveiros, M., Leandro, C., and Amaral, L. (2003) Mycobacterial efflux pumps and chemotherapeutic implications, *Int. J. Antimicrob. Agents 22*, 274–278.

121. Siddiqi, N., Das, R., Pathak, N., Banerjee, S., Ahmed, N., Katoch, V. M., and Hasnain, S. E. (2004) *Mycobacterium tuberculosis* isolate with a distinct genomic identity over-expresses a tap-like efflux pump, *Infection 32*, 109–111.

122. Silva, P. E., Bigi, F., Santangelo, M. P., Romano, M. I., Martin, C., Cataldi, A., and Ainsa, J. A. (2001) Characterization of P55, a multidrug efflux pump in *Mycobacterium bovis* and *Mycobacterium tuberculosis, Antimicrob. Agents Chemother. 45*, 800–804.

123. Goldberg, M., Pribyl, T., Juhnke, S., and Nies, D. H. (1999) Energetics and topology of CzcA, a cation/proton antiporter of the resistance–nodulation–cell division protein family, *J. Biol. Chem. 274*, 26065–26070.

124. Tseng, T. T., Gratwick, K. S., Kollman, J., Park, D., Niles, D. H., Goffeau, A., and Saier, M. H., Jr. (2003) The RND permease family: an ancient, ubiquitus and diverse family that includes human diseases and development proteins, *Mol. Microbiol. Biotechnol. 1*, 107–125.

125. Elkins, C. A., and Nikaido, H. (2003) 3D structure of AcrB: the archetypal multidrug efflux transporter of *Escherichia coli* likely captures substrates from periplasm, *Drug Resist. Update 6*, 9–13.

126. Mao, W., Warren, M. S., Black, D. S., Satou, T., Murata, T., Nishino, T., Gotoh, N., and Lomovskaya, O. (2002) On the mechanism of substrate specificity by resistance nodulation division (RND)-type multidrug resistance pumps: the large periplasmic loops of MexD from *Pseudomonas aeruginosa* are involved in substrate recognition, *Mol. Microbiol. 46*, 889–901.

127. Pos, K. M., and Diederichs, K. (2002) Purification, crystallization and preliminary diffraction studies of AcrB, an inner-membrane multi-drug efflux protein, *Acta Crystallogr. 58*, 1865–1867.

128. Pos, K. M., Schiefner, A., Seeger, M. A., and Diederichs, K. (2004) Crystallographic analysis of AcrB, *FEBS Lett. 564*, 333–339.

129. Murakami, S., Nakashima, R., Yamashita, E., and Yamaguchi, A. (2002) Crystal structure of bacterial multidrug efflux transporter AcrB, *Nature 419*, 587–593.

130. Elkins, C. A., and Nikaido, H. (2002) Substrate specificity of the RND-type multidrug efflux pumps AcrB and AcrD of *Escherichia coli* is determined predominantly by two large periplasmic loops, *J. Bacteriol. 184*, 6490–6498.

131. Murakami, S., Nakashima, R., Yamashita, E., Matsumoto, T., and Yamaguchi, A. (2006) Crystal structures of a multidrug transporter reveal a functionally rotating mechanism, *Nature 443*, 173–179.

132. Seeger, M. A., Schiefner, A., Eicher, T., Verrey, F., Diederichs, K., and Pos, K. M. (2006) Structural asymmetry of AcrB trimer suggests a peristaltic pump mechanism, *Science 313*, 1295–1298.

133. Seeger, M. A., Diederichs, K., Eicher, T., Brandstatter, L., Schiefner, A., Verrey, F., and Pos, K. M. (2008) The AcrB efflux pump: conformational cycling and peristalsis lead to multidrug resistance, *Curr. Drug Targets 9*, 729–749.

134. Murakami, S. (2008) Multidrug efflux transporter, AcrB—the pumping mechanism, *Curr. Opin. Struct. Biol. 18*, 459–465.

135. Pos, K. M. (2009) Drug transport mechanism of the AcrB efflux pump, *Biochim. Biophys. Acta 1794*, 782–793.

136. Eicher, T., Brandstatter, L., and Pos, K. M. (2009) Structural and functional aspects of the multidrug efflux pump AcrB, *Biol. Chem. 390*, 693–699.

137. Nikaido, H. (2003) Molecular basis of bacterial outer membrane permeability revisited, *Microbiol. Mol. Biol. Rev. 67*, 593–656.

138. Nikaido, H., and Takatsuka, Y. (2009) Mechanisms of RND multidrug efflux pumps, *Biochim. Biophys. Acta 1794*, 769–781.

139. Fralick, J. A., and Burns-Keliher, L. L. (1994) Additive effect of *tolC* and *rfa* mutations on the hydrophobic barrier of the outer membrane of *Escherichia coli* K-12, *J. Bacteriol. 176*, 6404–6406.

140. Higgins, M. K., Bokma, E., Koronakis, E., Hughes, C., and Koronakis, V. (2004) Structure of the periplasmic component of a bacterial drug efflux pump, *Proc. Natl. Acad. Sci. USA 101*, 9994–9999.

141. Mikolosko, J., Bobyk, K., Zgurskaya, H. I., and Ghosh, P. (2006) Conformational flexibility in the multidrug efflux system protein AcrA, *Structure 14*, 577–587.

142. Piao, S., Xu, Y., and Ha, N. C. (2008) Crystallization and preliminary X-ray crystallographic analysis of MacA from *Actinobacillus actinomycetemcomitans*, *Acta Crystallogr. 64*, 391–393.

143. Lobedanz, S., Bokma, E., Symmons, M. F., Koronakis, E., Hughes, C., and Koronakis, V. (2007) A periplasmic coiled-coil interface underlying TolC recruitment and the assembly of bacterial drug efflux pumps, *Proc. Natl. Acad. Sci. USA 104*, 4612–4617.

144. Stegmeier, J. F., Polleichtner, G., Brandes, N., Hotz, C., and Andersen, C. (2006) Importance of the adaptor (membrane fusion) protein hairpin domain for the functionality of multidrug efflux pumps, *Biochemistry 45*, 10303–10312.

145. Zgurskaya, H. I., Yamada, Y., Tikhonova, E. B., Ge, Q., and Krishnamoorthy, G. (2009) Structural and functional diversity of bacterial membrane fusion proteins, *Biochim. Biophys. Acta 1794*, 794–807.

146. Koronakis, V., Sharff, A., Koronakis, E., Luisi, B., and Hughes, C. (2000) Crystal structure of the bacterial membrane protein TolC central to multidrug efflux and protein export, *Nature 405*, 914–919.

147. Akama, H., Kanemaki, M., Tsukihara, T., Nakagawa, A., and Nakae, T. (2005) Preliminary crystallographic analysis of the antibiotic discharge outer membrane lipoprotein OprM of *Pseudomonas aeruginosa* with an exceptionally long unit cell and complex lattice structure, *Acta Crystallogr. 61*, 131–133.

148. Federici, L., Du, D., Walas, F., Matsumura, H., Fernández-Recio, J., McKeegan, K. S., Borges-Walmsley, M. I., Luisi, B. F., and Walmsley, A. R. (2005) The crystal structure of the outer membrane protein VceC from the bacterial pathogen *Vibrio cholerae* at 1.8 Å resolution, *J. Biol. Chem. 280*, 15307–15314.

149. Vediyappan, G., and Fralick, J. A. (2009) Unpublished results.

150. Husain, F., Humbard, M., and Misra, R. (2004) Interaction between the TolC and AcrA proteins of a multidrug efflux system of *Escherichia coli*, *J. Bacteriol. 186*, 8533–8536.

151. Tikhonova, E. B., and Zgurskaya, H. I. (2004) AcrA, AcrB, and TolC of *Escherichia coli* form a stable intermembrane multidrug efflux complex, *J. Biol. Chem. 279*, 32116–32124.

152. Touzé, T., Eswaran, J., Bokma, E., Koronakis, E., Hughes, C., and Koronakis, V. (2004) Interactions underlying assembly of the *Escherichia coli* AcrAB–TolC multidrug efflux system, *Mol. Microbiol. 53*, 697–706.

153. Aires, J. R., and Nikaido, H. (2005) Aminoglycosides are captured from both periplasm and cytoplasm by the AcrD multidrug efflux transporter of *Escherichia coli, J. Bacteriol. 187*, 1923–1929.

154. Fernández-Recio, J., Totrov, M., and Abagyan, R. (2004) Identification of protein–protein interaction sites from docking energy landscapes, *J. Mol. Biol. 335*, 843–865.

155. Misra, R., and Bavro, V. N. (2009) Assembly and transport mechanism of tripartite drug efflux systems, *Biochim. Biophys. Acta 1794*, 817–825.

156. Eswaran, J., Koronakis, E., Higgins, M. K., Hughes, C., and Koronakis, V. (2004) Three's company: component structures bring a closer view of tripartite drug efflux pumps, *Curr. Opin. Struct. Biol. 14*, 741–747.

157. Bavro, V. N., Pietras, Z., Furnham, N., Perez-Cano, L., Fernández-Recio, J., Pei, X. Y., Misra, R., and Luisi, B. (2008) Assembly and channel opening in a bacterial drug efflux machine, *Mol. Cell 30*, 114–121.

158. Nehme, D., and Poole, K. (2007) Assembly of the MexAB–OprM multidrug pump of *Pseudomonas aeruginosa*: component interactions defined by the study of pump mutant suppressors, *J. Bacteriol. 189*, 6118–6127.

159. Koronakis, V., Eswaran, J., and Hughes, C. (2004) Structure and function of TolC: the bacterial exit duct for proteins and drugs, *Annu. Rev. Biochem. 73*, 467–489.

160. Andersen, C., Koronakis, E., Bokma, E., Eswaran, J., Humphreys, D., Hughes, C., and Koronakis, V. (2002) Transition to the open state of the TolC periplasmic tunnel entrance, *Proc. Natl. Acad. Sci. USA 99*, 11103–11108.

161. Chollet, R., Chevalier, J., Bryskier, A., and Pages, J. M. (2004) The AcrAB–TolC pump is involved in macrolide resistance but not in telithromycin efflux in *Enterobacter aerogenes* and *Escherichia coli, Antimicrob. Agents Chemother. 48*, 3621–3624.

162. Ma, D., Cook, D. N., Alberti, M., Pon, N. G., Nikaido, H., and Hearst, J. E. (1995) Genes *acrA* and *acrB* encode a stress-induced efflux system of *Escherichia coli, Mol. Microbiol. 16*, 45–55.

163. Wandersman, C., and Delepelaire, P. (1990) TolC, an *Escherichia coli* outer membrane protein required for hemolysin secretion, *Proc. Natl. Acad. Sci. USA 87*, 4776–4780.

164. Blair, J. M., and Piddock, L. J. (2009) Structure, function and inhibition of RND efflux pumps in gram-negative bacteria: an update, *Curr. Opin. Microbiol. 12*, 512–519.

165. Fralick, J. A. (1996) Evidence that TolC is required for functioning of the Mar/AcrAB efflux pump of *Escherichia coli, J. Bacteriol. 178*, 5803–5805.

166. Baucheron, S., Tyler, S., Boyd, D., Mulvey, M. R., Chaslus-Dancla, E., and Cloeckaert, A. (2004) AcrAB–TolC directs efflux-mediated multidrug resistance in *Salmonella enterica* serovar *typhimurium* DT104, *Antimicrob. Agents Chemother. 48*, 3729–3735.

167. Jellen-Ritter, A. S., and Kern, W. V. (2001) Enhanced expression of the multidrug efflux pumps AcrAB and AcrEF associated with insertion element transposition in *Escherichia*

coli mutants selected with a fluoroquinolone, *Antimicrob. Agents Chemother.* *45*, 1467–1472.

168. Schweizer, H. P. (1998) Intrinsic resistance to inhibitors of fatty acid biosynthesis in *Pseudomonas aeruginosa* is due to efflux: application of a novel technique for generation of unmarked chromosomal mutations for the study of efflux systems, *Antimicrob. Agents Chemother.* *42*, 394–398.

169. Chuanchuen, R., Beinlich, K., Hoang, T. T., Becher, A., Karkhoff-Schweizer, R. R., and Schweizer, H. P. (2001) Cross-resistance between triclosan and antibiotics in *Pseudomonas aeruginosa* is mediated by multidrug efflux pumps: exposure of a susceptible mutant strain to triclosan selects *nfxB* mutants overexpressing MexCD–OprJ, *Antimicrob. Agents Chemother.* *45*, 428–432.

170. Chuanchuen, R., Narasaki, C. T., and Schweizer, H. P. (2002) The MexJK efflux pump of *Pseudomonas aeruginosa* requires OprM for antibiotic efflux but not for efflux of triclosan, *J. Bacteriol.* *184*, 5036–5044.

171. Morita, Y., Sobel, M. L., and Poole, K. (2006) Antibiotic inducibility of the MexXY multidrug efflux system of *Pseudomonas aeruginosa*: involvement of the antibiotic-inducible PA5471 gene product, *J. Bacteriol.* *188*, 1847–1855.

172. Sekiya, H., Mima, T., Morita, Y., Kuroda, T., Mizushima, T., and Tsuchiya, T. (2003) Functional cloning and characterization of a multidrug efflux pump, MexHI–OpmD, from a *Pseudomonas aeruginosa* mutant, *Antimicrob. Agents Chemother.* *47*, 2990–2992.

173. Li, Y., Mima, T., Komori, Y., Morita, Y., Kuroda, T., Mizushima, T., and Tsuchiya, T. (2003) A new member of the tripartite multidrug efflux pumps, MexVW–OprM, in *Pseudomonas aeruginosa*, *J. Antimicrob. Chemother.* *52*, 572–575.

174. Chang, L. L., Chen, H. F., Chang, C. Y., Lee, T. M., and Wu, W. J. (2004) Contribution of integrons and SmeABC and SmeDEF efflux pumps to multidrug resistance in clinical isolates of *Stenotrophomonas maltophilia*, *J. Antimicrob. Chemother.* *53*, 518–521.

175. Pumbwe, L., Randall, L. P., Woodward, M. J., and Piddock, L. J. (2004) Expression of the efflux pump genes *cmeB*, *cmeF* and the porin gene *porA* in multiple-antibiotic-resistant *Campylobacter jejuni*, *J. Antimicrob. Chemother.* *54*, 341–347.

176. Kumar, A., and Worobec, E. A. (2005) Cloning, sequencing, and characterization of the SdeAB multidrug efflux pump of *Serratia marcescens*, *Antimicrob. Agents Chemother.* *49*, 1495–1501.

177. Chen, J., Kuroda, T., Huda, M. N., Mizushima, T., and Tsuchiya, T. (2003) An RND-type multidrug efflux pump SdeXY from *Serratia marcescens*, *J. Antimicrob. Chemother.* *52*, 176–179.

178. Zarantonelli, L., Borthagaray, G., Lee, E. H., and Shafer, W. M. (1999) Decreased azithromycin susceptibility of *Neisseria gonorrhoeae* due to *mtrR* mutations, *Antimicrob. Agents Chemother.* *43*, 2468–2472.

179. Cousin, S. L., Jr., Whittington, W. L., and Roberts, M. C. (2003) Acquired macrolide resistance genes and the 1 bp deletion in the *mtrR* promoter in *Neisseria gonorrhoeae*, *J. Antimicrob. Chemother.* *51*, 131–133.

180. Dewi, B. E., Akira, S., Hayashi, H., and Ba-Thein, W. (2004) High occurrence of simultaneous mutations in target enzymes and MtrRCDE efflux system in quinolone-resistant *Neisseria gonorrhoeae*, *Sex. Transm. Dis.* *31*, 353–359.

181. Nair, B. M., Cheung, K. J., Jr., Griffith, A., and Burns, J. L. (2004) Salicylate induces an antibiotic efflux pump in *Burkholderia cepacia* complex genomovar III (*B. cenocepacia*), *J. Clin. Invest. 113*, 464–473.

182. Kaczmarek, F. S., Gootz, T. D., Dib-Hajj, F., Shang, W., Hallowell, S., and Cronan, M. (2004) Genetic and molecular characterization of beta-lactamase-negative ampicillin-resistant *Haemophilus influenzae* with unusually high resistance to ampicillin, *Antimicrob. Agents Chemother. 48*, 1630–1639.

183. Magnet, S., Courvalin, P., and Lambert, T. (2001) Resistance–nodulation–cell division–type efflux pump involved in aminoglycoside resistance in *Acinetobacter baumannii* strain BM4454, *Antimicrob. Agents Chemother. 45*, 3375–3380.

184. Saier, M. H., Jr., Tran, C., and Barabote, R. D. (2006) TCDB: The transporter classification database for membrane transport protein analyses and information, *Nucleic Acids Res. 34*, D191–D186.

185. Barabote, R. D., Rendulic, S., Schuster, S. C., and Saier, M. H., Jr. (2007) Comprehensive analysis of transport proteins encoded within the genome of *Bdellovibrio bacteriovorus*, *Genomics 90*, 424–446.

186. Barabote, R. D., and Saier, M. H., Jr. (2005) Comparative genomic analyses of the bacterial phosphotransferase system, *Microbiol. Mol. Biol. Rev. 69*, 608–634.

187. Schneiker, S., Perlova, O., Kaiser, O., Gerth, K., Alici, A., Altmeyer, M.O., Bartels, D., Bekel, T., Beyer, S., Bode, E., et al. (2007) Complete genome sequence of the *Myxobacterium sorangium cellulosum*, *Nat. Biotechnol. 25*, 1281–1289.

188. McCutcheon, J. P., and Moran, N. A. (2007) Parallel genomic evolution and metabolic interdependence in an ancient symbiosis, *Proc. Natl. Acad. Sci. USA 104*, 19392–19397.

189. Lorca, G. L., Barabote, R. D., Zlotopolski, V., Tran, C., Winnen, B., Hvorup, R. N., Stonestrom, A. J., Nguyen, E., Huang, L. W., Kim, D. S., and Saier, M. H., Jr. (2007) Transport capabilities of eleven gram-positive bacteria: comparative genomic analyses, *Biochim. Biophys. Acta, 1768*, 1342–1366.

190. Nagata, Y., Matsuda, M., Komatsu, H., Imura, Y., Sawada, H., Ohtsubo, Y., and Tsuda, M. (2005) Organization and localization of the *dnaA* and *dnaK* gene regions on themultichromosomal genome of *Burkholderia multivorans* ATCC 17616, *J. Biosci. Bioeng. 99*, 603–610.

191. Zahariadis, G., Levy, M. H., and Burns, J. L. (2003) Cepacia-like syndrome caused by *Burkholderia multivorans*, *Can. J. Infect. Dis., 14*, 123–125.

192. Horn, M., Harzenetter, M. D., Linner, T., Schmid, E. N., Müller, K. D., Michel, R., and Wagner, M. (2001) Members of the *Cytophaga–Flavobacterium–Bacteroides* phylum as intracellularbacteria of acanthamoebae: proposal of "*Candidatus amoebophilus asiaticus*," *Environ. Microbiol. 3*, 440–449.

193. Wu, D., Daugherty, S., Van Aken, S. E., Pai, G. H., Watkins, K. L., Khouri, H., Tallon, L. J., Zaborsky, J. M., Dunbar, H. E., Tran, P. L., Moran, N. A., and Eisen, J. (2006) Metabolic complementarity and genomics of the dual bacterial symbiosis of sharp-shooters, *PLoS Biol. 4*, e188.

194. Raymond, J. C., and Sistrom, W. R. (1969) *Ectothiorhodospirahalophila*: a new species of the genus, *Ectothiorhodospira Arch. Mikrobiol. 69*, 121–126.

195. Kapatral, V., Anderson, I., Ivanova, N., Reznik, G., Los, T., Lykidis, A., Bhattacharyya, A., Bartman, A., Gardner, W., Grechkin, G., et al. (2002) Genome sequence and analysis of the oralbacterium *Fusobacterium nucleatum* strain ATCC 25586, *J. Bacteriol. 184*, 2005–2018.

196. Albertson, G. D., Niimi, M., Cannon, R. D., and Jenkinson, H. F. (1996) Multiple efflux mechanisms are involved in *Candida albicans* fluconazole resistance, *Antimicrob. Agents Chemother. 40*, 2835–2841.

197. Lamping, E., Monk, B. C., Niimi, K., Holmes, A. R., Tsao, S., Tanabe, K., Niimi, M., Uehara, Y., and Cannon, R. D. (2007) Characterization of three classes of membrane proteins involved in fungal azole resistance by functional hyperexpression in *Saccharomyces cerevisiae, Eukaryot. Cell 6*, 1150–1165.

198. Griffin, D. H. (1994) *Fungal Physiology*, 2nd ed., Wiley-Liss, New York.

199. Saville, S. P., Lazzell, A. L., Monteagudo, C., and Lopez-Ribot, J. L. (2003) Engineered control of cell morphology in vivo reveals distinct roles for yeast and filamentous forms of *Candida albicans* during infection, *Eukaryot. Cell 2*, 1053–1060.

200. Kim, S., Kim, E., Shin, D. S., Kang, H., and Oh, K. B. (2002) Evaluation of morphogenic regulatory activity of farnesoic acid and its derivatives against *Candida albicans* dimorphism, *Bioorg. Med. Chem. Lett. 12*, 895–898.

201. Nunes, L. R., Costa de Oliveira, R., Leite, D. B., da Silva, V. S., dos Reis Marques, E., da Silva Ferreira, M. E., Ribeiro, D. C., de Souza Bernardes, L. A., Goldman, M. H., Puccia, R., et al. (2005) Transcriptome analysis of *Paracoccidioides brasiliensis* cells undergoing mycelium-to-yeast transition, *Eukaryot. Cell 4*, 2115–2128.

202. Campos, C. B., Di Benedette, J. P., Morais, F. V., Ovalle, R., and Nobrega, M. P. (2008) Evidence for the role of calcineurin in morphogenesis and calcium homeostasis during mycelium-to-yeast dimorphism of *Paracoccidioides brasiliensis, Eukaryot. Cell 7*, 1856–1864.

203. Moss, A. S., Reddy, N. S., Dortaj, I. M., and San Francisco, M. J. (2008) Chemotaxis of the amphibian pathogen *Batrachochytrium dendrobatidis* and its response to a variety of attractants, *Mycologia 100*, 1–5.

204. Pfaller, M. A., and Diekema, D. J. (2004) Rare and emerging opportunistic fungal pathogens: concern for resistance beyond *Candida albicans* and *Aspergillus fumigatus, J. Clin. Microbiol. 42*, 4419–4431.

205. Blehert, D. S., Hicks, A. C., Behr, M., Meteyer, C. U., Berlowski-Zier, B. M., Buckles, E. L., Coleman, J. T., Darling, S. R., Gargas, A., Niver, R., et al. (2009) Bat white-nose syndrome: an emerging fungal pathogen? *Science 323*, 227.

206. Longcore, J. E., Pessier, A. P., and Nichols, D. K. (1999) *Batrachochytrium dendrobatidis* gen et sp nov, a chytrid pathogenic to amphibians, *Mycologia 91*, 219–227.

207. Gulshan, K., and Moye-Rowley, W. S. (2007) Multidrug resistance in fungi, *Eukaryot. Cell 6*, 1933–1942.

208. Anderson, J. B. (2005) Evolution of antifungal-drug resistance: mechanisms and pathogen fitness, *Nat. Rev. Microbiol. 3*, 547–556.

209. Podust, L. M., von Kries, J. P., Eddine, A. N., Kim, Y., Yermalitskaya, L. V., Kuehne, R., Ouellet, H., Warrier, T., Altekoster, M., Lee, J. S., Rademann, J., Oschkinat, H.,

Kaufmann, S. H., and Waterman, M. R. (2007) Small-molecule scaffolds for CYP51 inhibitors identified by high-throughput screening and defined by x-ray crystallography, *Antimicrob. Agents Chemother. 51*, 3915–3923.

210. Talele, T. T., and Kulkarni, V. M. (1999) Three-dimensional quantitative structure–activity relationship (QSAR) and receptor mapping of cytochrome P-450 (14 alpha DM) inhibiting azole antifungal agents, *J. Chem. Inf. Comput. Sci. 39*, 204–210.

211. Che, X., Sheng, C., Wang, W., Cao, Y., Xu, Y., Ji, H., Dong, G., Miao, Z., Yao, J., and Zhang, W. (2009) New azoles with potent antifungal activity: design, synthesis and molecular docking, *Eur. J. Med. Chem. 44*, 4218–4226.

212. Mota, C. R., Miranda, K. C., Lemos J. de, A., Costa, C. R., Hasimoto e Souza, L. K., Passos, X. S., Meneses e Silva, H., and Silva M. do, R. (2009) Comparison of in vitro activity of five antifungal agents against dermatophytes, using the agar dilution and broth microdilution methods, *Rev. Soc. Bras. Med. Trop. 42*, 250–254.

213. Cowen, L. E., and Steinbach, W. J. (2008) Stress, drugs, and evolution: the role of cellular signaling in fungal drug resistance, *Eukaryot. Cell 7*, 747–764.

214. Kauffman, C. A. (2006) Clinical efficacy of new antifungal agents, *Curr. Opin. Microbiol. 9*, 483–488.

215. Odds, F. C., Brown, A. J., and Gow, N. A. (2003) Antifungal agents: mechanisms of action, *Trends Microbiol. 11*, 272–279.

216. Chapman, S. W., Sullivan, D. C., and Cleary, J. D. (2008) In search of the holy grail of antifungal therapy, *Trans. Am. Clin. Climatol. Assoc. 119*, 197–215; discussion 215–196.

217. Tincu, J. A., and Taylor, S. W. (2004) Antimicrobial peptides from marine invertebrates, *Antimicrob. Agents Chemother. 48*, 3645–3654.

218. Jenssen, H., Hamill, P., and Hancock, R. E. (2006) Peptide antimicrobial agents. *Clin. Microbiol. Rev. 19*, 491–511.

219. Ghannoum, M. A., and Rice, L. B. (1999) Antifungal agents: mode of action, mechanisms of resistance, and correlation of these mechanisms with bacterial resistance, *Clin. Microbiol. Rev. 12*, 501–517.

220. Ren, Q., Chen, K., and Paulsen, I. T. (2007) TransportDB: a comprehensive database resource for cytoplasmic membrane transport systems and outer membrane channels, *Nucleic Acids Res. 35*, D274–D279.

221. Morrissey, J. P., and Osbourn, A. E. (1999) Fungal resistance to plant antibiotics as a mechanism of pathogenesis, *Microbiol. Mol. Biol. Rev. 63*, 708–724.

222. Saier, M. H., Jr. (1998) Molecular phylogeny as a basis for the classification of transport proteins from bacteria, archaea and eukarya, *Adv. Microb. Physiol. 40*, 81–136.

223. Souid, A. K., Gao, C., Wang, L., Milgrom, E., and Shen, W. C. (2006) ELM1 is required for multidrug resistance in *Saccharomyces cerevisiae*, *Genetics 173*, 1919–1937.

224. Schmitt, L., and Tampe, R. (2002) Structure and mechanism of ABC transporters, *Curr. Opin. Struct. Biol. 12*, 754–760.

225. Jungwirth, H., and Kuchler, K. (2006) Yeast ABC transporters—a tale of sex, stress, drugs and aging, *FEBS Lett. 580*, 1131–1138.

226. Piddock, L. J. (2006) Multidrug-resistance efflux pumps—not just for resistance, *Nat. Rev. Microbiol. 4*, 629–636.

227. Walker, J. E., Saraste, M., Runswick, M. J., and Gay, N. J. (1982) Distantly related sequences in the alpha- and beta-subunits of ATP synthase, myosin, kinases and other ATP-requiring enzymes and a common nucleotide binding fold, *EMBO J. 1*, 945–951.

228. Decottignies, A., and Goffeau, A. (1997) Complete inventory of the yeast ABC proteins, *Nat. Genet. 15*, 137–145.

229. Coleman, J. J., and Mylonakis, E. (2009) Efflux in fungi: la pièce de résistance, *PLoS Pathog. 5*, e1000486.

230. Sipos, G., and Kuchler, K. (2006) Fungal ATP-binding cassette (ABC) transporters in drug resistance and detoxification, *Curr. Drug Targets 7*, 471–481.

231. Rogers, B. D. A., Kolaczkowski, M., Carvajal, E., Balzi, E., and Goffeau, A. (2003) The pleitropic drug ABC transporters from *Saccharomyces cerevisiae*, *J. Mol. Microbiol. Biotechnol. 3*, 207–214.

232. Iwaki, T., Giga-Hama, Y., and Takegawa, K. (2006) A survey of all 11 ABC transporters in fission yeast: two novel ABC transporters are required for red pigment accumulation in a *Schizosaccharomyces pombe* adenine biosynthetic mutant, *Microbiology 152*, 2309–2321.

233. Rea, P. A. (2007) Plant ATP-binding cassette transporters, *Annu. Rev. Plant Biol. 58*, 347–375.

234. Bauer, B. E., Wolfger, H., and Kuchler, K. (1999) Inventory and function of yeast ABC proteins: about sex, stress, pleiotropic drug and heavy metal resistance, *Biochim. Biophys. Acta 1461*, 217–236.

235. Rogers, B., Decottignies, A., Kolaczkowski, M., Carvajal, E., Balzi, E., and Goffeau, A. (2001) The pleitropic drug ABC transporters from *Saccharomyces cerevisiae*, *J. Mol. Microbiol. Biotechnol. 3*, 207–214.

236. Kralli, A., Bohen, S. P., and Yamamoto, K. R. (1995) LEM1, an ATP-binding-cassette transporter, selectively modulates the biological potency of steroid hormones, *Proc. Natl. Acad. Sci. USA 92*, 4701–4705.

237. Bissinger, P. H., and Kuchler, K. (1994) Molecular cloning and expression of the *Saccharomyces cerevisiae* STS1 gene product: a yeast ABC transporter conferring mycotoxin resistance, *J. Biol. Chem. 269*, 4180–4186.

238. Garcia-Effron, G., Dilger, A., Alcazar-Fuoli, L., Park, S., Mellado, E., and Perlin, D. S. (2008) Rapid detection of triazole antifungal resistance in *Aspergillus fumigatus*, *J. Clin. Microbiol. 46*, 1200–1206.

239. da Silva Ferreira, M. E., Capellaro, J. L., dos Reis Marques, E., Malavazi, I., Perlin, D., Park, S., Anderson, J. B., Colombo, A. L., Arthington-Skaggs, B. A., Goldman, M. H., and Goldman, G. H. (2004) In vitro evolution of itraconazole resistance in *Aspergillus fumigatus* involves multiple mechanisms of resistance, *Antimicrob. Agents Chemother. 48*, 4405–4413.

240. Tobin, M. B., Peery, R. B., and Skatrud, P. L. (1997) Genes encoding multiple drug resistance-like proteins in *Aspergillus fumigatus* and *Aspergillus flavus*, *Gene 200*, 11–23.

241. Del Sorbo, G., Andrade, A. C., Van Nistelrooy, J. G., Van Kan, J. A., Balzi, E., and De Waard, M. A. (1997) Multidrug resistance in *Aspergillus nidulans* involves novel ATP-binding cassette transporters, *Mol. Gen. Genet. 254*, 417–426.

242. Semighini, C. P., Marins, M., Goldman, M. H., and Goldman, G. H. (2002) Quantitative analysis of the relative transcript levels of ABC transporter Atr genes in *Aspergillus nidulans* by real-time reverse transcription-PCR assay, *Appl. Environ. Microbiol. 68*, 1351–1357.

243. Pappas, P. G., Rex, J. H., Lee, J., Hamill, R. J., Larsen, R. A., Powderly, W., Kauffman, C. A., Hyslop, N., Mangino, J. E., Chapman, S., et al., and NIAID Mycoses Study Group (2003) A prospective observational study of candidemia: epidemiology, therapy, and influences on mortality in hospitalized adult and pediatric patients, *Clin. Infect. Dis. 37*, 634–643.

244. Wisplinghoff, H., Bischoff, T., Tallent, S. M., Seifert, H., Wenzel, R. P., and Edmond, M. B. (2009) Nosocomial bloodstream infections in US hospitals: analysis of 24,179 cases from a prospective nationwide surveillance study, *Clin. Infect. Dis. 49*, 699–701.

245. Tsao, S., Rahkhoodaee, F., and Raymond, M. (2009) Relative contributions of the *Candida albicans* ABC transporters Cdr1p and Cdr2p to clinical azole resistance, *Antimicrob. Agents Chemother. 53*, 1344–1352.

246. Holmes, A. R., Lin, Y. H., Niimi, K., Lamping, E., Keniya, M., Niimi, M., Tanabe, K., Monk, B. C., and Cannon, R. D. (2008) ABC transporter Cdr1p contributes more than Cdr2p does to fluconazole efflux in fluconazole-resistant *Candida albicans* clinical isolates, *Antimicrob. Agents and Chemother. 52*, 3851–3862.

247. Prasad, R. K. S., Prasad, R., Gupta, V., and Lata, S. (1996) Multidrug resistance: an emerging threat, *Curr. Sci. 71*, 205–213.

248. Shukla, S., Saini, P., Smriti, Jha, S., Ambudkar, S. V., and Prasad, R. (2003) Functional characterization of *Candida albicans*. ABC transporter Cdr1p, *Eukaryot. Cell 2*, 1361–1375.

249. Jha, S., Karnani, N., Lynn, A. M., and Prasad, R. (2003) Covalent modification of cysteine 193 impairs ATPase function of nucleotide-binding domain of a *Candida* drug efflux pump, *Biochem. Biophys. Res. Commun. 310*, 869–875.

250. Jha, S., Dabas, N., Karnani, N., Saini, P., and Prasad, R. (2004) ABC multidrug transporter Cdr1p of *Candida albicans* has divergent nucleotide-binding domains which display functional asymmetry, *FEMS Yeast Res. 5*, 63–72.

251. Prasad, R., Gaur, N. A., Gaur, M., and Komath, S. S. (2006) Efflux pumps in drug resistance of *Candida*, *Infect. Disord. Drug Targets 6*, 69–83.

252. Shukla, S., Rai, V., Saini, P., Banerjee, D., Menon, A. K., and Prasad, R. (2007) *Candida* drug resistance protein 1, a major multidrug ATP binding cassette transporter of *Candida albicans*, translocates fluorescent phospholipids in a reconstituted system, *Biochemistry 46*, 12081–12090.

253. Clark, F. S., Parkinson, T., Hitchcock, C. A., and Gow, N. A. (1996) Correlation between rhodamine 123 accumulation and azole sensitivity in *Candida* species: possible role for drug efflux in drug resistance, *Antimicrob. Agents Chemother. 40*, 419–425.

254. Niimi, K., Maki, K., Ikeda, F., Holmes, A. R., Lamping, E., Niimi, M., Monk, B. C., and Cannon, R. D. (2006) Overexpression of *Candida albicans* CDR1, CDR2, or MDR1 does not produce significant changes in echinocandin susceptibility, *Antimicrob. Agents Chemother. 50*, 1148–1155.

255. Moran, G., Sullivan, D., Morschhauser, J., and Coleman, D. (2002) The *Candida dubliniensis* CdCDR1 gene is not essential for fluconazole resistance, *Antimicrob. Agents Chemother. 46*, 2829–2841.

256. Chen, W. J., Delmotte, F., Richard-Cervera, S., Douence, L., Greif, C., and Corio-Costet, M. F. (2007) At least two origins of fungicide resistance in grapevine downy mildew populations, *Appl. Environ. Microbiol. 73*, 5162–5172.

257. Reimann, S., and Deising, H. B. (2005) Inhibition of efflux transporter-mediated fungicide resistance in *Pyrenophora tritici-repentis* by a derivative of 4′-hydroxyflavone and enhancement of fungicide activity, *Appl. Environ. Microbiol. 71*, 3269–3275.

258. Leonard, K. J., Anikster, Y., and Manisterski, J. (2005) Virulence associations in oat crown rust, *Phytopathology 95*, 53–61.

259. Urban, M., Bhargava, T., and Hamer, J. E. (1999) An ATP-driven efflux pump is a novel pathogenicity factor in rice blast disease, *EMBO J. 18*, 512–521.

260. Sun, C. B., Suresh, A., Deng, Y. Z., and Naqvi, N. I. (2006) A multidrug resistance transporter in *Magnaporthe* is required for host penetration and for survival during oxidative stress, *Plant Cell 18*, 3686–3705.

261. Chen, C. G., Yang, Y. L., Tseng, K. Y., Shih, H. I., Liou, C. H., Lin, C. C., and Lo, H. J. (2009) Rep1p negatively regulating MDR1 efflux pump involved in drug resistance in *Candida albicans*, *Fungal Genet. Biol. 46*, 714–720.

262. Zwiers, L. H., Stergiopoulos, I., Van Nistelrooy, J. G., and De Waard, M. A. (2002) ABC transporters and azole susceptibility in laboratory strains of the wheat pathogen *Mycosphaerella graminicola*, *Antimicrob. Agents Chemother. 46*, 3900–3906.

263. Vermeulen, T., Schoonbeek, H., and De Waard, M. A. (2001) The ABC transporter BcatrB from Botrytis cinerea is a determinant of the activity of the phenylpyrrole fungicide fludioxonil. *Pest Mang. Sci. 57*, 393–402.

264. Hof, H. (2001) Critical annotations to the use of azole antifungals for plant protection, *Antimicrob. Agents Chemother. 45*, 2987–2990.

265. Berger, L., Speare, R., and Kent, A. (1999) Diagnosis of chytridiomycosis in amphibians by histologic examination. *Proc. Frog Symposium: Frogs in the Community*, Queensland Museum, Brisbane, Australia.

266. Harvey, P. H., and Pagel, M. D. (1991). *The Comparative Method in Evolutionary Biology*, Oxford University Press, Oxford, UK.

267. Larson, A., and Losos, J. B. (1996) Phylogenetic systematics of adaptation, in *Adaptation*, Rose, M. R., and Lauder, G. V., Eds., Academic Press, San Diego, CA, pp. 187–220.

268. Martins, E. (2000) Adaptation and the comparative method, *Trends Ecol. Evol. 15*, 296–299.

269. Robbertse, B., Reeves, J. B., Schoch, C. L., and Spatafora, J. W. (2006) A phylogenomic analysis of the Ascomycota, *Fungal Genet. Biol. 43*, 715–725.

270. Suh, S. O., Blackwell, M., Kurtzman, C. P., and Lachance, M. A. (2006) Phylogenetics of Saccharomycetales, the ascomycete yeasts, *Mycologia 98*, 1006–1017.

271. Geiser, D. M., Gueidan, C., Miadlikowska, J., Lutzoni, F., Kauff, F., Hofstetter, V., Fraker, E., Schoch, C. L., Tibell, L., Untereiner, W. A., and Aptroot, A. (2006) Eurotiomycetes: Eurotiomycetidae and Chaetothyriomycetidae, *Mycologia 98*, 1053–1064.

272. Fitzpatrick, D. A., Logue, M. E., Stajich, J. E., and Butler, G. (2006) A fungal phylogeny based on 42 complete genomes derived from supertree and combined gene analysis, *BMC Evol. Biol. 6*, 99.

273. Soanes, D. M., Richards, T. A., and Talbot, N. J. (2007) Insights from sequencing fungal and oomycete genomes: What can we learn about plant disease and the evolution of pathogenicity? *Plant Cell 19*, 3318–3326.

274. Podani, J., Engloner, A., and Major, A. (2009) Multilevel comparison of dendrograms: a new method with an application for genetic classifications, *Stat. Appl. Genet. Mol. Biol. 8*, 1–14.

275. Thompson, J. D., Higgins, D. G., and Gibson, T. J. (1994) CLUSTAL W: improving the sensitivity of progressive multiple sequence alignment through sequence weighting, position-specific gap penalties and weight matrix choice, *Nucleic Acids Res. 22*, 4673–4680.

276. Nei, M., and Kumar, S. (2000) *Molecular Evolution and Phylogenetics*, Oxford University Press, New York.

277. Felsenstein, J. (1985) Confidence limits on phylogenies: an approach using the bootstrap, *Evolution 39*, 783–791.

278. Eck, R. V. a. M. O. D. (1966) *Atlas of Protein Sequences and Structure*, National Biomedical Research Foundation, Silver Springs, MD.

279. Tamura, K., Dudley, J., Nei, M., and Kumar, S. (2007) MEGA4: Molecular Evolutionary genetics analysis (MEGA), software version 4.0, *Mol. Biol. Evol. 24*, 1596–1599.

280. Sa-Correia, I., dos Santos, S. C., Teixeira, M. C., Cabrito, T. R., and Mira, N. P. (2009) Drug:H^+ antiporters in chemical stress response in yeast, *Trends Microbiol. 17*, 22–31.

281. Ehrenhofer-Murray, A. E., Wurgler, F. E., and Sengstag, C. (1994) The *Saccharomyces cerevisiae* SGE1 gene product: a novel drug-resistance protein within the major facilitator superfamily, *Mol. Gen. Genet. 244*, 287–294.

282. Felder, T., Bogengruber, E., Tenreiro, S., Ellinger, A., Sa-Correia, I., and Briza, P. (2002) Dtrlp, a multidrug resistance transporter of the major facilitator superfamily, plays an essential role in spore wall maturation in *Saccharomyces cerevisiae*, *Eukaryot. Cell 1*, 799–810.

283. Jenkins, C. C., Mata, J., Crane, R. F., Thomas, B., Akoulitchev, A., Bahler, J., and Norbury, C. J. (2005) Activation of AP-1-dependent transcription by a truncated translation initiation factor, *Eukaryot. Cell 4*, 1840–1850.

284. Calvo, I. A., Gabrielli, N. G., Iglesias-Baena, I., Garcia-Santamarina, S., Hoe, K. L., Kim D. U., Sanso' M., M. Zuin A., Pérez, P., Ayté, J., and Hidalgo, E. (2009) Genome-wide screen of genes required for caffeine tolerance in fission yeast, *PLoS One 4*, e6619.

285. Nascimento, A. M., Goldman, G. H., Park, S., Marras, S. A., Delmas, G., Oza, U., Lolans, K., Dudley, M. N., Mann, P. A., and Perlin, D. S. (2003) Multiple resistance mechanisms among *Aspergillus fumigatus* mutants with high-level resistance to itraconazole, *Antimicrob. Agents Chemother. 47*, 1719–1726.

286. Lechenne, B., Reichard, U., Zaugg, C., Fratti, M., Kunert, J., Boulat, O., and Monod, M. (2007) Sulphite efflux pumps in *Aspergillus fumigatus* and dermatophytes, *Microbiology 153*, 905–913.

287. Gaur, M., Puri, N., Manoharlal, R., Rai, V., Mukhopadhayay, G., Choudhury, D., and Prasad, R. (2008) MFS transportome of the human pathogenic yeast *Candida albicans*, *BMC Genom. 9*, 579.

288. Fling, M. E., Kopf, J., Tamarkin, A., Gorman, J. A., Smith, H. A., and Koltin, Y. (1991) Analysis of a *Candida albicans* gene that encodes a novel mechanism for resistance to benomyl and methotrexate, *Mol. Gen. Genet. 227*, 318–329.

289. Ben-Yaacov, R., Knoller, S., Caldwell, G. A., Becker, J. M., and Koltin, Y. (1994) *Candida albicans* gene encoding resistance to benomyl and methotrexate is a multidrug resistance gene, *Antimicrob. Agents Chemother. 38*, 648–652.

290. Goldway, M., Teff, D., Schmidt, R., Oppenheim, A. B., and Koltin, Y. (1995) Multidrug resistance in *Candida albicans*: disruption of the BENr gene, *Antimicrob. Agents Chemother. 39*, 422–426.

291. Cheng, S., Clancy, C. J., Nguyen, K. T., Clapp, W., and Nguyen, M. H. (2007) A *Candida albicans* petite mutant strain with uncoupled oxidative phosphorylation overexpresses MDR1 and has diminished susceptibility to fluconazole and voriconazole, *Antimicrob. Agents Chemother. 51*, 1855–1858.

292. Becker, J. M., Henry, L. K., Jiang, W., and Koltin, Y. (1995) Reduced virulence of *Candida albicans* mutants affected in multidrug resistance, *Infect. Immun. 63*, 4515–4518.

293. Pasrija, R., Banerjee, D., and Prasad, R. (2007) Structure and function analysis of CaMdr1p, a major facilitator superfamily antifungal efflux transporter protein of *Candida albicans*: identification of amino acid residues critical for drug/H$^+$ transport, *Eukaryot. Cell 6*, 443–453.

294. Calabrese, D., Bille, J., and Sanglard, D. (2000) A novel multidrug efflux transporter gene of the major facilitator superfamily from *Candida albicans* (FLU1) conferring resistance to fluconazole, *Microbiology 146*(Pt. 11), 2743–2754.

295. Sengupta, M., and Datta, A. (2003) Two membrane proteins located in the Nag regulon of *Candida albicans* confer multidrug resistance, *Biochem. Biophys. Res. Commun. 301*, 1099–1108.

296. White, T. C., Holleman, S., Dy, F., Mirels, L. F., and Stevens, D. A. (2002) Resistance mechanisms in clinical isolates of *Candida albicans*, *Antimicrob. Agents Chemother. 46*, 1704–1713.

297. Hayashi, K., Schoonbeek, H. J., and De Waard, M. A. (2002) Bcmfs1, a novel major facilitator superfamily transporter from *Botrytis cinerea*, provides tolerance towards the natural toxic compounds camptothecin and cercosporin and towards fungicides, *Appl. Environ. Microbiol. 68*, 4996–5004.

298. Roohparvar, R., De Waard, M. A., Kema, G. H., and Zwiers, L. H. (2007) MgMfs1, a major facilitator superfamily transporter from the fungal wheat pathogen *Mycosphaerella graminicola*, is a strong protectant against natural toxic compounds and fungicides, *Fungal Genet. Biol. 44*, 378–388.

299. Roohparvar, R., Mehrabi, R., Van Nistelrooy, J. G., Zwiers, L. H., and De Waard, M. A. (2008) The drug transporter MgMfs1 can modulate sensitivity of field strains of the fungal wheat pathogen *Mycosphaerella graminicola* to the strobilurin fungicide trifloxystrobin, *Pest Mang. Sci. 64*, 685–693.

300. Coste, A., Turner, V., Ischer, F., Morschhauser, J., Forche, A., Selmecki, A., Berman, J., Bille, J., and Sanglard, D. (2006) A mutation in Tac1p, a transcription factor regulating CDR1 and CDR2, is coupled with loss of heterozygosity at chromosome 5 to mediate antifungal resistance in *Candida albicans, Genetics 172*, 2139–2156.

301. Liu, T. T., Znaidi, S., Barker, K. S., Xu, L., Homayouni, R., Saidane, S., Morschhauser, J., Nantel, A., Raymond, M., and Rogers, P. D. (2007) Genome-wide expression and location analyses of the *Candida albicans* Tac1p regulon, *Eukaryot. Cell 6*, 2122–2138.

302. Sellam, A., Tebbji, F., and Nantel, A. (2009) Role of Ndt80p in sterol metabolism regulation and azole resistance in *Candida albicans, Eukaryot. Cell 8*, 1174–1183.

303. Morschhauser, J., Barker, K. S., Liu, T. T., Bla, B. W. J., Homayouni, R., and Rogers, P. D. (2007) The transcription factor Mrr1p controls expression of the MDR1 efflux pump and mediates multidrug resistance in *Candida albicans, PLoS Pathog. 3*, e164.

304. Harry, J. B., Oliver, B. G., Song, J. L., Silver, P. M., Little, J. T., Choiniere, J., and White, T. C. (2005) Drug-induced regulation of the MDR1 promoter in *Candida albicans, Antimicrob. Agents Chemother. 49*, 2785–2792.

305. Riggle, P. J., and Kumamoto, C. A. (2006) Transcriptional regulation of MDR1, encoding a drug efflux determinant, in fluconazole-resistant *Candida albicans* strains through an Mcm1p binding site, *Eukaryot. Cell 5*, 1957–1968.

306. Znaidi, S., Weber, S., Al-Abdin, O. Z., Bomme, P., Saidane, S., Drouin, S., Lemieux, S., De Deken, X., Robert, F., and Raymond, M. (2008) Genomewide location analysis of *Candida albicans* Upc2p, a regulator of sterol metabolism and azole drug resistance, *Eukaryot. Cell 7*, 836–847.

307. Ferrari, S., Ischer, F., Calabrese, D., Posteraro, B., Sanguinetti, M., Fadda, G., Rohde, B., Bauser, C., Bader, O., and Sanglard, D. (2009) Gain of function mutations in CgPDR1 of *Candida glabrata* not only mediate antifungal resistance but also enhance virulence, *PLoS Pathog. 5*, e1000268.

308. Kolaczkowska, A., Kolaczkowski, M., Goffeau, A., and Moye-Rowley, W. S. (2008) Compensatory activation of the multidrug transporters Pdr5p, Snq2p, and Yor1p by Pdr1p in *Saccharomyces cerevisiae, FEBS Lett. 582*, 977–983.

309. Niimi, K., Harding, D. R., Parshot, R., King, A., Lun, D. J., Decottignies, A., Niimi, M., Lin, S., Cannon, R. D., Goffeau, A., and Monk, B. C. (2004) Chemosensitization of fluconazole resistance in *Saccharomyces cerevisiae* and pathogenic fungi by a D-octapeptide derivative, *Antimicrob. Agents Chemother. 48*, 1256–1271.

310. Zhang, H., Gao, A., Li, F., Zhang, G., Ho, H. I., and Liao, W. (2009) Mechanism of action of tetrandrine, a natural inhibitor of *Candida albicans* drug efflux pumps, *Yakugaku Zasshi 129*, 623–630.

311. Diwischek, F., Morschhauser, J., and Holzgrabe, U. (2009) Cerulenin analogues as inhibitors of efflux pumps in drug-resistant *Candida albicans, Arch. Pharm. (Weinheim) 342*, 150–164.

312. Sharma, M., Manoharlal, R., Shukla, S., Puri, N., Prasad, T., Ambudkar, S. V., and Prasad, R. (2009) Curcumin modulates efflux mediated by yeast ABC multidrug transporters and is synergistic with antifungals, *Antimicrob. Agents Chemother. 53*, 3256–3265.

313. Costerton, J. W., Lewandowski, Z., Caldwell, D. E., Korber, D. R., and Lappin-Scott, H. M. (1995) Microbial biofilms, *Annu. Rev. Microbiol. 49*, 711–745.

314. Seidler, M. J., Salvenmoser, S., and Muller, F. M. (2008) *Aspergillus fumigatus* forms biofilms with reduced antifungal drug susceptibility on bronchial epithelial cells, *Antimicrob. Agents Chemother. 52*, 4130–4136.

315. Jabra-Rizk, M. A., Falkler, W. A., and Meiller, T. F. (2004) Fungal biofilms and drug resistance, *Emerg. Infect. Dis. 10*, 14–19.

316. Kumamoto, C. A. (2002) *Candida* biofilms, *Curr. Opin. Microbiol. 5*, 608–611.

317. d'Enfert, C. (2006) Biofilms and their role in the resistance of pathogenic *Candida* to antifungal agents, *Curr. Drug Targets 7*, 465–670.

318. Mukherjee, P.K., Chandra, J., Kuhn, D. M., and Ghannoum, M. A. (2003) Mechanism of fluconazole resistance in *Candida albicans* biofilms: phase-specific role of efflux pumps and membrane sterols, *Infect. Immun. 71*, 4333–4340.

319. Ramage, G., Bachmann, S., Patterson, T. F., Wickes, B. L., and López-Ribot, J. L. (2002) Investigation of multidrug efflux pumps in relation to fluconazole resistance in *Candida albicans* biofilms, *J. Antimicrob. Chemother. 49*, 973–980.

320. Perumal, P., Mekala, S., and Chaffin, W. L. (2007) Role for cell density in antifungal drug resistance in *Candida albicans* biofilms, *Antimicrob. Agents Chemother. 51*, 2454–2463.

321. Chaffin, W. L. (2008) *Candida albicans* cell wall proteins, *Microbiol. Mol. Biol. Rev. 72*, 495–544.

A SURVEY OF OXIDATIVE PARACATALYTIC REACTIONS CATALYZED BY ENZYMES THAT GENERATE CARBANIONIC INTERMEDIATES: IMPLICATIONS FOR ROS PRODUCTION, CANCER ETIOLOGY, AND NEURODEGENERATIVE DISEASES

By VICTORIA I. BUNIK, *School of Bioinformatics and Bioengineering, and Belozersky Institute of Physico-Chemical Biology, Moscow Lomonosov State University, Moscow, Russian Federation,* JOHN V. SCHLOSS, *Department of Pharmaceutical Sciences, University of New England, Portland, Maine,* JOHN T. PINTO, *Department of Biochemistry and Molecular Biology, New York Medical College, Valhalla, New York,* NATALIA DUDAREVA, *Department of Horticulture and Landscape Architecture, Purdue University, West Lafayette, Indiana,* and ARTHUR J. L. COOPER, *Department of Biochemistry and Molecular Biology, New York Medical College, Valhalla, New York*

CONTENTS

Advances in Enzymology and Related Areas of Molecular Biology, Volume 77
Edited by Eric J. Toone   Copyright © 2011 John Wiley & Sons, Inc.

I. Overview

Enzymes that generate carbanionic intermediates often catalyze paracatalytic reactions with O_2 and other electrophiles not considered "normal" reactants. For example, pyridoxal 5'-phosphate (PLP)–containing pig kidney dopa decarboxylase oxidizes dopamine with molecular O_2 to 3,4-dihydroxyphenylacetaldehyde at about 1% of the rate at which it catalyzes nonoxidative dopa decarboxylation. The mutant Y332F enzyme, however, catalyzes stoichiometric conversion of dopa to 3,4-dihydroxyphenylacetaldehyde, suggesting that even minor structural changes may alter or initiate paracatalytic reactions catalyzed by certain enzymes. Carbanions generated by several thiamine diphosphate (ThDP)–dependent enzymes react with different electrophiles, transforming some xenobiotics and endogenous compounds into potentially biologically hazardous products. The detrimental effects of paracatalytic reactions may be greatly increased by cellular compartmentation of enzymes and intermediates. For example, in two of the the three multienzyme complexes involved in oxidative α-keto acid decarboxylation, paracatalytic reactions of the third component inactivate the first carbanion-generating component.

In this review we provide an outline of carbanion-generating enzymes known to catalyze paracatalytic reactions. We also discuss the potential of some of these reactions to contribute to irreversible damage in cancer and neurodegeneration through disease-induced alterations in the metabolic state and/or protein structure.

II. Paracatalytic Reactions: Definition

In a series of seminal articles published more than 25 years ago, Philipp Christen and co-workers showed that the carbanionic intermediates of several enzymes [e.g., class I and class II fructose 1,6-bisphosphate aldolase, 6-phosphogluconate dehydrogenase, cytosolic aspartate amino-transferase, and pyruvate decarboxylase (PDC)] were accessible to various oxidants (electrophiles). Depending on the enzyme, these oxidants included, for example, 2,6-dichloroindophenol (DCIP), hexacyanoferrate(III), porphyrindin, tetranitromethane, and H_2O_2 (1–8). Christen coined the word *paracatalytic* to describe enzyme-catalyzed interactions between substrate and a reagent (especially an oxidant) not generally considered to be a physiological reactant (4, 5, 7). Reactions of enzyme-generated carbanions with O_2 were not included in Christen's original definition of paracatalytic reactions. However, several enzymes have been shown to catalyze oxidative side reactions with O_2 as the oxidant.[1] Therefore, in the present chapter we include enzyme-catalyzed oxidation side reactions with O_2 as a type of paracatalytic reaction. In many cases, the side products, whether generated from an artificial electron acceptor or from O_2, are reactive and can lead to inactivation of the enzyme catalyzing the reaction or result in other types of

---

[1] In this chapter we use the Enzyme Commission guidelines in defining oxygenases and oxidases. Oxygenases catalyze reactions in which at least one O atom derived from molecular oxygen (O_2) is incorporated into the main substrate. A monooxygenase catalyzes the incorporation of one O atom into the main substrate (AH), generating AOH; the other O atom is reduced to water, with concomitant oxidation of the co-substrate (BH_2) to B. The reaction may be written formally as $AH + BH_2 + O_2 \rightarrow AOH + B + H_2O$. Dioxygenases catalyze the incorporation of two O atoms into a substrate. In some cases, a single oxidized product retains both O atoms derived from O_2. In other cases, however, oxidation may generate two identical products, each containing an O atom derived from O_2. In yet other cases, oxidation may yield two dissimilar products from a single substrate, each containing an O atom derived from O_2. Oxidases catalyze oxidation reactions in which electrons are removed from the substrate and are used to reduce an electron acceptor, usually, but not always, O_2. A one-electron reduction of O_2 will generate superoxide anion radical ($O_2 \cdot^-$). More common is a two-electron reduction of O_2 to hydrogen peroxide (H_2O_2). In other cases, a four-electron reduction of O_2 yields water. In none of these three cases is O derived from O_2 incorporated into the oxidized product.

oxidative damage. These reactive species might also be released from the active site of one enzyme to affect neighboring enzymes and macromolecules within the cellular milieu. As discussed below, these paracatalytic side reactions may contribute to disease processes.

III. Discovery of Enzyme-Catalyzed Oxygenase Side Reactions: Rubisco

Originally it was not suspected that enzymes catalyzing reactions with carbanionic intermediates might also catalyze reactions with O_2. Therefore, it came as a surprise to enzymologists when it was shown that ribulose 1,5-bisphosphate carboxylase is capable of catalyzing an oxygenase reaction [see the article by Bowes and Ogren (9) for the original discovery]. Since ribulose 1,5-bisphosphate carboxylase was the first enzyme shown to catalyze an unexpected oxygenase side reaction, we include a discussion of this enzyme here for historical reasons and to use this enzyme as a prototype in our discussion of enzyme-catalyzed oxidase/oxygenase side reactions.

Ribulose 1,5-bisphosphate carboxylase, which is probably the most abundant enzyme on earth, catalyzes the first and controlling step of carbon fixation in photosynthesis (capture of atmospheric CO_2 to form 3-phosphoglycerate). However, ribulose 1,5-bisphosphate carboxylase also catalyzes carbon photorespiration (consumption of O_2 with concomitant production of 2-phosphoglycolate) (9–12). Because of this finding, the enzyme is now generally referred to as ribulose 1,5-bisphosphate carboxylase/oxygenase (Rubisco). Photosynthesis results in a net gain of carbon biomass, whereas photorespiration does not. Photorespiration is energetically inefficient because it does not capture released energy as ATP or as a reduced nicotinamide adenine dinucleotide (phosphate). Thus, the oxygenase activity of Rubisco impedes the carbon biomass productivity of plants (9).

During CO_2 fixation catalyzed by Rubisco, a carbanionic intermediate (enediolate) derived from ribulose 1,5-bisphosphate reacts with CO_2 to form a β-keto acid:

$$^{-2}O_3POCH_2C(OH) = C(O^-)CH(OH)CH_2OPO_3{}^{2-} + CO_2 \rightarrow$$
$$\text{enediolate carbanion intermediate}$$

$$^{-2}O_3POCH_2C(CO_2{}^-)(OH)C(O)CH(OH)CH_2OPO_3{}^{2-}$$
$$\text{β-keto acid intermediate}$$

(1)

which then undergoes hydrolysis at the active site to yield two equivalents of D-3-phosphoglycerate:

$$^{-2}O_3POCH_2C(CO_2^-)(OH)C(O)C(OH)CH_2OPO_3^{2-} + H_2O \rightarrow$$

$$2\ ^-O_2CCH(OH)CH_2OPO_3^{2-} + H^+ \tag{2}$$
D-3-phosphoglycerate

The net carbon gain is 1 (i.e., 5C + 1C → 3C + 3C). In the oxygenase reaction catalyzed by Rubisco, the electrophilic O_2 attacks the enediolate intermediate, generating a peroxide intermediate, which then fragments into phosphoglycolate and D-3-phosphoglycerate. The net reaction is

$$^{-2}O_3POCH_2C(OH) = C(O^-)CH(OH)CH_2OPO_3^{2-} + O_2 \rightarrow$$
enediolate carbanion intermediate

$$^-O_2CCH_2OPO_3^{2-} + \ ^-O_2CCH(OH)CH_2OPO_3^{2-} + H^+ \tag{3}$$
phosphoglycolate   D-3-phosphoglycerate

There is no carbon gain (i.e., 5C → 2C + 3C). These and other reactions catalyzed by wild-type and mutant forms of Rubisco are discussed in more detail by Schloss and Hixon (13).

Although the oxygenase reaction of Rubisco may be regarded as a side reaction, the maximal oxygenase rate at atmospheric levels of O_2 and CO_2 may reach about 50% of the CO_2 fixation rate (13). Because the ratio of CO_2/O_2 utilization varies markedly among CO_2-fixing higher plants, algae, and cyanobacteria, much effort has been devoted to understanding the topology of the active site that controls the CO_2/O_2 utilization ratio in the various Rubisco enzymes (14). It has been hoped by some that by suitable molecular engineering, the CO_2/O_2 utilization ratio in Rubisco of commercially important crops may be increased. If successful, the increased biomass yields would be of enormous benefit to the burgeoning human population (12). At the present time, however, it appears to be extraordinary difficult, if not impossible, to divorce the O_2-consuming reaction of Rubisco from its CO_2-consuming reaction.

Possibly, the oxygenase reaction catalyzed by Rubisco has an important physiological role in its own right and was therefore not eliminated during evolution. Remarkably, the intermediate peroxide in this reaction does not leave the active site, there is no release of reactive oxygen species (ROS), and O_2 is consumed. Therefore, the oxygenase reaction catalyzed by

Rubisco may be a mechanism to lower the concentration of potentially harmful O_2. However, many other enzyme-catalyzed oxidative side reactions may on occasion be at best neutral, or at worst detrimental. Indeed, we suggest later that such side reactions may play roles in cancer and neurodegenerative diseases. As we discuss below, it is likely that even relatively slow oxidative side reactions may be physiologically important in some cases. Thus, although we use the term *side reaction* liberally here, it should be borne in mind that such reactions are not necessarily merely enzymological curiosities and may have biological and clinical relevance.

IV. Reaction of Carbanions with Oxygen Catalyzed by Selected Enzymes (Excluding Pyridoxal 5′-Phosphate–Containing Enzymes)

A. BACKGROUND

By the early 1990s it was well known that Rubisco catalyzes an oxygenase side reaction and that several enzymes catalyze the transfer of electrons from substrate to artificial oxidants. However, the full extent to which carbanion-forming enzymes catalyze side reactions directing the transfer of electrons from substrate to ambient O_2 was not appreciated at that time. In 1991, Abell and Schloss pointed out that any enzyme that generates a carbanionic intermediate is *theoretically* capable of reacting not only with electrophiles (oxidants) other than O_2 but also with O_2 itself (15). Accordingly, these authors investigated several enzymes that catalyze reactions in which carbanionic species are intermediates for their ability to catalyze oxygenase/oxidase side reactions. Carbanion-generating enzymes reported to catalyze oxygenase/oxidase side reactions included *Salmonella typhimurium* acetolactate synthase isozyme II (ALS II), brewer's yeast PDC, *Escherichia coli* glutamate decarboxylase (GAD), *Staphylococcus aureus* fructose 1,6-bisphosphate aldolase, and *E. coli* L-rhamnulose 1-phosphate aldolase (15–17). On the other hand, no oxygenase/oxidase side reactions could be detected for several other carbanion-forming enzymes, including rabbit muscle fructose 1,6-bisphosphate aldolase, *E. coli* [(phosphoribosyl)amino]imidazole carboxylase, *Torula* yeast 6-phosphogluconate dehydrogenase, pig heart isocitrate dehydrogenase ($NADP^+$), baker's yeast triose phosphate isomerase, and *E. coli* L-fuculose 1-phosphate aldolase (15–17).

Four well-studied examples of enzymes that catalyze a paracatalytic side reaction with molecular O_2 are discussed in detail below. Two of these enzymes (ALS II and PDC) are thiamine diphosphate (ThDP)-dependent. The third and fourth examples are fructose-1,6-bisphosphate aldolase and L-rhamnulose-1-phosphate aldolase. These aldolases do not contain an organic cofactor. Several pyridoxal 5'-phosphate (PLP)–containing enzymes that catalyze paracatalytic reactions with O_2 are discussed in Section V. The α-ketoglutarate dehydrogenase complex (KGDHC), which contains ThDP in one of its component enzymes, catalyzes paracatalytic reactions with electron acceptors. Under certain circumstances, KGDHC also catalyzes the formation of $O_2^{\bullet-}$ (dismutating to H_2O_2). Because of the likely biological importance of these reactions catalyzed by KGDHC, they are discussed separately (Sections VII and VIIIC).

B. ACETOLACTATE SYNTHASE

ALSs are enzymes that catalyze important steps in the biosynthetic pathway to the branched-chain amino acids (18). When two equivalents of pyruvate are used as a substrate pair, the product is acetolactate: a precursor of valine and leucine:

$$2 \underset{\text{pyruvate}}{CH_3C(O)CO_2^-} + H^+ \rightarrow \underset{\text{acetolactate}}{CH_3C(O)C(CH_3)(OH)CO_2^-} + CO_2 \quad (4)$$

On the other hand, when pyruvate and α-ketobutyrate are used as a substrate pair, the product is α-aceto-α-hydroxybutyrate:

$$CH_3C(O)CO_2^- + CH_3CH_2C(O)CO_2^- + H^+ \rightarrow$$
$$\underset{\text{α-aceto-α-hydroxybutyrate}}{CH_3C(O)C(CH_2CH_3)(OH)CO_2^-} + CO_2 \quad (5)$$

Hence, the alternative name for ALS is acetohydroxy acid synthase (AHAS). α-Aceto-α-hydroxybutyrate is a precursor of isoleucine.

As noted above, ALS II catalyzes a paracatalytic side reaction with O_2. The ultimate product of the reaction is acetate. The mechanism of the reaction catalyzed by S. typhimurium ALS II has been studied extensively. The enzyme contains both flavin adenine dinucleotide (FAD) and ThDP within the active site. 4,5-Deaza-FAD (a substitute for FAD that is not competent for one-electron chemistry) and added superoxide dismutase had no effect on O_2 consumption, indicating that the FAD in the active site and adventitious metal ions do not play a role in the oxygenase reaction

[(15); see also below]. These data, together with the labeling patterns obtained with $^{18}O_2$, [2-^{13}C]pyruvate, and $H_2^{18}O$ were consistent with the oxygenase reaction:

$$CH_3C(O)CO_2^- + O_2 \rightarrow CH_3CO_3^- + CO_2 \qquad (6)$$
$$\text{pyruvate} \qquad\qquad\qquad \text{peracetate}$$

in which FAD plays mostly a structural role (15).

Taken together, the data suggested that a (hydroxyethyl)thiamine diphosphate hydroperoxide intermediate is converted to peracetate [$CH_3C(O)O_2^-$] and ThDP [reaction (6)]; the peracetate then reacts with one equivalent of pyruvate, forming two equivalents of acetate (19):

$$CH_3C(O)OOH + CH_3C(O)CO_2^- \rightarrow 2\,CH_3CO_2^- + H^+ + CO_2 \qquad (7)$$

The 2 : 1 stoichiometry between acetate produced and O_2 consumed was confirmed in careful labeling studies with [^{18}O]O_2 (15, 19). In the normal condensation reaction, the second pyruvate (or α-ketobutyrate) competes with O_2 for reaction with a common intermediate (20). It was noted that the enzyme was eventually inactivated by the products of the reaction (15). This inactivation is presumably paracatalytic, as noted previously, for example, with rabbit muscle fructose 1,6-bisphosphate aldolase during trapping experiments with the external oxidant hexacyanoferrate(III) (21).

Note that the product of the oxygenase reaction catalyzed by S. typhimurium ALS II has one less carbon than in the substrate (pyruvate). There is no H_2O_2 produced from the peroxide intermediate in the oxygenase side reaction catalyzed by ALS II [reaction (6)], although a peracid is produced as an intermediate. This situation is similar to that observed for the oxygenase reaction catalyzed by Rubisco, where no H_2O_2 is produced [reaction (3)].

S. typhimurium ALS II was previously reported to either reduce the enzyme-bound FAD or to support attack of the carbanionic intermediate on the isoalloxazine ring of the flavin (22, 23). Recently, it was proposed that E. coli ALS II (AHAS) transfers electrons to the enzyme-bound FAD (i.e., reduces the flavin) and produces H_2O_2 as one of the products of the O_2-consuming reaction (24). However, no evidence was provided for H_2O_2 production. Further, as noted above, S. typhimurium ALS II does not produce H_2O_2 under aerobic conditions (15, 25), but does produce peracetate (25). ALS II (ilvG) is thought to have evolved from pyruvate oxidase (*poxB*) or from a common ancestor that used the tightly bound FAD cofactor to convert pyruvate and O_2 to acetate, CO_2, and H_2O_2 (26). Rather than reducing O_2 to H_2O_2, pyruvate oxidase transfers electrons in vivo to a lipophilic quinone (Q_{10}).

A number of commercial herbicides, exemplified by sulfonylureas and imidazolinones, are thought to bind to the evolutionary vestige of the quinone-binding site in ALS (27). Consistent with this notion, water-soluble quinones (Q_0 and Q_1) compete with these herbicides for binding to ALS II and can reversibly inhibit the enzyme in a noncompetitive fashion with respect to pyruvate (27). Similar to the binding of quinones to pyruvate oxidase, the herbicides bind most tightly to ALS II after addition of pyruvate (27). In contrast to pyruvate oxidase, however, *S. typhimurium* ALS II does not reduce Q_0 in the presence of pyruvate (23). On the other hand, if FAD at the active site of ALS II is photoreduced, the enzyme is able to reduce Q_0. However, the enzyme does not catalyze this reaction. Half of the enzyme-bound photoreduced flavin is oxidized by Q_0 at a rate similar to the rate at which free FAD is oxidized by Q_0 ($>10^8\,M^{-1}\,s^{-1}$). The other half of the enzyme-bound photoreduced flavin is oxidized more slowly ($28,200\,M^{-1}\,s^{-1}$) (23).

These data suggest that the flavin quenching observed for *E. coli* ALS II (24) and *S. typhimurium* ALS II (22, 23) is not due to a reduction of the flavin to $FADH_2$, but rather to a reversible attack of the carbanion on the flavin isoalloxazine ring. Furthermore, ALS II, which has been stripped of its tightly bound flavin and reconstituted with FAD, no longer exhibits quenching of flavin absorbance upon addition of pyruvate (28). Substitution of FAD with flavins that are incompetent for one electron transfer (e.g., 8-demethyl-8-chloro-FAD, 5-deaza-5-carba-FAD, or 8-demethyl-8-hydroxy-5-deaza-5-carba-FAD) produces a catalytically competent ALS II, exhibiting 84, 100, or 57% specific activity, respectively, relative to that of enzyme containing FAD (23). Similarly, substitution of the FAD in the active sites of ALS I or ALS III (two additional isozymes of ALS) with 8-demethyl-8-hydroxy-5-deaza-5-carba-FAD has only modest effects on the specific activity of their synthase reactions [2 pyruvates $\rightarrow$ acetolactate + CO_2; reaction (4)], exhibiting 98 and 39% specific activity, respectively, relative to that of enzyme containing FAD (23). By contrast, substitution of the FAD in pyruvate oxidase with 8-demethyl-8-hydroxy-5-deaza-5-carba-FAD results in a catalytically incompetent enzyme (<2% activity) (23).

The accumulated evidence strongly suggests that O_2 reacts directly with the carbanionic intermediate (radical form) generated by the ALS

II–catalyzed decarboxylation of pyruvate. The commercial herbicides that inhibit ALS II (e.g., sulfometuron methyl, a sulfonylurea) trap an O_2-sensitive reaction intermediate on the enzyme, thereby inducing para-catalytic inactivation of the enzyme by O_2 (29). The structures of several ALS–sulfonylurea complexes have been reported (18, 30–32). The crystal structures and structures deduced by homology modeling are consistent with the herbicides closing the active site. Paracatalytic inactivation of ALS, induced by interaction with commercial herbicides, is an important aspect that dramatically enhances herbicidal activity at the level of the whole plant. The peracetate produced in the oxygenase reaction of ALS II is a highly reactive oxidant which will react readily with thiols $(6000 \, M^{-1} \, s^{-1})$, pyruvate $(20 \, M^{-1} \, s^{-1})$, and various assay components (e.g., buffers, cofactors, and other enzymes) (25, 33). Peracetate also reacts with various metals, such as Co^{2+} $(2000 \, M^{-1} \, s^{-1})$, Mn^{2+}, Pb^{2+}, and Fe^{3+}, to produce molecular O_2, light, and other products (e.g., CO_2 and meth-ane) (13, 19, 25, 33). The light produced by peracetate is likely to be a consequence of radiation produced by an activated metal, as it does not have the proper spectral characteristics to be produced by decay of singlet oxygen to its triplet state (13, 19, 25).

C. PYRUVATE DECARBOXYLASE

Yeast PDC, which catalyzes the decarboxylation of pyruvate to acetal-dehyde, utilizes ThDP as a cofactor. The enzyme catalyzes formation of the same (hydroxyethyl)thiamine diphosphate intermediate as ALS II, but lacks the flavin in the active site (15, 25). This enzyme was shown to catalyze an oxygenase side reaction, the end product of which is acetate (15, 25). The reaction proceeds through the oxidation of pyruvate with peracetate as shown for ALS II [reactions (6) and (7)]. The corresponding PDC from *Zymomonas mobilis* supports a somewhat higher level of oxygenase activity than the yeast enzyme (25). The *Z. mobilis* PDC oxygenase activity is about twice that of yeast PDC in the presence of 50 mM 2-(*N*-morpholino)ethanesulfonic acid (MES)–NaOH (pH 6.0), 10 mM $MgCl_2$, 0.1 mM ThDP, and 25 mM sodium pyruvate (33 versus 16 nmol min^{-1} mg^{-1}, respectively) (25).

Yeast PDC, *Z. mobilis* PDC, and ALS II were confirmed to produce a peracid as their initial oxygenase product by relying on the more rapid interaction of peracetate with 5-thionitrobenzoate (TNB) than with pyru-vate (25). By using TNB, a continuous assay for the formation of peracetate was devised for both ALS II and the *Z. mobilis* PDC. The activity of this assay was dependent on the respective substrate and O_2. This assay could also distinguish between the production of H_2O_2 and peracetate, due to the much greater reactivity of peracetate toward TNB (25). Under the conditions used for assay of ALS II, for example, TNB reacted about 3000 times faster with peracetate ($6000\,M^{-1}\,s^{-1}$) than with H_2O_2 ($2\,M^{-1}\,s^{-1}$).

D. ALDOLASES

S. aureus fructose 1,6-bisphosphate aldolase is a class I aldolase. Class I aldolases do not require a cofactor but make use of Schiff-base chemistry involving a lysine residue during the catalytic cycle. *S. aureus* fructose 1,6-bisphosphate aldolase catalyzes the slow uptake of O_2 in the presence of substrate (15). For every equivalent of O_2 taken up, 0.25 equivalent of hydroxypyruvaldehyde phosphate and 0.25 equivalent of 3-phosphoglycerate are generated. Evidently, the hydroperoxide formed from the carbanionic intermediate (1,3-dihydroxyacetone phosphate, enolate form) plus O_2:

$$
\begin{aligned}
&{}^{-2}_{3}OPOCH_2C(O^-) = CHOH + O_2 + H^+ \rightarrow \\
&\quad\text{1,3-dihydroxyacetone phosphate enolate} \\[4pt]
&{}^{-2}_{3}OPOCH_2C(O)CH(OOH)OH \\
&\quad\text{peroxide intermediate}
\end{aligned}
\tag{8}
$$

can break down to form either hydroxypyruvaldehyde phosphate and H_2O_2:

$$
{}^{-2}_{3}OPOCH_2C(O)CH(OOH)OH \rightarrow \underset{\text{hydroxypyruvaldehyde phosphate}}{{}^{-2}_{3}OPOCH_2C(O)C(O)H} + H_2O_2
\tag{9}
$$

or 3-phosphoglycerate and H_2O_2:

$$
\begin{aligned}
&{}^{-2}_{3}OPOCH_2C(O)CH(OOH)OH + H_2O \\
&\rightarrow \underset{\text{3-phosphoglycerate}}{{}^{-2}_{3}OPOCH_2CH(OH)CO_2{}^-} + H_2O_2 + H^+
\end{aligned}
\tag{10}
$$

Interestingly, the enzyme undergoes inactivation during the course of the reaction (15). It should be pointed out that hydroxypyruvaldehyde phosphate in the form of its hydrate (gem-diol) can isomerize to 3-phosphoglycerate under alkaline conditions (19).

In later work, the ability of two metal-containing class II aldolases to catalyze oxygenase-type side reactions was investigated (16, 17). It was found that *E. coli* L-rhamnulose 1-phosphate aldolase catalyzes an O_2-consuming side reaction. In a fashion similar to Rubisco, this enzyme catalyzes the formation of an ene-diolate intermediate. However, the Rubisco oxygenase reaction is unlike that of L-rhamnulose 1-phosphate aldolase, in that the former reaction results in scission of the intermediate with no production of H_2O_2 [reaction (3)], whereas in the side reaction catalyzed by L-rhamnulose 1-phosphate aldolase, H_2O_2 is generated in amounts roughly stoichiometric with O_2 consumption (16).

Unlike *E. coli* L-rhamnulose 1-phosphate aldolase, *E. coli* L-fuculose 1-phosphate aldolase is unable to support an oxygenase/oxidase side reaction (16). Despite limited sequence homology, L-rhamnulose 1-phosphate aldolase and L-fuculose 1-phosphate aldolase are structurally and mechanistically similar (34–36). Thus, some carbanion-forming enzymes have evolved the ability to either fully protect intermediates from attack by O_2 whereas others have not (15) or have substantially more radical-forming character than others have (13).

E. MECHANISMS

The question arises as to why some carbanion-forming enzymes catalyze oxygenase/oxidase side reactions, whereas others do not. In the absence of stabilizing factors, any peroxide intermediate formed between carbanion and O_2 could decompose back to carbanion and O_2. Stabilizing factors could include protonation of the peroxide anion and/or metal coordination (15). These factors may differ even within conserved active sites in closely related enzymes, due to variations in the conformational mobility of the protein. For example, the difference observed in the oxygenase activity between yeast PDC and *Z. mobilis* PDC may correlate with differences in the role of conformational changes in regulation of the enzymes. Whereas the yeast enzyme is subject to allosteric activation by pyruvate, the *Z. mobilis* enzyme is not (37).

It is interesting to note that a metal cofactor is not an absolute requirement for an enzyme to catalyze a side reaction with molecular O_2. For example, *S. aureus* fructose 1,6-bisphosphate aldolase, but not the rabbit muscle fructose 1,6-bisphosphate aldolase, catalyzes an oxygenase side reaction (15). Although the former enzyme was initially misidentified as a class II (metal-dependent) aldolase (15–17), neither of these enzymes possesses a metal cofactor. Thus, with *S. aureus* fructose 1,6-bisphosphate aldolase and rabbit muscle fructose 1,6-bisphosphate aldolase, the difference in reactivity toward molecular O_2 is not related to the presence of metal cofactor.

Nevertheless, in other cases a metal cofactor may be essential for an enzyme to carry out side reactions with molecular O_2. For example, both L-rhamnulose 1-phosphate aldolase and L-fuculose 1-phosphate aldolase contain Zn^{2+} as a cofactor. The substrates for these two enzymes are epimeric at a single position, and both enzymes catalyze similar reactions (35, 36). However, as noted above, only the former enzyme catalyzes an oxidase reaction. Interestingly, L-rhamnulose 1-phosphate aldolase has substantially greater oxidase activity with Co^{2+} or Mn^{2+} as the activating metal than with the natural cofactor Zn^{2+}. Similar to the oxygenase activity of the *Rhodospirillum rubrum* Rubisco, where replacement of active site Mg^{2+} with Co^{2+} results in an enzyme with exclusive oxygenase activity (38), various metals support different relative levels of oxidase activities with the rhamnulose 1-phosphate aldolase. Although different metal ions cause different relative amounts of "normal" versus oxidase/oxygenase activities for both rhamnulose 1-phosphate aldolase and Rubisco, the rank order in which metals support these two activities varies. Moreover, substitution of various metals for Mg^{2+} in the oxygenase reaction of ALS II has only a modest effect on the relative levels of synthase and oxygenase activity (20). Clearly, the relative oxygenase/oxidase activities are not solely a feature of the activating metal ion, but also depend on the structure of the active site.

A summary of the carbanion-generating enzymes (excluding PLP-containing enzymes) currently known to catalyze oxygenase/oxidase side reactions is given in Table 1.

V. Oxidation Reactions Catalyzed by PLP-Containing Amino Acid Decarboxylases

A. BACKGROUND

Healy and Christen in 1973 described the first paracatalytic reaction catalyzed by a PLP-dependent enzyme (2). These authors showed that in the presence of the amino acid substrate analog D,L-*erythro*-β-hydroxy-

TABLE 1
Carbanion-Forming Enzymes (Excluding PLP-Containing Enzymes) That Catalyze
Oxygenase/Oxidase Side Reactions

Enzyme[a]	Oxygenase or Oxidase	Reaction Catalyzed
Ribulose 1,5-bisphosphate carboxylase/oxygenase (Rubisco)	Oxygenase	Ribulose 1,5-bisphosphate + O_2 → phosphoglycolate + D-3-phosphoglycerate
Acetolactate synthase	Oxygenase	Pyruvate + O_2 → peracetate + CO_2 nonenzymatic: pyruvate + peracetic acid → 2 acetate + CO_2
Pyruvate decarboxylase	Oxygenase	Pyruvate + O_2 → peracetate + CO_2 non-enzymatic: pyruvate + peracetic acid → 2 acetate + CO_2
Rhamnulose 1-phosphate aldolase	Oxidase	1,3-Dihydroxyacetone phosphate + O_2 → hydroxypyruvaldehyde phosphate + H_2O_2
Fructose 1,6-bisphosphate aldolase	Oxidase	1,3-Dihydroxyacetone phosphate + O_2 → hydroxypyruvaldehyde phosphate (or 3-phosphoglycerate) + H_2O_2
α-Ketoglutarate dehydrogenase	Oxidase	α-Ketoglutarate → succinate + CO_2 + $2e^-$ (with various electron acceptors)

[a]The enzymes in this list have been purified from various sources. For simplicity the source is not identified. The only case where species differences appear to make a difference is fructose 1,6-bisphosphate aldolase. The enzyme isolated from *Staphylococcus aureus* catalyzes a side reaction with O_2, whereas the enzyme isolated from rabbit muscle does not. For original references, see the text.

aspartate, cytosolic pig heart aspartate aminotransferase catalyzes reduction of the artificial electron acceptor (oxidant) hexacyanoferrate(III). No enzyme-catalyzed reduction was observed in the absence of amino acid (2). At that time, however, no PLP-containing enzyme was known to interact paracatalytically with O_2. Subsequently, several PLP-dependent decarboxylases were shown to react directly with O_2, and these enzymes are described individually below. In three cases [glutamate decarboxylase (GAD), dopa decarboxylase (DDC), ornithine decarboxylase (ODC)], the enzyme-catalyzed oxidative reaction with O_2 is slower than the enzyme-catalyzed "natural" nonoxidative decarboxylation reaction. However, in one case [phenylacetaldehyde synthase (PAAS)], evolutionary pressure has converted an ancestral decarboxylase that presumably catalyzed a nonoxidative amino acid decarboxylation to an enzyme that catalyzes the stoichiometric oxidative decarboxylation of its amino acid substrate. The considerations above raise the intriguing question of why PLP-dependent decarboxylases, among the many disparate classes of PLP-containing enzymes, are apparently unique in their ability to catalyze paracatalytic reactions with O_2.

B. GLUTAMATE DECARBOXYLASE

E. coli GAD catalyzes O_2 uptake in the presence of glutamate (15). The nonoxidative and oxidative decarboxylase reactions are

$$^-O_2CCH_2CH_2C(H)(NH_3{}^+)CO_2{}^- + H^+ \rightarrow$$
$$\text{L-glutamate}$$
$$^-O_2CCH_2CH_2CH_2NH_3{}^+ + CO_2 \tag{11}$$
$$\text{GABA}$$

and

$$^-O_2CCH_2CH_2C(H)(NH_3{}^+)CO_2{}^- + O_2 + H_2O + H^+ \rightarrow$$
$$\text{L-glutamate}$$
$$^-O_2CCH_2CH_2C(O)H + NH_4{}^+ + CO_2 + H_2O_2 \tag{12}$$
$$\text{succinic semialdehyde}$$

respectively. Bertoldi et al. showed that *E. coli* GAD oxidizes α-methyl-L-glutamate to ammonium and levulinate. Only half an equivalent of O_2 was shown to be consumed for every equivalent of ammonium and levulinate produced; no H_2O_2 was detected (39):

$$^-O_2CCH_2CH_2C(CH_3)(NH_3{}^+)CO_2{}^- + \tfrac{1}{2}O_2 + H^+ \rightarrow$$
$$\text{α-methyl-L-glutamate}$$
$$^-O_2CCH_2CH_2C(O)CH_3 + NH_4{}^+ + CO_2 \tag{13}$$
$$\text{levulinate}$$

Bertoldi et al. stated that GABA is also an oxygenase substrate of *E. coli* GAD (39). Since other authors reported that this enzyme could not support an oxidation reaction with GABA (15, 40), the level of oxidase activity with GABA must be quite low.

More recent work has established that the *E. coli* GAD oxidase reaction has a rapid equilibrium ordered kinetic mechanism in which O_2 traps glutamate in a ternary complex with enzyme that turns over to products very slowly (40). Partitioning between the "normal" GABA-forming reaction and the oxidase reaction is shifted dramatically in ^2H_2O relative to that in ^1H_2O. The addition rate of O_2 (k_{cat}/K_m oxygen) and the fraction of glutamate converted to oxidase-specific products (H_2O_2, ammonium, and succinic semialdehyde) are approximately 10-fold greater in ^2H_2O than in ^1H_2O. This appears to be a consequence, under steady-state conditions, of a larger fraction of the enzyme intermediate being present as a radical form in ^2H_2O than in ^1H_2O. It is the radical intermediate that reacts with O_2.

The recombinant isozymes of human GAD that are either GABA-vesicle associated (GAD65) or localized in the cytosol (GAD67) are sensitive to reversible inhibition by molecular O_2 or by nitric oxide (NO) under anaerobic conditions (41). Two other PLP-dependent decarboxylases, that is, two isozymes of porcine cysteine sulfinic acid decarboxylase (CSAD I and CSAD II), are also sensitive to reversible inhibition by O_2 and NO (41). Unlike molecular O_2, which is a biradical (i.e., contains two unpaired electrons), NO has only a single unpaired electron. It is important to specify that experiments with NO were conducted in the absence of O_2, since NO will react rapidly with O_2 to form other products. NO is known to bind to various paramagnetic substances and is a commonly used paramagnetic probe.

E. coli GAD, recombinant human GAD65, recombinant human GAD67, recombinant porcine CSAD I, and recombinant porcine CSAD II are remarkably similar with respect to their sensitivity to reversible inhibition by O_2 ($K_I = 0.2$ to 0.5 mM) or NO ($K_I = 0.2$ to 0.5 mM) (41). However, the inhibition does not reflect the ability of the enzymes to use O_2 as a substrate. (The ability of porcine CSAD I and CSAD II to catalyze an oxidative decarboxylation reaction has not been studied.) Human GAD65 does not consume O_2 (as assessed with an O_2 electrode) or produce H_2O_2 (as assessed by the glutathione–glutathione peroxidase coupled reaction) at levels comparable to those observed for the *E. coli* enzyme (42). Also, the rate at which the *E. coli* enzyme utilizes O_2 in the H_2O_2/NH_4^+/succinic semialdehyde-forming reaction [reaction (12)] is disproportionate to the effect that O_2 has on the GABA-forming reaction (41). Under normal atmospheric conditions, the oxidase reaction catalyzed by *E. coli* GAD proceeds at approximately 0.2% of the rate of the GABA-forming reaction. However, under the same conditions, O_2 inhibits the GABA-forming reaction by 30% relative to the reaction rate observed under anaerobic conditions. In the presence of 1 atm of pure O_2, the GABA-forming reaction is inhibited by 70%. However, under these conditions, the rate of the oxidase reaction is 1.2% of the rate of the GABA-forming reaction.

Clearly, there is more O_2-induced inhibition of the GABA-forming reaction than can be accounted for by diversion of glutamate into the oxidase reaction. Inhibition of the GABA-forming reaction correlates with saturation of the H_2O_2/succinic semialdehyde/NH_4^+-forming reaction. Oxygen ($K_m = 0.56$ mM) traps glutamate ($K_{IA} = 13$ mM) into a complex that turns over to products (H_2O_2, succinic semialdehyde, and NH_4^+) very slowly (0.9 min^{-1}). Although the complex formed between O_2 and human GAD65 does not turn over at a rate comparable to that of *E. coli* GAD, the rate of O_2 release is likely to be similar, since the amount of inhibition (uncompetitive, $K_{II} = 0.2$ mM) of the GABA-forming reaction by O_2 is comparable. In the presence of O_2, GAD should exhibit slow binding kinetics, due to the slow rate of O_2 release ($t_{1/2} = 0.8$ min). [See the article by Schloss (43) for a discussion of slow-binding enzyme inhibition and its relationship to reaction-intermediate analogs.] Further, assuming that O_2 actually binds rapidly to the radical enzyme form that is O_2 sensitive, the abundance of the radical form can be estimated. In ^1H_2O, k_{cat}/K_m for O_2 is 27 M^{-1} s^{-1} and increases to 175 M^{-1} s^{-1} in ^2H_2O (40). If O_2 actually binds to the radical at a diffusion-limited rate [about 2.5×10^9 M^{-1} s^{-1} (13)], only 0.000001% of the enzyme exists in a radical form in steady state. The fraction of GAD existing as a radical in steady state would increase to 0.000007% in ^2H_2O.

Unlike the *E. coli* GAD, human GAD65 is inactivated under aerobic assay conditions in the absence of various thiols. This has precluded the same type of detailed kinetic analysis of human GAD65 that has recently been completed for *E. coli* GAD. Attempts to determine whether human GAD65 can support an oxidase reaction under any conditions, similar to *E. coli* GAD, have met with mixed results (42). However, it has been reported that pig brain GAD produces succinic semialdehyde in the presence of glutamate, with concomitant inactivation of the enzyme by transamination of the PLP cofactor to PMP (44). A similar inactivation of pig brain GAD by aspartate has also been reported (45). It is unclear whether the inactivation of pig brain GAD by transamination is in addition to and independent of interaction of the enzyme with O_2 or whether this may also be an O_2-dependent process. However, inhibition of human GAD65 by O_2, which would interfere with loading of GABA vesicles and GABAergic synaptic function in the brain, is likely to be responsible for O_2-induced seizures (41) (see also Section XIII). Under conditions of hyperbaric oxygen, an unusual cyclic change in the concentration of GAD67 (cytoplasmic GAD), but not in GAD65 (GABA vesicle-bound GAD), has been observed (46). Presumably, this cyclic change in the intracellular level of GAD67 is related to the reversible inhibition of GAD by oxygen, since there was no irreversible loss of either GAD65 or GAD67 during the period of observation.

C. DOPA DECARBOXYLASE

Bertoldi et al. showed that native pig kidney DDC catalyzes oxidation of aromatic amines (39, 47–53). The pig kidney DDC-catalyzed conversion of L-dopa to dopamine and L-5-hydroxytryptophan to serotonin, however, is about 100 times faster than the enzyme-catalyzed conversion of dopamine to 3,4-dihydroxyphenylacetaldehyde and serotonin to 5-hydroxyindole-acetaldehyde, respectively (51). Ammonium production from the amine is stoichiometric with aldehyde formation and with consumption of half an equivalent of O_2. In initial reports, no H_2O_2 formation was detected (51). However, more recently the production of both H_2O_2 and superoxide was inferred, based on the slower rate of oxygen consumption observed in the presence of catalase or superoxide dismutase (53). The nonoxidative decarboxylation of aromatic amino acids is shown as:

$$ArCH_2CH(NH_3{}^+)CO_2{}^- + H^+ \rightarrow ArCH_2CH_2(NH_3{}^+) + CO_2 \qquad (14)$$

Oxidation of the corresponding amines apparently proceeds with the stoichiometry:

$$ArCH_2CH_2(NH_3{}^+) + \tfrac{1}{2}O_2 \rightarrow ArCH_2C(O)H + NH_4{}^+ \qquad (15)$$

Presumably, the oxygen or superoxide produced in the first round of catalysis is consumed as a substrate in a second round of catalysis (53).

Earlier reports that pig kidney DDC catalyzes an abortive half-transamination with L-dopa under aerobic conditions (54) may require reinterpretation, since it appears that abortive transamination competes with the oxygen-consuming reaction, depending on the availability of O_2 or other oxidants (e.g., H_2O_2 or superoxide) (51, 53). Nevertheless, pig kidney DDC can catalyze half-transamination reactions with (1) D-aromatic amino acids under aerobic conditions, (2) with L-aromatic amino acids, D-aromatic amino acids, and aromatic amines under anaerobic conditions, and (3) α-methyl L-dopa under aerobic conditions (48, 51, 53). With α-methyl L-dopa, the product of oxidative decarboxylation is 3,4-dihydroxy-phenylacetone (39, 48, 51, 53):

$$(OH)_2C_6H_3CH_2C(CH_3)(NH_3^+)CO_2^- + \tfrac{1}{2}O_2 + H^+ \rightarrow$$
$$\underset{\text{α-methyl L-dopa}}{}$$

$$\underset{\text{3,4-dihydroxyphenylacetone}}{(OH)_2C_6H_3CH_2C(O)CH_3} + NH_4^+ + CO_2 \tag{16}$$

The partitioning ratio between oxidative decarboxylation and abortive transamination with this substrate is about 100 (51). The oxidized products generated from serotonin (47) and dopamine (51) eventually cause inactivation of the enzyme by covalent modifications of the enzyme/cofactor. Thus, molecular O_2 seems to play a role as substrate and as an enzyme regulator.

Bertoldi et al. showed that mutation of tyrosine 332 to phenylalanine converts pig kidney DDC to a decarboxylation-dependent deaminase, in which decarboxylation of amino acid is *stoichiometric* with oxidation (50, 53). Unlike the reaction catalyzed by the native enzyme, the amine product cannot be detected in the reaction catalyzed by the mutated enzyme. However, under aerobic conditions, both native and mutated enzymes catalyze oxidation of dopamine at about the same rate. Under anaerobic conditions, the mutated enzyme exclusively catalyzes a decarboxylation-dependent half-transamination (50, 53).

Several cases of dopa decarboxylase deficiency have been reported [reviewed by Allen et al. (55)]. The defect is inherited in an autosomal recessive fashion. Although rare, it is suggested that the incidence might be underreported. Activity may be low to undetectable. Some patients, although not all, are helped by treatment with a monoamine oxidase inhibitor, dopamine antagonist, and pyridoxine. In those patients helped by pyridoxine it is possible that the mutation results in less effective binding of PLP in the active site. Since, as noted above, a relatively conservative change in a tyrosine residue to a phenylalanine residue at the active site converts DDC to an oxidative decarboxylase, it would be interesting to determine whether mutations of DDC that have arisen naturally in the human population may enhance oxidation reactions at the expense of nonoxidative decarboxylation.

D. ORNITHINE DECARBOXYLASE

Bertoldi et al. showed that ODC prepared from *Lactobacillus* 30a catalyzes the oxidative deamination/decarboxylation of L-α-methylornithine (39). The product of the reaction was identified as 2-methyl-Δ^1-pyrroline, which arises through cyclization/dehydration of the corresponding aldehyde (5-aminopentan-2-one). One equivalent of ammonium was generated at the expense of half an equivalent of O_2 consumed (39). The decarboxylation reaction with L-ornithine is:

$$^+H_3NCH_2CH_2CH_2C(H)(NH_3{}^+)CO_2{}^- + H^+ \rightarrow$$
L-ornithine

$$^+H_3NCH_2CH_2CH_2CH_2NH_3{}^+ + CO_2$$
putrescine

$$(17)$$

and the oxidative decarboxylation of α-methyl-L-ornithine is:

$$^+H_3NCH_2CH_2CH_2C(CH_3)(NH_3{}^+)CO_2{}^- + \tfrac{1}{2}O_2 + H^+ \rightarrow$$
α-methyl-L-ornithine

$$^+H_3NCH_2CH_2CH_2C(O)CH_3 + NH_4{}^+ + CO_2$$
5-aminopentan-2-one

$$(18)$$

The authors also found that putrescine is oxidatively deaminated by *Lactobacillus* ODC at a relatively slow rate. Half an equivalent of O_2 was consumed per equivalent of putrescine oxidized (39).

The *Lactobacillus* ODC was shown to catalyze a half-transamination reaction with α-methyl-L-ornithine (39). The partition ratio between oxidative deamination and half-transamination with this substrate was 12 : 1 (39). The *Lactobacillus* ODC is therefore remarkably similar to the *E. coli* GAD in its ability to oxidatively deaminate/decarboxylate α-methyl amino acid substrates and to catalyze a low-efficiency half-transamination of PLP with α-methyl amino acid substrate (39).

In earlier work (56), Sakai et al. showed that an ODC purified from *Hafnia alvei* catalyzes the oxidative decarboxylation of L-ornithine to 4-aminobutanal:

$$^+H_3NCH_2CH_2CH_2C(H)(NH_3{}^+)CO_2{}^- + O_2 + H_2O + H^+ \rightarrow$$
L-ornithine

$$^+H_3NCH_2CH_2CH_2C(O)H + NH_4{}^+ + CO_2 + H_2O_2$$
4-aminobutanal

$$(19)$$

which reversibly cyclizes and dehydrates nonenzymatically to Δ^1-pyrroline (not shown). The authors named this activity L-ornithine oxidase (decarboxylating) (OOD). Most turnover events at the active site lead to the release of CO_2 and to the formation of putrescine [reaction (17)]. Occasionally (about one turnover in 160), however, the enzyme converts L-ornithine and O_2 to 4-aminobutanal [reaction (19)]. Unlike DDC, which can oxidize dopamine (the end product of the nonoxidative decarboxylation of substrate L-dopa), ODC/OOD could not be shown to oxidize putrescine (the end product of the nonoxidative decarboxylation of substrate L-ornithine) (56).

E. PHENYLACETALDEHYDE SYNTHASE

Recent work has shown that petunia petal PAAS oxidatively converts L-phenylalanine stoichiometrically to phenylacetaldehyde, ammonium, H_2O_2, and CO_2 (57):

$$\underset{\text{L-phenylalanine}}{C_6H_5CH_2CH(NH_3{}^+)CO_2{}^-} + O_2 + H^+ + H_2O \rightarrow$$

$$\underset{\text{phenylacetaldehyde}}{C_6H_5CH_2C(O)H} + NH_4{}^+ + CO_2 + H_2O_2 \tag{20}$$

Phenylethylamine is neither a product nor a substrate (57). PAAS also converts α-methyl-L-phenylalanine to phenylacetone:

$$\underset{\alpha\text{-methyl-L-phenylalanine}}{C_6H_5CH_2C(CH_3)(NH_3{}^+)CO_2{}^-} + O_2 + H^+ + H_2O \rightarrow$$

$$\underset{\text{phenylacetone}}{C_6H_5CH_2C(O)CH_3} + NH_4{}^+ + CO_2 + H_2O_2 \tag{21}$$

In the presence of catalase, the enzyme can catalyze turnover of phenylalanine to phenylacetaldehyde for several hours without loss of activity. Interestingly, however, in the absence of catalase to trap H_2O_2, PAAS is slowly inactivated (57).

As noted above, native pig kidney DDC catalyzes the *nonoxidative* decarboxylation of dopa to dopamine. However, mutation of an active-site tyrosine residue to a phenylalanine converts pig kidney DDC to an enzyme that stoichiometrically catalyzes an *oxidative* decarboxylation of dopa to an aromatic aldehyde (50, 53). It is interesting to note that during evolution the corresponding residue in the ancestral dopa-decarboxylase-like enzyme was converted from a tyrosine to a phenylalanine in the case of rose PAAS and to a valine in the case of petunia PAAS (57).

An interesting analogous reaction to that catalyzed by PAAS has recently been described in which N-hydroxy-L-phenylalanine is oxidatively decarboxylated to Z-phenylacetaldoxime. The enzyme that catalyzes this reaction, tentatively named N-hydroxy-L-phenylalanine decarboxylase/oxidase, was isolated from a *Bacillus* strain and shown to contain only PLP as a cofactor (58). Although not measured, it is probable that H_2O_2 is generated in the reaction.

VI.   Mechanisms Contributing to the Oxidation Reactions Catalyzed by PLP-Containing Amino Acid Decarboxylases

The oxidation reactions catalyzed by various PLP-containing decarboxylases are shown in Table 2. *E. coli* GAD, *H. alvei* ODC, and petunia PAAS catalyze the oxidative decarboxylation of their respective amino acid substrates to aldehydes, ammonium, and CO_2 with the stoichiometric production of H_2O_2. Oxygen in the aldehyde moiety of the product is derived from water. These enzymes, therefore, clearly meet the definition of oxidases. In the case of the petunia PAAS reaction, net oxidation is stoichiometric with decarboxylation. However, in the case of the *E. coli* GAD or *H. alvei* ODC, oxidative decarboxylation occurs only about once in every 500 or 160 turnovers, respectively, under normal atmospheric conditions (0.24 mM dissolved O_2; under 21% oxygen at 1 atm).

The putative mechanism by which petunia PAAS catalyzes the oxidative decarboxylation of phenylalanine is shown in Figure 1. We suggest that PAAS catalyzes a biradical mechanism similar to that proposed for *E. coli* GAD (41). There are many possible radical mechanisms that could be written for this reaction. Normally, even for reactions that are commonly accepted to involve radical intermediates, such as those involving the isoalloxazine ring of flavin cofactors, the radical intermediates are not shown. This is because, in most cases, radical forms are short-lived, high-energy intermediates, which are usually difficult to observe directly, due to their low abundance in the steady state. However, in the hypothetical reaction sequence illustrated in Figure 1, radical intermediates are included for purposes of illustration only. These are not the only possible radical forms, nor are they likely to build up to detectable levels during steady state. The following mechanistic steps include those commonly accepted

TABLE 2
Carbanion-Forming PLP Enzymes That Catalyze Oxidation Reactions

Enzyme[a]	Oxygenase or Oxidase	Reaction Catalyzed
Escherichia coli glutamate decarboxylase (GAD)	Oxidase	Glutamate + O_2 → succinic semialdehyde + NH_4^+ + CO_2 + H_2O_2
E. coli glutamate decarboxylase (GAD)	Uncertain	α-Methyl-L-glutamate + $\frac{1}{2}O_2$ → levulinate + NH_4^+ + CO_2
Pig kidney dopa decarboxylase (DDC)	Uncertain	Aromatic amine + $\frac{1}{2}O_2$ → aromatic aldehyde + NH_4^+
Pig kidney dopa decarboxylase (DDC)	Uncertain	α-Methyl L-dopa + $\frac{1}{2}O_2$ → dihydroxyphenylacetone + NH_4^+ + CO_2
Lactobacillus ornithine decarboxylase (ODC)	Uncertain	α-Methyl L-ornithine + $\frac{1}{2}O_2$ → 2-methyl-Δ^1-pyrroline + NH_4^+ + CO_2 + H_2O
Lactobacillus ornithine decarboxylase (ODC)	Uncertain	Putrescine + $\frac{1}{2}O_2$ → Δ^1-pyrroline + NH_4^+ + H_2O
Hafnia alvei ornithine decarboxylase (ODC)	Oxidase	L-Ornithine + O_2 → Δ^1-pyrroline + NH_4^+ + CO_2 + H_2O_2
Petunia phenylacetaldehyde synthase (PAAS)	Oxidase	L-Phenylalanine + O_2 → phenylacetaldehyde + NH_4^+ + CO_2 + H_2O_2
Petunia phenylacetaldehyde synthase (PAAS)	Oxidase	α-Methyl-L-phenylalanine + O_2 → phenylacetone + NH_4^+ + CO_2 + H_2O_2

[a]In each case except PAAS, the reactions shown are side reactions (paracatalytic) with O_2. With PAAS the reaction shown is the dominant reaction. For *E. coli* GAD, *H. alvei* ODC, and petunia PAAS, oxidative decarboxylation results in stoichiometric formation of H_2O_2. The enzymes catalyzing these reactions are thus clearly oxidases. For pig heart DDC and *Lactobacillus* ODC, where no H_2O_2 has been directly detected in the oxidation reaction, the situation is not yet clear. For original references, see the text.

for PLP-dependent enzymes and additional steps unique to the oxygen-utilizing reactions. In the absence of substrate, the PLP coenzyme forms a Schiff base with the ε-amino group of a lysine residue at the enzyme active site (E-PLP aldimine). Binding of phenylalanine at the active site leads to formation of a new Schiff base (Phe-PLP aldimine) in which the α-amino group of phenylalanine (Phe) displaces the ε-amino group of the active site lysine. The double bond of the imine is converted from a singlet state (paired electrons) to a triplet state (unpaired electrons) (biradical intermediate 1). Subsequent decarboxylation generates biradical intermediate 2. Since O_2

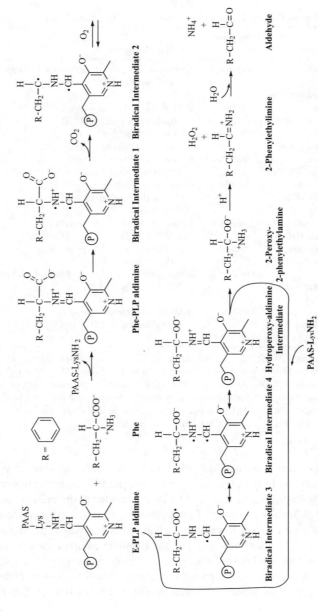

Figure 1. Proposed mechanism for the phenylacetaldehyde synthase (PAAS)-catalyzed reaction leading to biosynthesis of phenylacetaldehyde from L-phenylalanine. [Modified from (57).]

330

is a biradical, it should react extremely rapidly with biradical intermediate 2. Addition of O_2 could theoretically occur at either of the two unpaired electrons of the biradical 2 intermediate (41). Here we show the addition of O_2 at the aldehyde carbon of the intermediate, generating biradical intermediate 3. Transfer of an electron from the nitrogen base to the peroxy radical of biradical intermediate 3 will generate a zwitterionic structure containing a peroxy anion and a radical cation (biradical intermediate 4). This zwitterionic structure is expected to be short-lived, producing a hydroperoxy intermediate. Reaction of the ε-amino group of a lysine residue in the active site will generate the E-PLP aldimine with the liberation of 1-peroxy-2-phenylethylamine, which undergoes nonenzymatic elimination of H_2O_2 with the formation of 2-phenylethylimine. Hydrolysis of the imine will yield phenylacetaldehyde and ammonium. Generation of a biradical form of ribulose 1,5-bisphosphate from an ene-diol intermediate has been proposed as part of the oxygenase reaction of Rubisco (13). Quantum mechanical calculations have indicated that the energy level of the radical intermediate can be reduced substantially by appropriate adjustment of the molecular geometry (59–61). Since the PLP-imine and other PLP intermediates, especially the quinonoid intermediate, more closely resemble the structure of the flavin isoalloxazine ring, it is likely that these radical intermediates will be even lower in energy than those in the oxygenase reaction catalyzed by Rubisco.

The oxidative decarboxylation reactions catalyzed by *E. coli* GAD and *H. alvei* ODC presumably occur via a radical mechanism similar to that depicted in Figure 1 for petunia PAAS. However, there are differences among the three enzymes. Petunia PAAS has evolved to ensure maximal oxidation of substrate intermediate, so that the oxidation is stoichiometric with decarboxylation. On the other hand, the other two enzymes have either evolved to exclude O_2 from the active site or to reduce the steady-state levels of radical intermediates available for reaction with O_2. Whichever method is employed, the oxidase reaction is not totally eliminated, since the oxidative decarboxylation reaction still occurs at 0.2 and 0.6% of the nonoxidative reaction rates, respectively. It will be interesting to compare the active sites of these enzymes to determine those steric factors that may exclude O_2 from the active site or that affect the steady-state levels of radical intermediates.

PAAS, *E. coli* GAD, and *H. alvei* ODC clearly catalyze oxidase reactions as O_2 consumption in the presence of L-amino acid substrate is stoichiometrically coupled to ammonium and H_2O_2 production. However, the situation with the pig kidney DDC is less clear. As noted above, in the oxidation reactions catalyzed by DDC, H_2O_2 cannot be detected directly and only half an equivalent of O_2 is consumed per equivalent of substrate oxidized. There are several possible explanations for this observation. If DDC were to utilize H_2O_2 produced in one round of catalysis as an oxidant in a second round of oxidation, two equivalents of aldehyde or ketone would be produced for each equivalent of oxygen consumed. In this explanation, H_2O_2 would be expected to serve as a substrate for the oxidation reaction. Such an explanation could be consistent with the recent observation that the rate of oxygen consumption by DDC is reduced in the presence of catalase or superoxide dismutase (53). To the best of our knowledge, H_2O_2 has not been tested as an alternative substrate for any of the pig kidney DDC oxygen-consuming reactions. A second possible explanation for the inability to detect the H_2O_2 intermediate directly would be that the pig kidney DDC was prepared by including 0.1 mM dithiothreitol to preserve the enzyme activity (47). Any thiol present in the assay would effectively prevent detection of H_2O_2 produced by DDC (25). However, the interference by thiols in assaying for H_2O_2 would not explain the $1:2$ stoichiometry between O_2 consumed and aldehyde or ketone produced, which was observed for DDC (47, 48, 51), *Lactobacillus 30a* ODC (39), and *E. coli* GAD (39). Since a $1:1$ stoichiometry between O_2 consumed and H_2O_2/aldehyde produced has been observed by other investigators for one type of bacterial ODC (56), *E. coli* GAD (15, 40, 41), and PAAS (57), there are clearly problems in the literature that need to be resolved.

Another point of interest is that all the native decarboxylases listed in Table 3, except pig kidney DDC, can oxidatively decarboxylate their L-amino acid substrates with varying efficiencies, but cannot effectively oxidize the amine product when this is added directly to the enzyme. On the other hand, pig kidney DDC can oxidize aromatic amines, but not the L-amino acid substrate. It would seem, however, that not much perturbation (either of the active site itself or with a modified substrate) is required to enable pig kidney DDC to catalyze an oxidative decarboxylation. Thus, as noted above, a Tyr $\rightarrow$ Phe mutation at position 322 results in an enzyme that can effectively catalyze oxidative decarboxylation of dopa. Additionally, the native enzyme can catalyze the oxidative decarboxylation of L-α-methyldopa (51, 53).

Vitamin B_6 in its active form (usually, PLP, but occasionally, PMP) is a remarkably versatile biological catalyst. Enzyme-catalyzed reactions in

TABLE 3

Relative Nonoxidative Decarboxylation and Oxidation Reactions Catalyzed by Various PLP-Containing Amino Acid Decarboxylases[a]

| | Enzyme | | | | | |
Reaction	Escherichia coli GAD	Native Pig Kidney DDC	Pig Kidney DDC Y332F Mutant	Lactobacillus ODC	Hafnia alvei ODC	Petunia PAAS
Nonoxidative decarboxylation of L-amino acid substrate	+++	+++	–	+++	+++	–
Oxidative decarboxylation of L-amino acid substrate	+	–	+++	+	+	+++
Amine oxidation	+/–	+	+	–	–	–
Oxidative decarboxylation of α-methyl L-amino acid substrate	+	+	ND[b]	+	ND[a]	+++

[a]For details, see the text.
[b]Not determined.

333

which PLP is a cofactor include, but are not limited to, racemization, transamination, α-decarboxylation, β-decarboxylation, β-decarboxylation–addition, reverse aldol condensation, α,β-elimination, α,β-replacement, β,γ-elimination, and β,γ-replacement (62, 63). Most reactions catalyzed by vitamin B_6–containing enzymes are not generally considered to involve radical intermediates. The mechanism of almost all such enzymes is thought to be based on their ability to stabilize high-energy anionic intermediates in their reaction pathways by the pyridinium moiety of PLP/PMP (64). However, a few exceptions are known. One example is *Clostridium subterminale* lysine 2,3-aminomutase (65). This enzyme catalyzes the PLP-dependent interconversion of L-lysine and L-β-lysine. The reaction mechanism includes one-electron chemistry and uses a [4Fe–4S] cluster and *S*-adenosylmethionine (65). Another example of a PLP enzyme that catalyzes a radical mechanism is *Clostridium stricklandii* lysine 5,6-aminomutase. This enzyme catalyzes an adenosylcobalamin (vitamin B_{12})–dependent interconversion of D-lysine with 2,5-diaminohexanoate and of L-β-lysine with 3,5-diaminohexanoate (66). A third vitamin B_6–dependent enzyme that participates in a radical mechanism is CDP-6-deoxy-L-*threo*-D-glycero-4-hexulose-3-dehydrase. This enzyme catalyzes the C-3 deoxygenation in the biosynthesis of 3,6-dideoxyhexoses in *Yersinia pseudotuberculosis* and is a PMP-dependent enzyme that also contains a [2Fe–2S] center (67).

Based on the findings that several PLP-containing decarboxylases can catalyze oxidations with O_2 as side reactions (and in at least one case an oxidative decarboxylation that is the predominant pathway), it would seem that the potential for PLP to participate in radical mechanisms is more widespread than is generally appreciated.

VII. Paracatalytic Oxidation of the α-Ketoglutarate Dehydrogenase–Generated Carbanion

α-Ketoglutarate dehydrogenase (E_1k) is a member of a family of α-keto acid dehydrogenases that also includes pyruvate dehydrogenase (E_1p) and the branched-chain α-keto acid dehydrogenase (E_1b). The enzymes catalyze the first and rate-controlling steps of the overall reactions within the α-ketoglutarate dehydrogenase complex (KGDHC), pyruvate dehydrogenase complex (PDHC), and branched-chain α-keto acid dehydrogenase complex (BCDHC), respectively. These enzyme complexes are similar in their architecture and catalytic mechanism. Both PDHC and KGDHC have been shown to catalyze reactions with external oxidants, which can lead to paracatalytic inactivation. For example, pig heart PDHC was shown to be paracatalytically inactivated in the presence of [^{14}C]pyruvate, ThDP, and the oxidant DCIP (68). Inactivation was accompanied by incorporation of radioactivity into the PDHC complex, mostly into the E_2p component (68). However, most of the work on the paracatalytic reactions of α-keto acid dehydrogenases has been carried out with KGDHC. Therefore, most of the discussion in this section is devoted to this enzyme complex.

The α-ketoglutarate dehydrogenase complex (KGDHC) self-assembles from multiple copies of α-ketoglutarate dehydrogenase (E_1k), dihydrolipoamide succinyl transferase (E_2k), and dihydrolipoamide dehydrogenase (E_3). KGDHC catalyzes the oxidative decarboxylation of α-ketoglutarate (KG) with the concomitant formation of succinyl-CoA and reduction of NAD^+ to NADH (Figure 2). E_1k is a ThDP-dependent enzyme that generates a carbanionic intermediate after the ThDP-dependent decarboxylation of the substrate (Figure 2, reaction (1). In the course of the overall reaction this intermediate undergoes oxidation by the lipoyl moiety covalently bound to the second enzyme of the complex, E_2k (Figure 2, reaction (2)). Free lipoate and lipoamide are relatively poor substrates of the α-keto acid dehydrogenases compared to the protein-bound lipoyl group (69, 70). Nevertheless, free lipoate and lipoamide may be reductively acylated in the E_1k-catalyzed reaction. Also, a number of small-molecular-weight electron acceptors may participate in the E_1k-catalyzed reaction and can therefore be considered as paracatalytic substrates of E_1k. Several aspects of these paracatalytic reactions deserve attention regarding their potential pathophysiological significance.

First, the E_1k-catalyzed oxidation of α-keto acids is supported by structurally and chemically different oxidants (electrophiles) (Table 4). Reaction rates at saturating concentrations of different oxidants may vary by more than an order of magnitude, but similar saturation with KG is observed with these oxidants (Table 4). Thus, the K_m^{KG} values determined for the paracatalytic reactions are close to the K_d^{KG} values measured in independent experiments under equilibrium conditions (74). This finding shows that the E_1k-generated carbanion is formed rapidly compared to the subsequent oxidation step. That is, the oxidoreduction step, which is the slowest step of the overall reaction (reaction (2) in Figure 2), is also slow compared to carbanion formation during paracatalytic side reactions. Most of the paracatalytic substrates [hexacyanoferrate(III), DCIP, 4-chloro-7-nitro-2,1,3-benzoxadiazole (NBD–Cl), tetranitromethane] of E_1k act as oxidants. After their reduction by the carbanionic intermediate at the active

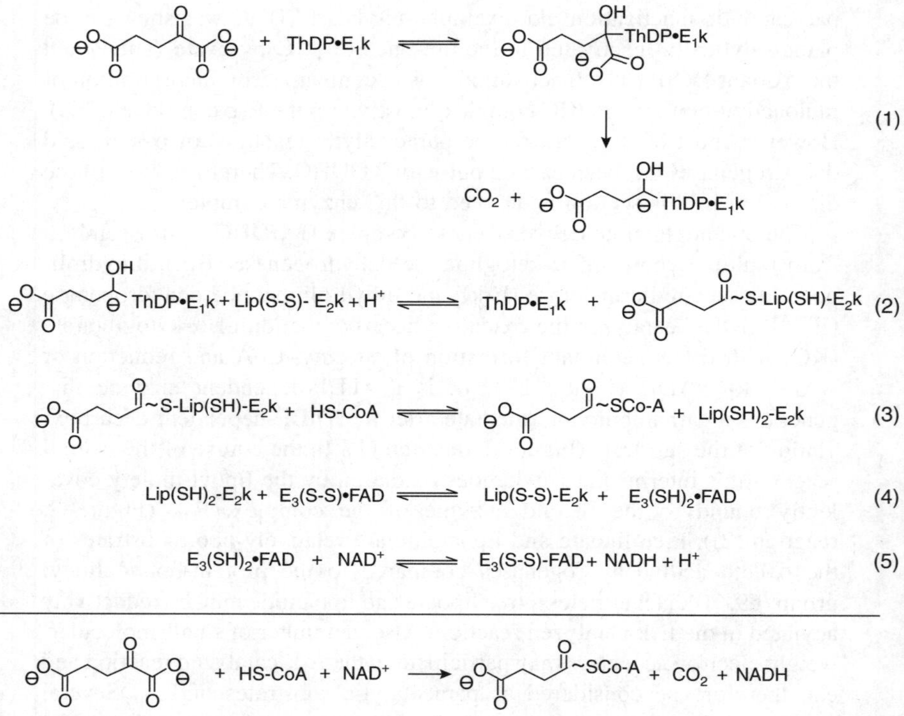

$$(1)$$

$$(2)$$

$$(3)$$

$$Lip(SH)_2\text{-}E_2k + E_3(S\text{-}S)\bullet FAD \rightleftharpoons Lip(S\text{-}S)\text{-}E_2k + E_3(SH)_2\bullet FAD \qquad (4)$$

$$E_3(SH)_2\bullet FAD + NAD^+ \rightleftharpoons E_3(S\text{-}S)\text{-}FAD + NADH + H^+ \qquad (5)$$

Figure 2. α-Ketoglutarate dehydrogenase complex (KGDHC)–catalyzed reaction. In the first step, KG reacts with the ThDP cofactor and is decarboxylated to generate CO_2 and active aldehyde (carbanion) at the active site of E_1k (step 1). In the second step, ThDP is regenerated in the E_1k-catalyzed reductive transacylation reaction with the lipoyl residue covalently bound to E_2k (step 2). This thio ester contains a high-energy bond as depicted by $\sim$. In the next step, at the active site of E_2k, coenzyme A reacts with the succinyl dihydrolipoyl thio ester to generate the high-energy compound succinyl CoA and a dihydrolipoyl residue (step 3). Subsequently, a disulfide linkage in the dihydrolipoyl residue is regenerated concomitant with the $2e^-$ reduction of $E_3(S\text{-}S)\bullet FAD$ (step 4), which in turn is reoxidized back with NAD^+ (step 5). [From (115), with permission.]

site, the resulting succinyl residue is transferred from succinyl-ThDP to water. However, the reaction with 4-chloronitrosobenzene includes both an oxidoreduction step and an acyl transfer step, producing an *N*-succinyl hydroxamic acid (72, 75–77). Analogous reactions are known for other

ThDP-dependent enzymes transforming the "active aldehyde" intermediate of the "natural" substrate in the presence of 4-chloronitrosobenzene to an N-acyl hydroxamic acid (75). The overall reaction catalyzed by such an enzyme (E) is depicted by:

$$\underset{\text{"active aldehyde" intermediate}}{\text{E–ThDP–C(OH)(H)R}} + \underset{\text{nitroso compound}}{\text{Ar–N} = \text{O}} \rightarrow$$

$$\underset{N\text{-acyl hydroxamic acid}}{\text{Ar–N(OH)C(O)R}} + \underset{\text{enzyme-bound coenzyme}}{\text{E–ThDP}} \tag{22}$$

where $R = -CH_3$, $-CH_2OH$, and $-CH_2CH_2CO_2H$ for reactions catalyzed by pyruvate dehydrogenase (E_1p), transketolase, and E_1k, respectively. This finding implicates ThDP-containing enzymes in the metabolic transformation of aromatic nitroso compounds.

A unique feature of E_1k in this paracatalytic reaction is that in addition to the N-acyl hydroxamic acid, E_1k also produces significant amounts of other products. These products include some highly polar unidentified material and N-succinyl 1-hydroxy-4-chloroaniline (75–77). The N-succinyl 1-hydroxy-4-chloroaniline presumably arises through a Bamberger rearrangement, where the –OH group migrates from the nitrogen to the aromatic ring (76). The distribution of the products among the different pathways depends on the reaction conditions (e.g., salts, pH) (77). The distribution also depends on the enzyme source, presumably due to subtle differences in active-site structures (72). Table 4 shows that compared to *E. coli* E_1k, both bovine and porcine E_1k exhibit high affinities and high rates for the paracatalytic reaction with 4-chloronitrosobenzene. This is accompanied by a significant decrease in N-succinyl hydroxamic acid production in favor of the other products originating from an electrophilic aromatic intermediate stabilized at the E_1k active site.

Second, the E_1k-catalyzed paracatalytic reactions occur not only with isolated E_1k, which is *not* the in vivo form of the enzyme, but also with the E_1k in its natural complex-bound state. Moreover, an increase in the maximal reaction rates with artificial electron acceptors in intact KGDHC compared to those with isolated E_1k is observed, probably due to the overall stabilization of E_1k when bound to the complex (Table 5). Thus, incorporation of E_1k into the native structure of the multienzyme complex does not prevent the paracatalytic reactions. These findings further support the potential significance of such reactions in vivo.

TABLE 4

Paracatalytic Reactions of KG Oxidation Catalyzed by E_1k Purified from Different Sources[a]

Enzyme Source	Artificial Electron Acceptor	Relative KG Oxidation Rates (%)	K_m^{KG} (mM)	$K_m^{acceptor}$ at Saturating KG (mM)	pH	Assay Conditions				Refs. for λ_{max} and ϵ
						Concentration of the Acceptor (mM)	Potassium Phosphate (M)	λ_{max} (nm)	ϵ (M⁻¹cm⁻¹)	
Pigeon breast muscle	Hexacyanoferrate(III)	100	0.046 ± 0.013	0.084 ± 0.009	6.3	0.8	0.05	420	1,000	(2)
	DCIP	10	0.050 ± 0.02	0.058 ± 0.012	6.8	0.1	0.1	600	20,800	(2)
	NBD-Cl[b]	3	—	0.3	6.8	1	0.2	540	12,500[c]	(73)
	Tetranitromethane	200	0.043 ± 0.017	0.6 ± 0.2	6.3	1	0.1	350	14,400	(2)
Escherichia coli	4-Chloronitrosobenzene	100	—	[d]	7.5	0.1	0.05	320	—	(72)
Bovine	4-Chloronitrosobenzene	230		0.1						
Porcine	4-Chloronitrosobenzene	380		0.1						

[a]The initial reaction rates were determined and corrected for the non-specific reduction of oxidants in the absence of KG. The data obtained with pigeon breast muscle E_1k are from (71). The data with 4-chlorinitrosobenzene are from (72). λ_{max} and ϵ are the absorbance maximum and extinction coefficient, respectively, of the reduced species obtained after reaction of the redox indicator (artificial electron acceptor) with carbanion intermediate.

[b]KG oxidation by NBD-Cl is highly dependent on conditions; in particular, it is stimulated by increasing concentration of potassium phosphate, is light-sensitive in MOPS buffer, and does not take place in imidazole buffer.

[c]Determined under the indicated conditions from the maximal ΔD_{540} with limiting KG at an excess of E_1k and NBD-Cl. Linear increase in ΔD_{540} was observed with 1 to 4 μM KG.

[d]Exceeds the solubility of 4-chloronitrosobenzene.

Third, an irreversible inactivation of E_1k occurs in the course of both paracatalytic and native reactions (78). This process is different from the inactivation resulting from the nonspecific action of oxidants per se. The catalysis-associated inactivation involves a reactive catalytic intermediate formed during the oxidoreduction step in the presence of both the α-keto acid and oxidant substrates of E_1k. The fate of this intermediate may differ depending on the surrounding medium, and in particular, on the substrates that are present. Table 5 shows that the hexacyanoferrate(III)-dependent oxidation of KG is characterized by an apparent rate constant for irreversible inactivation $(k_{in}) \sim 0.1 \, \text{min}^{-1}$. With another α-keto dicarboxylate substrate of E_1k, α-ketoadipate, the k_{in} value is increased up to fivefold, while dicarboxylates such as glutarate, succinate, and malonate protect E_1k from the inactivation (78).

What are the intermediates and/or side reactions leading to E_1k inactivation? As shown in Fig. 2, the E_1k carbanionic intermediate is normally oxidized by a two-electron acceptor, the E_2k-bound lipoyl residue (reaction (2). However, the intermediate can also be efficiently oxidized by a one-electron acceptor such as hexacyanoferrate(III) (Table 4). A radical of the carbanionic intermediate of isolated E_1k from *E. coli* was observed by EPR, presumably resulting from the one-electron oxidation of the intermediate by molecular oxygen (79). Similar one-electron oxidation reactions take place in pyruvate:ferridoxin oxidoreductases and in chemical models of key intermediates of thiamine catalysis [reviewed by Bunik and Sievers (80)]. A reactive intermediate containing an uncoupled electron generated in the course of a one-electron oxidation could induce a modification of the E_1k active site leading to KGDHC inactivation. This

TABLE 5

Maximal Reaction Rate (V_{max}) and Apparent Rate Constant for Irreversible Inactivation Associated with Turnover (k_{in}), Characteristic of the Paracatalytic Reaction of E_1k Oxidizing KG with Hexacyanoferrate(III)[a]

Enzyme Source	E_1k Preparation	KG: Hexacyanoferrate(III) Oxidoreduction V_{max} (μmol/min·mg)	k_{in} (min^{-1})
Pigeon breast muscle	Isolated E_1k	0.2	0.12
	Complex-bound E_1k	4.2	0.05
Pig heart	Complex-bound E_1k	3.0	0.04
Azotobacter vinelandii	Complex-bound E_1k	2.0	0.08

[a]The data for the mammalian and pigeon E_1k are from (88). The data for the *Azotobacter* E_1k are from (126).

would explain the first-order loss of activity of E_1k operating through a one-electron oxidation in the presence of hexacyanoferrate(III) (Table 4). A strong oxidant of the E_1k-generated carbanion is the stabilized thiyl radical of the complex-bound lipoyl residue. Experiments on the substrate-dependent reduction of the α-keto acid dehydrogenase complexes under anaerobic conditions showed that there is transient formation of complex-bound thiyl radicals of lipoyl residues, which arises from the semiquinone species of the E_3-bound FAD (80). In the course of the anaerobic reduction of KGDHC, a carbon-centered radical and a complex-bound thiyl radical were trapped with 5,5'-dimethyl-1-pyrroline-N-oxide (DMPO) and α-phenyl-*tert*-butylnitrone, respectively. Production of these radicals coincided with E_1k inactivation. These findings support the mechanism of E_1k inactivation during catalysis.

When terminal two-electron oxidation is perturbed or totally blocked (e.g., at limiting or no NAD$^+$), thiyl radicals of the complex-bound lipoyl moieties may be formed. As shown by Bunik and Sievers (80), these radicals react with the E_1k-generated carbanion through one-electron oxidation (reaction (23), with the resulting carbon-centered radical species inactivating E_1k according to the following reactions:

$$E_1k\cdot\text{hydroxybutyryl-ThDP} \quad + \quad \overset{\cdot}{S}\vee SH \quad \rightarrow$$
$$\underset{E_2k}{\big|}$$

$$\text{(23)}$$

$$E_1k\cdot \text{ hydroxybutyryl}^{\cdot}\text{-ThDP} \quad + \quad SH \vee SH$$
$$\underset{E_2k}{\big|}$$

$$E_1k \cdot \text{hydroxybutyryl}^{\bullet}\text{-ThDP} \rightarrow E_1k_{\text{inactive}} \qquad \text{(24)}$$

Further evidence for this mechanism of the E_1k inactivation is provided by protection from the inactivation by a known thiyl radical scavenger thioredoxin (81–83). Thioredoxin prevents the substrate-induced inactivation by catalyzing dismutation of the complex-bound thiyl radicals, which arise upon the KGDHC reduction with KG and CoA (Figure 3). Remarkably, the formation of the thiyl radicals is transient under anaerobic conditions, but becomes permanent in the presence of O_2 because O_2 stimulates formation of the semiquinone species of the complex-bound FADH$_2$ through its one-electron oxidation (80, 82). That is, at saturating concentrations of KG and CoA, the E_2k-bound dihydrolipoyl residues and

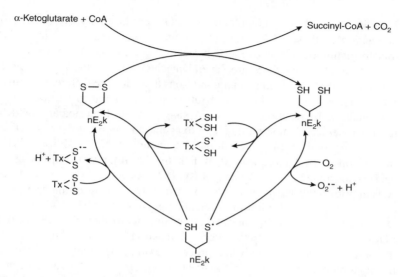

Figure 3.   Formation of KGDHC-dependent radical species and the thioredoxin-catalyzed dismutation of KGDHC-bound dihydrolipoyl radicals. For a detailed discussion, see the text. [Modified from (82).]

E_3-bound FAD of KGDHC are reduced (reactions (1) to (4) in Figure 2). At low or zero concentrations of NAD^+, O_2 competitively oxidizes E_3, with the concomitant formation of $O_2 \bullet^-$ and semiquinone form of the E_3-bound FAD. The semiquinone is in redox equilibrium with the catalytic thiol groups of KGDHC. This redox equilibrium gives rise to the permanent O_2-dependent generation of the complex-bound thiyl radicals of the lipoyl residues. As a result, the E_1k component is inactivated according to reactions (23) and (24) under both aerobic and anaerobic conditions. However, generation of the inactivating thiyl radicals under anaerobic conditions occurs by different mechanisms (incomplete reduction versus an oxidase reaction) and kinetics (transient versus permanent). The oxidative decarboxylation of KG may thus proceed under low concentrations of NAD^+ only when thioredoxin protects E_1k by dismutating the complex-bound thiyl radicals (Figure 3). Under these conditions, however, the electron flow to O_2 is not blocked by E_1k inactivation, leading to accumulation of the superoxide anion radical and the product of its dismutation, H_2O_2 (80). Studies with isolated mitochondria (84, 85) and cultivated neurons (86) indicated the significance of KGDHC for the production of

ROS in vivo. Thus, accumulation of ROS produced by KGDHC and interaction of these ROS with cellular metabolites may complicate the transformation pathways of the E_1k reactive intermediates in vivo.

It is possible that there are several mechanisms for the paracatalytic inactivation of E_1k, arising under conditions that stabilize different reactive intermediates. Some of these inactivation mechanisms may be due to the oxygenase side reaction of a catalysis-generated intermediate, which results in peracid formation. Such an oxygenase reaction at the E_1k active site is supported by the observation that under aerobic conditions, TNB formed upon thiol modification of E_1k by 5,5′-dithiobis(2-nitrobenzoic acid) is consumed in the presence of KG (87). As pointed out above, TNB consumption may be indicative of peracid formation resulting from the oxidase/oxygenase reaction of carbanions generated by ThDP-dependent enzymes (25). A peracid intermediate analogous to that observed in the PDC- and ALS II–catalyzed oxygenase reactions may well be involved, therefore, in the E_1k-catalyzed process, explaining why paracatalytic inactivation occurs not only in the reaction with the one-electron acceptor hexacyanoferrate(III), but also with the two-electron acceptor DCIP, and during the course of the native reaction with the lipoyl-bound moiety at saturating NAD^+. Moreover, an O_2-dependent reaction on E_1k would explain the rather strong α-keto acid–dependent inactivation observed with some preparations of E_1k (88). This inactivation varies with the α-keto acid and E_1k preparation. These variations may be due to differences in the stability of reactive intermediates and to structural differences among the isolated E_1k preparations, resulting, for example, from binding of regulatory metal ions to the enzyme and/or posttranslational modifications. As shown above with other enzymes, such factors greatly affect the paracatalytic oxidase/oxygenase reactions.

Recently, the isolated E_1k component of the KGDHC from *E. coli* was shown to catalyze an O_2-dependent reaction with KG, the product of which could be detected by the Amplex Red–peroxidase assay (79). Because formation of resorufin in the Amplex Red assay is known to be catalyzed by peroxidase in the presence of H_2O_2, the authors concluded that H_2O_2 is the product of the E_1k-catalyzed reaction. However, although the Amplex Red–peroxidase kit is used to assay H_2O_2-generating enzymes, it cannot be employed as an analytical tool to prove H_2O_2 production. In particular, N–C bond scission, which in Amplex Red leads to resorufin, is known to be catalyzed by peroxidase in the presence not only of H_2O_2 but also of peracids (89, 90). Moreover, peracids may generate hydrogen peroxide hydrolytically (90). Taken together, formation of resorufin, interpreted by Frank et al. (79) as H_2O_2 generation by E_1k, is also consistent with an E_1k-dependent oxygenase reaction, resulting in peracid formation. EPR detection of the carbanion/enamine–ThDP radical intermediate in the presence of KG and oxygen (79) led Frank et al. to assume that the peroxide production occurs through two successive $1e^-$ oxidations of the carbanion/enamine–ThDP intermediate by two molecules of O_2, with the resulting superoxide anion radical dismutated to the peroxide. However, the radical intermediate detected could arise from secondary reactions in the presence of peracid or its decomposition product H_2O_2, or in a side reaction concomitant with the peracid formation. In addition, the suggested ping-pong mechanism, in which the ThDP radical intermediate is stable enough to let one superoxide anion radical leave the E_1k active site, and the other oxygen molecule arrive to accept the second electron, seems much less probable than the reaction sequence proposed for other ThDP-dependent enzymes in which the carbanionic intermediate reacts with O_2, resulting in peracid formation (25). Remarkably, the resorufin accumulation curves at 1 mM KG do not show any significant inactivation of the enzyme in the course of this reaction (79). This is in accord with our study of pig heart KGDHC, in which inactivation in the presence of KG alone occurred at a rate one order of magnitude lower ($k = 0.01\,\text{min}^{-1}$) than under conditions in which the thiyl radical of the complex-bound lipoate was efficiently generated ($k = 0.1\,\text{min}^{-1}$) (80). Thus, both the isolated E_1k and complex-bound E_1k generate reactive species by oxidizing α-ketoglutarate with O_2 (79, 80) without significant inactivation. The isolated E_1p component also generates reactive species by oxidizing pyruvate with O_2 (79). However, under normal physiological conditions (i.e., when the components are integrated within the multienzyme complexes and the substrates of other component enzymes are present as well), side reactions may be blocked rapidly by self-inactivation (80–82), which is subject to metabolic and redox regulation by substrate structural analogs (78) and by thioredoxin and/or $NADH/NAD^+$ (81, 82). These data indicate that natural systems may use paracatalytic enzyme inactivation for limiting side reactions in vivo. Moreover, for enzymes that produce highly reactive compounds, self-inactivation in the course of catalysis was shown to be a feasible means of preventing cellular accumulation of such reactive compounds (91, 92). KGDHC may be classified as one such enzyme, as it produces ROS, and in vivo ROSs are known to be signal transducers that become toxic at excessively high levels.

VIII. Potential Role of Paracatalytic Side Reactions in Diseases

A. BACKGROUND

During the course of evolution, nature has made numerous efforts to prevent O_2 and other electrophiles from reacting at the active sites of enzymes that generate carbanionic and/or radical intermediates. For some enzyme reactions that do not benefit from radical chemistry, carbanionic intermediates may not require sequestration from molecular O_2, but may still need to be protected from other possible reactants. Those enzymes that do support reactions with substantial radical character must somehow prevent reaction with molecular O_2 and other radical species that may be present. With the luxury of hindsight, we now know that enzymes are only as "perfect" as required by natural selection (13). In other words, excluding O_2 may or may not be desirable, but will be decided upon by selection pressure.

Tipping the balance from an inefficient to an efficient oxygenase/ oxidase in an enzyme that catalyzes the formation of a carbanionic intermediate may only require surprisingly minor alteration in an enzyme's architecture. There are a number of examples when post-translational modification of amino acid residues of proteins, which typically arise from the oxygenation of aromatic residues, covalent cross-linking of amino acid residues, or cyclization or cleavage of internal amino acid residues, may give rise to altered or even new catalytic functions (93). Thus, as noted above, native pig kidney DDC does not oxidatively decarboxylate a significant fraction of L-dopa, although it can oxidize dopamine at a relatively low rate (51, 53). A Y332F mutation results in an enzyme that retains the ability to catalyze the oxidation of dopamine at a relatively low rate. However, this mutated enzyme catalyzes a very active oxidative decarboxylation of L-dopa. The V_{max} of this reaction is at least as great as, and possibly greater than, that of the nonoxidative decarboxylation of L-dopa catalyzed by the native enzyme (50, 53). The affinity of the mutated enzyme for L-dopa is, however, somewhat lower than that exhibited by the native enzyme (50). These findings may have important ramifications. It is conceivable, for example, that mutation or chemical modification could give rise to a modified form of DDC with substantial oxidase/oxygenase activity. Such modifications of DDC could convert an enzyme that catalyzes a purely nonoxidative dopa decarboxylation reaction to an enzyme that catalyzes an exclusive or partial oxidative decarboxylation of L-dopa to an aromatic aldehyde (i.e., 3,4-dihydroxyphenylacetaldehyde;

dopal). Aromatic aldehydes are generated from the action of DDC on dopamine and serotonin (47), which can then inactivate the enzyme. A damaged or mutant DDC may generate a greater amount of reactive aromatic aldehyde than native DDC, resulting in pathology. It is interesting to note that dopal is far more toxic to the substantia nigra than is dopamine (94). It has been suggested that dopal is a major neurotoxin contributing to the death of neurons in the substantia nigra (94). Although dopal is generally regarded as arising via the action of monoamine oxidases on dopamine, it is conceivable that this reactive aldehyde may also arise as a new activity of a damaged DDC to catalyze oxidative decarboxylation of L-dopa or to increased ability of the enzyme to oxidize dopamine.

In this regard, as noted above, recent findings suggest that defects in DCC, although rare, occur in human populations (55). As summarized by Allen et al. (55), children with DCC deficiency present early in life with axial hypotonia, hypokinesia, choreoathetosis, developmental delay and episodes of dystonia, limb hypertonia, and oculogyric crises. Autonomic symptoms are common, including nasal congestion, temperature instability, and excessive sweating. Presumably, the symptoms are related to low levels of dopamine and serotonin (and products derived therefrom). However, the possibility should also be considered that a mutated enzyme can convert dopa to neurotoxic dopal.

As noted above, petunia PAAS has strong sequence homology to aromatic amino acid decarboxylases (57). Based on this finding and the finding that only a relatively simple mutation can convert DDC from a nonoxidative decarboxylase to an oxidative decarboxylase (50), it is reasonable to assume that nature readily exploited mutations in the basic structure of L-aromatic amino acid decarboxylases to maximize interaction of carbanionic intermediates with O_2 in the petunia PAAS. On the other hand, other PLP-dependent decarboxylases, where oxidative decarboxylation of the L-amino acid to an aldehyde is not quantitatively an important metabolic process, have evolved to minimize the interaction of carbanionic intermediates with molecular O_2. The fact that the reaction with O_2 has not been eliminated altogether suggests that due to the topology of the active site, some oxidation with O_2 is unavoidable if nonoxidative decarboxylation is to proceed at a reasonable rate. However, it is also possible that nature may use these side reactions with O_2 to regulate decarboxylase activity or to balance normal physiological function at various oxygen tensions. For example, inhibition of the human vesicle-associated GAD (GAD65) and porcine CSAD I/II (enzymes that catalyze an important step in the taurine biosynthesis pathway) by molecular O_2 may be involved in maintaining the balance between excitatory and inhibitory signaling during changes in atmospheric pressure. Failure to maintain balance due to excessive inhibition of GAD65 is thought to contribute to the genesis of oxygen-induced seizures (e.g., seizures that result from extended underwater dives on compressed air) (41, 46).

Certain enzymes can catalyze new or increased paracatalytic reactions if the natural metal ion cofactor is replaced by an inappropriate metal ion. For example, as noted above, replacement of active site Mg^{2+} with Co^{2+} in *Rhodospirillum rubrum* Rubisco converts the enzyme from a partial oxygenase into an exclusive oxygenase (38). We suggest that enhancement of oxygenase/oxidase activity in some carbanion-forming enzymes by alterations of the topology of the active site and/or exposure to an inappropriate metal may also occur in mammalian enzymes and contribute to metabolic stress in various diseases. For example, zinc homeostasis is thought to be perturbed in Alzheimer disease (AD) brain (95). Even sub-μM amounts of Zn^{2+} can promote, by as much as fivefold, the NADH oxidation by O_2 catalyzed by the E_3 component of KGDHC (96).

B. POSSIBLE ROLE OF ThDP IN THE PARACATALYTIC BIOACTIVATION OF *N*-NITROSO COMPOUNDS AND RELEVANCE TO CANCER INDUCTION

As discussed above, some nitroso compounds are converted to *N*-acyl hydroxamic acids in the presence of "active aldehyde" [reaction (22)] at the active sites of E_1p, transketolase, and E_1k (97). Due to its ability to accommodate bulky substrates at the active site, the BCDHC can participate in the biotransformation of *N*-nitroso aromatic compounds producing *N*-hydroxyarylamines (98). In a similar fashion, a transketolase-mediated reaction of nitroso aromatic compounds with fructose-6-phosphate can produce *N*-hydroxy-*N*-arylglycolamides (98). The formation of *N*-hydroxy-*N*-acetamide and *N*-hydroxyarylamines is facilitated by PDHC and BCDHC, respectively (98, 99). The formation of these derivatives is enhanced by cofactors (Mg^{2+} and ThDP) and inhibited by thiamine thiazolone pyrophosphate, a specific inhibitor of ThDP-dependent enzymes (98). These reactions may be very important in the bioactivation (toxification) of *N*-nitroso compounds.

Humans are exposed to *N*-nitroso compounds by exogenous routes (from processed meats in particular) and by endogenous routes. A variety of *N*-nitroso compounds are derived primarily from reaction of amides, substituted ureas, or aromatic amines within the stomach and intestines as a result of oral and gastric bacterial degradation of nitrates to nitrites, both common food additives used in the processing of ham, sausages, and other meat products to prevent the colonization of *Clostridia* bacteria. Red meat may contribute to the formation of colorectal cancers by exposure to *N*-nitroso compounds, and endogenous exposure to *N*-nitroso compounds is dose-dependently related to the amount of red meat in the diet (100). The polycyclic aromatic hydrocarbons (PAHs) contained in smoked and charcoal-broiled foods can be converted to *N*-substituted aromatic compounds and direct-acting mutagens by treatment with nitrite in acidic media. Thus, simultaneous consumption of charcoal-broiled and smoked foods with nitrite-containing food may contribute to esophageal, stomach, and colorectal tumors (101). Similarly, intragastric nitrosation of 4-chloroindole and of 4-chloro-6-methoxyindole in fava beans (*Vicia faba*) with nitrite yields a direct-acting mutagen, 4-chloro-2-hydroxy-N^1-nitroso-indolin-3-one oxime, a stable α-hydroxy *N*-nitrosylated compound (102). Other naturally occurring *N*-substituted aromatic compounds have been identified in a widely consumed tea from India (Kashmir), which are responsible for a high incidence of esophageal and stomach cancer in that area. Mononitrosocaffeidine and dinitrosocaffeidine are obtained from in vitro nitrosation of caffeidine, a hydrolysis product of caffeine present in the Kashmir tea (103). In addition, after nitrosation, naturally occurring aromatic amines, especially 2-carboline derivatives such as harman, norharman, harmaline, harmalol, harmine, and harmol, become more mutagenic to *Salmonella typhimurium* (104). Under gastric conditions, piperidine, detected in popular white and black pepper powders, reacts readily with nitrite to form carcinogenic *N*-nitroso-piperidine (105).

N-Substituted aromatic compounds have been implicated in induction of cancers and DNA mutations due to metabolic transformation into *N*-hydroxyarylamines and *N*-acyl hydroxamic acids, which can react with DNA after bioactivation into reactive electrophiles. As noted above, both of these classes of compounds can arise through paracatalytic reactions of aromatic *N*-nitroso compounds with aldehydes at the active sites of a number of ThDP-containing enzymes. The mechanism by which hydroxylamines and *N*-acyl hydroxamic acids are bioactivated is beyond the scope of this review, but it is worth mentioning that the bioactivation of the latter group of compounds appears to involve *N*-acyltransferase reactions (106). These reactions show certain degrees of similarity to *S*-acyltransferase reactions catalyzed by the ThDP-dependent E_1p and E_1k (99).

C.   CRITICAL INVOLVEMENT OF KGDHC IN NEURODEGENERATIVE
DISEASES

Considerable evidence has accumulated during the past two decades that reduction of brain KGDHC activity is inherent in the pathophysiology of thiamine deficiency and neurodegenerative diseases, such as AD, Parkinson disease, Huntington disease, Wernicke–Korsakoff disease, and progressive supranuclear palsy [for a review, see the article by Gibson et al. (107)]. Whether the neurodegeneration-associated inactivation of KGDHC is primary or secondary, the loss of activity would strongly aggravate the disease state, due to the key position of KGDHC at a branch point of the central metabolism of mitochondria, organelles known to play an important role in cell death. In particular, AD is associated with loss of neurons with a high KGDHC content and with a specific decrease of KGDHC activity in cortex and hippocampus (108). Cortex and hippocampus provide trophic support for cholinergic neurons of the forebrain basal nucleus and produce nerve growth factor (NGF), a deficit of which occurs in AD brain (109). Impairment of energy metabolism due to diminished KGDHC activity will disturb the levels of trophic factors.

Although the mechanisms by which KGDHC is inactivated in diseased brain are not clear, it is thought that metabolic imbalance and oxidative stress under disease conditions may contribute to the reduced KGDHC activity. The findings summarized in the present review may suggest potential mechanisms. For example, KGDHC activity is diminished in Wernicke–Korsakoff syndrome, a neuropsychiatric disorder that is induced in humans by thiamine deficiency as a result of chronic alcoholism (110) and in some cases by hyperemesis gravidarum. Curiously, however, the content of the thiamine-dependent E_1k is increased by as much as threefold in rats exposed to ethanol (111). Taken together, these data suggest that increased E_1k expression may be a compensatory response to E_1k-dependent KGDHC inactivation resulting from alcohol-induced stress. It is well established that excessive alcohol consumption increases the NADH/NAD^+ ratio in liver. This increased NADH/NAD^+ ratio should promote thiyl radical formation by KGDHC, thereby increasing E_1k inactivation according to reactions (23) and (24). The possibility of paracatalytic inactivation of E_1k under pathological conditions is supported by our finding that the glutamate excitotoxicity toward neurons is significantly alleviated by synthetic analogs of KG, which protect KGDHC from paracatalytic inactivation (112). Because glutamate excitotoxicity is known to contribute greatly to hypoxic–ischemic damage to both adult and developing brain, we tested analogs of KG in vivo as neuroprotective agents, using a model of prenatal hypoxia. Exposure of pregnant rats to KG analogs was shown to alleviate hypoxic damage to their offspring (113). It is worth noting that the glutamate neurotoxicity observed in neurodegenerative diseases is associated with increased ROS production in overstimulated neurons (114). Using specific inhibitors of cellular E_1k (115), we showed that neuronal KGDHC contributes significantly to the ROS production in response to glutamate overstimulation (86). The biological significance of the KGDHC-dependent paracatalytic reactions involving E_1k-E_3 interaction is supported further by the fact that KGDHC from brain possesses a specific isoform of E_1k and a lower E_3 content than that of the heart complex (116). Structural differences between the two E_1k isoforms (117) and the role of E_1k in the E_3 binding to the complex (118) suggest that the lower E_3 content of brain KGDHC is determined by the structure of the brain-specific isoform of E_1k. The decreased content of E_3 in the brain KGDHC may thus represent an additional means by which the brain is protected from excessive ROS generation.

On the basis of the observed diversity of the oxidants accepted at the E_1k active site (Table 4) and diverse chemistry of the interaction of these oxidants with the E_1k active site as discussed in the present review, we wish to emphasize the possibility that *some xenobiotics and some metabolites not on the main catalytic sequence shown in Figure 2 may be paracatalytic substrates of E_1k in vivo*, oxidizing the carbanionic intermediate formed in reaction (1). These paracatalytic reactions may result in products detrimental to KGDHC itself and/or to the mitochondrial milieu. For example, E_1k paracatalytic substrates might very well include cellular quinones, which are well known to be highly electrophilic. This possibility is supported by study of the cellular toxicity of coenzyme Q_0. KGDHC interacts with this quinone in vivo as documented by adduction of coenzyme Q_0 to E_2k and protection from this modification by coenzymes Q_1 and Q_2 upon incubation of cells with the various quinones (119).

An important factor in the KG plus CoA-dependent generation of ROS by KGDHC is that the ROS are produced concomitantly with, and in the vicinity of, CO_2/HCO_3^-, which are known to greatly increase the damaging potential of ROS (120, 121) and peroxynitrite (122). For example, bicarbonate enhances the radical-dependent oligomerization and nitration of the neuronal presynaptic protein α-synuclein, which is known to be a component of potentially toxic Lewy bodies in neurodegenerative disorders (122).

Furthermore, CO_2/HCO_3^- mediates the Mn^{2+}-catalyzed decomposition of H_2O_2, initiating a chain of biologically hazardous processes (121, 123). The structural and functional interaction of E_1k and E_3 within KGDHC is especially important in this regard because (1) the E_3-produced ROS may be coupled to KG oxidative decarboxylation, resulting in CO_2 formation, and (2) some mammalian E_1k subunits are known to bind Mn^{2+} and other metal ions in addition to the catalytically essential Mg^{2+} (124, 125). We conclude that changes in the cellular proteome and metabolome that occur as a result of neurodegenerative diseases may affect paracatalytic reactions of E_1k. Stimulation of such reactions may not only decrease the energy production by KGDHC but also damage the cellular milieu through generation of reactive metabolites by KGDHC. Further study of the mechanism of the E_1k catalysis and biologically occurring paracatalytic substrates will lead to a greater understanding of the complex processes which take place in diseased brain. Such an understanding may suggest new avenues for safe and effective treatment modalities against devastating neurodegenerative diseases.

IX. Conclusions

Many enzymes possess the ability to catalyze paracatalytic (including oxygenase/oxidase) side reactions with carbanionic intermediates. In some cases, this may be the result of an evolutionary balance between generating a biologically acceptable level of the "natural" product and a biologically acceptable level of a toxic or "nonproductive" side product. These side reactions may have additional consequences. For example, they may be used to regulate O_2 levels or enzyme activities. However, under pathological conditions, these side reactions may contribute to cancer induction and to the dysfunctions of cerebral metabolism during aging or in neurodegenerative diseases. The possibility exists that products resulting from enzyme-catalyzed paracatalytic reactions, including oxygenase/oxidase side reactions, may be elevated in aging and neurodegenerative diseases by (a) provision of an inappropriate metal, (b) increased availability of reactive electrophile/oxidant (e.g., quinones), and/or (c) oxidative stress–induced modification of an enzyme. Inhibition of PLP-dependent decarboxylases, such as human GAD65, by molecular O_2 could be responsible for acute manifestations of oxygen toxicity, such as O_2-induced seizures during diving. Disruption of glutamate and dopamine levels as a result of dysregulation of GAD and DDC may also have strong implications for neurodegenerative diseases. Finally, decreased NADH production by KGDHC, which is a robust finding for several neurodegenerative diseases, no doubt contributes to the decline in cerebral energy metabolism in many neurodegenerative diseases. However, paracatalytic activities of KGDHC may be activated under these conditions and aggravate the disease state due to other factors as well. The disease-induced modifications of enzymes and cellular milieu (in particular, perturbed thiol redox status, an increased $NADH/NAD^+$ ratio, and a perturbed metal ion balance) may increase oxidative stress in AD brain by stimulating production of $O_2 \bullet^-$ and H_2O_2. Pharmacological treatments may have side effects through the paracatalytic generation of hazardous compounds. Together, these factors could lead to a vicious cycle in which cellular metabolism is compromised.

Acknowledgements

V.I.B. greatly acknowledges support from Russian Foundation of Basic Research (grants 09-04-90473 and 10-04-90007) and long-standing support of her work on the mechanism of the KGDHC catalysis and regulation by the Alexander von Humboldt Foundation (Bonn, Germany). J.V.S. gratefully acknowledges support by the Office of Naval Research (N00014-94-1-0457; N00014-00-1-01-02; N00014-03-1-0450) and NIH (RO1GM48568), U.S.-Israel Binational Science Foundation, and the Gustavus and Louise Pfeiffer Research Foundation. Work cited from the A.J.L.C. laboratory was supported by NIH grants PO1 AG14930 and RO1 AG19589, and work from the N.D. laboratory was supported in part by the U.S.–Israel Binational Agriculture Research and Development funds, grant US-3437-03.

References

1. Healy, M. J., and Christen, P. (1972) Reaction of the carbanionic aldolase-substrate intermediate with tetranitromethane: identification of the products, hydroxypyruvaldehyde phosphate and D-5-ketofructose 1,6-diphosphate, *J. Am. Chem. Soc. 94*, 7911–7916.

2. Healy, M. J., and Christen, P. (1973) Mechanistic probes for enzymatic reactions: oxidation–reduction indicators as oxidants of intermediary carbanions (studies with aldolase, aspartate aminotransferase, pyruvate decarboxylase, and 6-phosphogluconate dehydrogenase), *Biochemistry 12*, 35–41.

3. Christen, P., Anderson, T. K., and Healy, M. J. (1974) H_2O_2 oxidizes an aldolase dihydroxyacetone phosphate intermediate to hydroxymethylglyoxal phosphate, *Experientia 30*, 603–605.

4. Christen, P., and Gasser, A. (1976) Oxidation of the carbanion intermediate of trans-aldolase by hexacyanoferrate(III), *J. Biol. Chem. 251*, 4220–4223.

5. Christen, P., Cogoli-Greuter, M., Healy, M. J., and Lubini, D. (1976) Specific irreversible inhibition of enzymes concomitant to the oxidation of carbanionic enzyme– substrate intermediates by hexacyanoferrate(III), *Eur. J. Biochem. 63*, 223–231.

6. Christen, P. (1977) Paracatalytic enzyme modification by oxidation of enzyme– substrate carbanion intermediates, *Methods Enzymol. 46*, 48–54.

7. Cogoli-Greuter, M., Hausner, U., and Christen, P. (1979) Irreversible inactivation of pyruvate decarboxylase in the presence of substrate and an oxidant: an example of paracatalytic enzyme inactivation, *Eur. J. Biochem. 100*, 295–300.

8. Christen, P., and Gasser, A. (1980) Production of glycolate by oxidation of the 1,2-dihydroxyethyl-thamin-diphosphate intermediate of transketolase with hexacyanoferrate (III) or H_2O_2, *Eur. J. Biochem. 107*, 73–77.

9. Bowes, G., and Ogren, W. L. (1972) Oxygen inhibition and other properties of soybean ribulose 1,5-diphosphate carboxylase, *J. Biol. Chem. 247*, 2171–2176.

10. Ryan, F. J., and Tolbert, N. E. (1975) Ribulose diphosphate carboxylase/oxygenase: III. Isolation and properties, *J. Biol. Chem. 250*, 4229–4233.

11. Chen, Z. X., Chastain, C. J., Al-Abed, S. R., Chollet, R., and Spreitzer, R. J. (1988) Reduced CO_2/O_2 specificity of ribulose-bisphosphate carboxylase/oxygenase in a temperature-sensitive chloroplast mutant of *Chlamydomonas*, *Proc. Natl. Acad. Sci. USA 85*, 4696–4699.

12. Yu, G.-X., Park, B.-H., Chandramohan, P., Geist, A., and Samatova, N. F. (2005) An evolution-based analysis scheme to identify CO_2/O_2 specificity-determining factors for ribulose 1,5-bisphosphate carboxylase/oxygenase, *Protein Eng. Des. Sel. 18*, 589–596.

13. Schloss, J. V., and Hixon, M. S. (1998) Enol chemistry and enzymology, in *Comprehensive Biological Catalysis*, Vol. 2, *Reactions of Nucleophilic/Carbanionoid Carbon*, Sinnod, M., Ed., Academic Press, London, pp. 43–114.

14. Ogren, W. L. (2003) Affixing the O to Rubisco: discovering the source of photorespiratory glycolate and its regulation, *Photosynth. Res. 76*, 53–63.

15. Abell, L. M., and Schloss, J. V. (1991) Oxygenase side reactions of acetolactate synthase and other carbanion-forming enzymes, *Biochemistry 30*, 7883–7887.

16. Hixon, M., Sinerius, G., Schneider, A., Walter, C., Fessner, W.-D., and Schloss, J. V. (1996) Quo vadis photorespiration: a tale of two aldolases, *FEBS Lett. 392*, 281–284.

17. Fessner, W.-D., Schneider, A., Held, H., Sinerius, G., Walter, C., Hixon, M., and Schloss, J. V. (1996) The mechanism of class II, metal-dependent aldolases, *Angew. Chem. Int. Ed. Engl. 35*, 2219–2221.

18. Duggleby, R. G., McCourt, J. A., and Guddat, L. W. (2008) Structure and mechanism of inhibition of plant acetohydroxyacid synthase, *Plant Physiol. Biochem. 46*, 309–324.

19. Hixon, M. S. (1997) Oxygen consuming side reactions of carbanion forming enzymes: the accidental oxygenases, Ph.D. dissertation, University of Kansas.

20. Tse, J. M.-T., and Schloss, J. V. (1993) The oxygenase reaction of acetolactate synthase, *Biochemistry 32*, 10398–10403.

21. Lubini, D. G. E., and Christen, P. (1979) Paracatalytic modification of aldolase: a side reaction of the catalytic cycle resulting in irreversible blocking of two active-site lysyl residues, *Proc. Natl. Acad. Sci. USA 76*, 2527–2531.

22. Schloss, J. V. (1984) Interaction of the herbicide sulfometuron methyl with acetolactate synthase: a slow-binding inhibitor, in *Flavins and Flavoproteins*, Bray, R. C., Engel, P. C., and Mayhew, S. G., Eds., Walter de Gruyter, New York, pp. 737–740.

23. Schloss, J. V., Ciskanik, L., Pai, E. F., and Thorpe, C. (1991) Acetolactate synthase: a deviant flavoprotein, in *Flavins and Flavoproteins*, Curti, B., Ronchi, S., and Zanetti, G., Eds., Walter de Gruyter, New York, pp. 907–914.

24. Tittmann, K., Schröder, K., Golbik, R., McCourt, J., Kaplun, A., Duggleby, R. G., Barak, Z., Chipman, D. M., and Hübner, G. (2004) Electron transfer in acetohydroxy acid synthase as a side reaction of catalysis. Implications for the reactivity and partitioning of the carbanion/enamine form of (α-hydroxyethyl)thiamin diphosphate in a "nonredox" flavoenzyme, *Biochemistry 43*, 8652–8661.

25. Schloss, J. V., Hixon, M. S., Chu, F., Chang, S., and Duggleby, R. G. (1996) Products formed in the oxygen-consuming reactions of acetolactate synthase and pyruvate decarboxylase, in *Biochemistry and Physiology of Thiamin Diphosphate Enzymes*, Bisswanger, H., and Schellenberger, A., Eds., Intemann, Prien, Germany, pp. 580–585.

26. Grabau, C., and Cronan, J. E., Jr. (1986) Nucleotide sequence and deduced amino acid sequence of *Escherichia coli* pyruvate oxidase, a lipid-activated flavoprotein, *Nucleic Acids Res. 14*, 5449–5460.

27. Schloss, J. V., Ciskanik, L. M., and Van Dyk, D. E. (1988) Origin of the herbicide binding site of acetolactate synthase, *Nature 331*, 360–362.

28. J. V. Schloss (E.I. Du Pont de Nemours & Co.), unpublished work, 1984.

29. Schloss, J. V. (1994) Recent advances in understanding the mechanism and inhibition of acetolactate synthase, in *Chemistry of Plant Protection*, Vol. 10, Stetter, J., Ed., Springer-Verlag, Berlin, pp. 3–14.

30. Ibdah, M., Bar-Ilan, A., Livnah, O., Schloss, J. V., Barak, Z., and Chipman, D. M. (1996) Homology modeling of the structure of bacterial acetohydroxy acid synthase and examination of the active site by site-directed mutagenesis, *Biochemistry 35*, 16282–16291.

31. McCourt, J. A., Pang, S. S., Guddat, L. W., and Duggleby, R. G. (2005) Elucidating the specificity of binding of sulfonylurea herbicides to acetohydroxyacid synthase, *Biochemistry 44*, 2330–2338.

32. McCourt, J. A., Pang, S. S., King-Scott, J., Guddat, L. W., and Duggleby, R. G. (2006) Herbicide-binding sites revealed in the structure of plant acetohydroxyacid synthase, *Proc. Natl. Acad. Sci. USA 103*, 569–573.

33. Chipman, D., Barak, Z., and Schloss, J. V. (1998) Biosynthesis of 2-aceto-2-hydroxy acids: acetolactate synthases and acetohydroxyacid synthases, *Biochim. Biophys. Acta 1385*, 401–419.

34. Grueninger, D., and Schulz, G. E. (2008) Antenna domain mobility and enzymatic reaction of L-rhamnulose-1-phosphate aldolase, *Biochemistry 47*, 607–614.

35. Kroemer, M., and Schulz, G. E. (2002) The structure of L-rhamnulose-1-phosphate aldolase (class II) solved by low-resolution SIR phasing and 20-fold NCS averaging, *Acta Crystallogr. D 58*, 824–832.

36. Kroemer, M., Merkel, I., and Schulz, G. E. (2003) Structure and catalytic mechanism of L-rhamnulose-1-phosphate aldolase, *Biochemistry 42*, 10560–10568.

37. Hong, J., Sun, S., Derrick, T., Larive, C., Schowen, K. B., and Schowen, R. L. (1998) Transition-state theoretical interpretation of the catalytic power of pyruvate decarboxylases: the roles of static and dynamical considerations, *Biochim. Biophys. Acta 1385*, 187–200.

38. Christeller, J. T. (1981) The effects of bivalent cations on ribulose bisphosphate carboxylase/oxygenase, *Biochem. J. 193*, 839–844.

39. Bertoldi, M., Carbone, V., and Borri Voltattorni, C. (1999) Ornithine and glutamate decarboxylases catalyse an oxidative deamination of their α-methyl substrates, *Biochem. J. 342*, 509–512.

40. Chang, S. (2005) On the oxygenase activity of bacterial glutamate decarboxylase: an examination of the kinetic mechanism, Ph.D. dissertation, University of Kansas.

41. Davis, K., Foos, T., Wu, J.-Y., and Schloss, J. V. (2001) Oxygen-induced seizures and inhibition of human glutamate decarboxylase and porcine cysteine sulfinic acid decarboxylase by oxygen and nitric oxide, *J. Biomed. Sci. 8*, 359–364.

42. Davis, K., and Schloss, J. V. (1999) Unpublished work, University of Kansas, Lawrence, KS.

43. Schloss, J. V. (1988) Significance of slow binding enzyme inhibition and its relationship to reaction-intermediate analogues, *Acc. Chem. Res. 21*, 348–353.

44. Choi, S. Y., and Churchich, J. E. (1986) Glutamate decarboxylase side reactions catalyzed by the enzyme, *Eur. J. Biochem. 160*, 515–520.

45. Porter, T. G., and Martin, D. L. (1987) Rapid inactivation of brain glutamate decarboxylase by aspartate, *J. Neurochem. 48*, 67–72.

46. Li, Q., Guo, M., Xu, X., Xiao, X., Xu, W., Sun, X., Tao, H., and Li, R. (2008) Rapid decrease of GAD 67 content before the convulsion induced by hyperbaric oxygen exposure, *Neurochem. Res. 33*, 185–193.

47. Bertoldi, M., Moore, P. S., Maras, B., Dominici, P., and Borri Voltattorni, C. B. (1996) Mechanism-based inactivation of dopa decarboxylase by serotonin, *J. Biol. Chem. 271*, 23954–23959.

48. Bertoldi, M., Dominici, P., Moore, P. S., Maras, B., and Borri Voltattorni, C. B. (1998) Reaction of dopa decarboxylase with α-methyldopa leads to an oxidative deamination producing 3,4-dihydroxyphenylacetone, an active site directed affinity label, *Biochemistry 37*, 6552–6561.

49. Bertoldi, M., Frigeri, P., Paci, M., and Borri Voltattorni, C. B. (1999) Reaction specificity of native and nicked 3,4-dihydroxyphenylalanine decarboxylase, *J. Biol. Chem. 274*, 5514–5521.

50. Bertoldi, M., Gonsalvi, M., Contestabile, R., and Borri Voltattorni C. B. (2002) Mutation of tyrosine 332 to phenylalanine converts dopa decarboxylase into a decarboxylation-dependent oxidative deaminase, *J. Biol. Chem. 277*, 36357–36362.

51. Bertoldi, M., and Borri Voltattorni, C. (2003) Reaction and substrate specificity of recombinant pig kidney dopa decarboxylase under aerobic and anaerobic conditions, *Biochim. Biophys. Acta 1647*, 42–47.

52. Bertoldi, M., Cellini, B., Maras, B., and Borri Voltattorni, C. (2005) A quinonoid is an intermediate of oxidative deamination reaction catalyzed by dopa decarboxylase, *FEBS Lett. 579*, 5175–5180.

53. Bertoldi, M., Cellini, B., Montioli, R., and Borri Voltattorni, C. (2008) Insights into the mechanism of oxidative deamination catalyzed by dopa decarboxylase, *Biochemistry 47*, 7187–7195.

54. O'Leary, M. H., and Baughn, R. L. (1977) Decarboxylation-dependent transamination catalyzed by mammalian 3,4-dihydroxyphenylalanine decarboxylase, *J. Biol. Chem. 252*, 7168–7173.

55. Allen, G. F., Land, J. M., and Heales, S. J. (2009) A new perspective on the treatment of aromatic L-amino acid decarboxylase deficiency, *Mol. Genet. Metab. 97*, 6–14.

56. Sakai, K., Miyasako, Y., Nagatomo, H., Watanabe, H., Wakayama, M., and Moriguchi, M. (1997) L-Ornithine decarboxylase from *Hafnia alvei* has a novel L-ornithine oxidase activity, *J. Biochem. 122*, 961–968.

57. Kaminaga, Y., Schnepp, J., Peel, G., Kish, C. M., Ben-Nissan, G., Weiss, D., Orlova, I., Lavie, O., Rhodes, D., Wood, K., et al. (2006) Plant phenylacetaldehyde synthase is a bifunctional homotetrameric enzyme that catalyzes phenylalanine decarboxylation and oxidation, *J. Biol. Chem. 281*, 23357–23399.

58. Kato, Y., Tsuda, Y., and Asano, Y. (2007) Purification and partial characterization of N-hydroxy-L-phenylalanine decarboxylase/oxidase from *Bacillus* sp. strain Ox-B1, an enzyme involved in the "aldoximine–nitrile" pathway, *Biochim. Biophys. Acta 1774*, 856–965.

59. Tapia, O., and Andrés, J. (1992) Towards an explanation of carboxylation/oxygenation bifunctionality in Rubisco: transition structure for the carboxylation reaction of 2,3,4-pentanetriol, *J. Mol. Eng. 2*, 37–41.

60. Andrés, J., Safont, V. S., and Tapia, O. (1992) Straining the double bond in 1,2-dihydroxyethylene: a simple theoretical model for the enediol moiety in Rubisco's substrate and analogs, *Chem. Phys. Lett. 198*, 515–520.

61. Andrés, J., Safont, V. S., Queralt, J., and Tapia, O. (1993) A theoretical study of the singlet-triplet energy gap dependence upon rotation and pyramidalization for 1,2-dihydroxyethylene: a simple model to study the enediol moiety in Rubisco's substrate, *J. Phys. Chem. 97*, 7888–7893.

62. Dolphin D., Poulson, R., and Avramović, O., Eds. (1986) *Vitamin B₆ Pyridoxal Phosphate*, Wiley, New York.

63. Amadasi, A., Bertoldi, M., Contestabile, R., Bettati, S., Cellini, B., di Salvo, M. L., Borri Voltattorni, C., Bossa, F., and Mozzarelli, A. (2007) Pyridoxal 5′-phosphate enzymes as targets for therapeutic agents, *Curr. Med. Chem. 14*, 1291–1324.

64. Agnihotri, G., and Liu, H. W. (2001) PLP and PMP radicals: a new paradigm in coenzyme B₆ chemistry, *Bioorg. Chem. 29*, 234–257.

65. Lepore, B. W., Ruzicka, F. J., Frey, P. A., and Ringe, D. T. (2005) The x-ray crystal structure of lysine-2,3-aminomutase from *Clostridium subterminale*, *Proc. Natl. Acad. Sci. USA 102*, 13819–13824.

66. Berkovitch, F., Behshad, E., Tang, K. H., Enns, E. A., Frey, P. A., and Drennan, C. L. (2004) A locking mechanism preventing radical damage in the absence of substrate, as revealed by the x-ray structure of lysine 5,6-aminomutase, *Proc. Natl. Acad. Sci. USA 101*, 15870–15875.

67. Agnihotri, G., Liu, Y. N., Paschal, B. M., and Liu, H. W. (2004) Identification of an unusual [2Fe–2S]-binding motif in the CDP-6-deoxy-D-glycero-l-threo-4-hexulose-3-dehydrase from *Yersinia pseudotuberculosis*: implication for C-3 deoxygenation in the biosynthesis of 3,6-dideoxyhexoses, *Biochemistry 43*, 14265–14274.

68. Sümegi, B., and Alkonyi, I. (1983) Paracatalytic inactivation of pig heart pyruvate dehydrogenase complex, *Arch. Biochem. Biophys. 223*, 417–424.

69. Frey, P. A., Flournoy, D. S., Gruys, K., and Yang, Y.-S. (1989) Intermediates in reductive transacetylation catalyzed by pyruvate dehydrogenase complex, *Ann. NY Acad. Sci. 573*, 21–35.

70. Graham, L. D., Packman, L. C., and Perham, R. N. (1989) Kinetics and specificity of reductive acylation of lipoyl domains from 2-oxo acid dehydrogenase multienzyme complexes, *Biochemistry 28*, 1574–1581.

71. Bunik, V. I. (1987) Unpublished work, Lomonosov Moscow State University, Moscow, Russia.

72. Doerge, D. R., and Corbett, M. D. (1985) The action of α-ketoglutarate dehydrogenase on 4-chloronitrosobenzene: evidence for species-dependent differences in active site properties, *Comp. Biochem. Physiol. 80C*, 161–165.

73. Carlberg, I., and Mannervik, B. (1980) Interaction of 2,4,6-trinitrobenzenesulfonate and 4-chloro-7-nitrobenzo-2-oxa-1,3-diazole with the active sites of glutathione reductase and lipoamide dehydrogenase, *Acta Chem. Scand. B 34*, 144–146.

74. Bunik, FV. I., and Gomazkova, V. S. (1996) Study of 2-oxoglutarate dehydrogenase by the method of chemical modification of amino acid residues, in *Chemical Modification of Enzymes*, Kurganov, B. I., Nagradova, N. K., and Lavrik, O. I., Eds., Nova Science Publishers, New York, pp. 479–521.

75. Corbett, M. D., and Chipko, B. R. (1980) Comparative aspects of hydroxamic acid production by thiamine-dependent enzymes, *Bioorg. Chem. 9*, 273–287.

76. Corbett, M. D., Corbett, B. R., and Doerge, D. R. (1982) Hydroxamic production and active-site induced Bamberger rearrangement from the action of α-ketoglutarate dehydrogenase on 4-chloronitrosobenzene, *J. Chem. Soc. Perkin Trans. I* 345–350.

77. Corbett, M. D., Doerge, D. R., and Corbett, B. R. (1983) Hydroxamic acid production by α-ketoglutarate dehydrogenase: 2. Evidence for an electrophilic reaction intermediate at the enzyme active site, *J. Chem. Soc. Perkin Trans. I* 765–769.

78. Bunik, V. I., and Pavlova, O. G. (1997) Inactivation of α-ketoglutarate dehydrogenase during its enzymatic reaction, *Biochemistry (Moscow) 62*, 973–982.

79. Frank, R. A. W., Kay, C. W. M., Hirst, J., and Luisis, B. F. (2008) Off-pathway oxygen-dependent thiamine radical in the Krebs cycle, *J. Am. Chem. Soc. 130*, 1662–1668.

80. Bunik, V. I., and Sievers, C. (2002) Inactivation of the 2-oxo acid dehydrogenase complexes upon generation of intrinsic radical species, *Eur. J. Biochem.* 269, 5004–5015.

81. Bunik, V. (2000) Increased catalytic performance of the 2-oxoacid dehydrogenase complexes in the presence of thioredoxin, a thiol-disulfide oxidoreductase, *J. Mol. Catal. B* 8, 165–174.

82. Bunik, V. I. (2003) 2-Oxo acid dehydrogenase complexes in redox regulation, *Eur. J. Biochem.* 270, 1036–1042.

83. Bunik, V., Raddatz, G., Lemaire, S., Meyer, Y., Jacquot, J.-P., and Bisswanger, H. (1999) Interaction of thioredoxins with target proteins: role of particular structural elements and electrostatic properties of thioredoxins in their interplay with 2-oxoacid dehydrogenase complexes, *Protein Sci.* 8, 65–74.

84. Starkov, A. A., Fiskum, G., Chinopoulos, C., Lorenzo, B. J., Browne, S. E., Patel, M. S., and Beal, M. F. (2004) Mitochondrial α-ketoglutarate dehydrogenase complex generates reactive oxygen species, *J. Neurosci.* 24, 7779–7788.

85. Tretter, L., and Adam-Vizi, V. (2004) Generation of reactive oxygen species in the reaction catalyzed by α-ketoglutarate dehydrogenase, *J. Neurosci.* 24, 7771–7778.

86. Zündorf, G., Kahlert, S., Bunik, V. I., and Reiser, G. (2009) α-Ketoglutarate dehydrogenase contributes to production of reactive oxygen species in glutamate-stimulated hippocampal neurons in situ, *Neuroscience* 158, 610–616.

87. Bunik, V. I. (1989) Unpublished work, Lomonosov Moscow State University, Moscow, Russia.

88. Pavlova, O. G. (1996) Interaction of α-ketoglutarate dehydrogenase from pigeon breast muscle with the α-keto substrate and its structural analogs, Ph.D. dissertation, Lomonosov Moscow State University, Moscow, Russia.

89. Kedderis, G. L., and Hollenberg, P. F. (1983) Characterization of the N-demethylation reactions catalyzed by horseradish peroxidase, *J. Biol. Chem.* 258, 8129–8138.

90. Schonbaum, G. R., and Lo, S. (1972) Interaction of peroxidases with aromatic peracids and alkyl peroxides. Product analysis, *J. Biol. Chem.* 247, 3353–3360.

91. Varfolomeev, S. D. (1984) Enzyme inactivation in the reaction process. Regulatory role [Russian], *Biokhimiia* 49, 723–735.

92. Sud'ina, G. F., Kobel'kov, G. M., and Varfolomeev, S. D. (1987) The macrokinetic behavior of an enzymatic system with an enzyme inactivated in the reaction, *Biotechnol. Bioeng.* 29, 625–632.

93. Davidson, V. L. (2007) Protein-derived cofactors. Expanding the scope of post-translational modifications, *Biochemistry* 46, 5283–5292.

94. Burke, W. J., Li, S. W., Williams, E. A., Nonneman, R., and Zahm, D. S. (2003) 3,4-Dihydroxyphenylacetaldehyde is the toxic dopamine metabolite in vivo: implications for Parkinson's disease pathogenesis, *Brain Res.* 989, 205–213.

95. Finefrock, A. E., Bush, A. I., and Doraiswamy, P. M. (2003) Current status of metals as therapeutic targets in Alzheimer's disease, *J. Am. Geriatr. Soc.* 51, 1143–1148.

96. Gazaryan, I. G., Krasnikov, B. F., Ashby, G. A., Thorneley, R. N. F., Kristal, B. S., and Brown, A. M. (2002) Zinc is a potent inhibitor of thiol oxidoreductase activity and stimulates reactive oxygen species production by lipoamide dehydrogenase, *J. Biol. Chem.* 277, 10064–10072.

97. Corbett, M. D., and Corbett, B. R. (1986) Effect of ring substituents on the transketolase-catalyzed conversion of nitroso aromatics to hydroxamic acids, *Biochem. Pharmacol. 35*, 3613–3621.

98. Yoshioka, T., and Uematsu, T. (1998) Biotransformation of nitroso aromatic compounds and 2-oxo acids to *N*-hydroxy-*N*-arylacylamides by thiamine-dependent enzymes in rat liver, *Drug Metab. Dispos. 26*, 705–710.

99. Yoshioka, T., Ohno, H., and Uematsu, T. (1996) Pyruvate dehydrogenase complex-catalyzed formation of *N*-arylacetohydroxamic acids from nitroso aromatic compounds in rat isolated cells and perfused organs, *J. Pharmacol. Exp. Ther. 279*, 1282–1289.

100. Ferguson, L. R. (2002) Natural and human-made mutagens and carcinogens in the human diet, *Toxicology 181–182*, 79–82.

101. Kangsadalampai, K., Butryee, C., and Manoonphol, K. (1997) Direct mutagenicity of the polycylic aromatic hydrocarbon-containing fraction of smoked and charcoal-broiled foods treated with nitrite in acid solution, *Food Chem. Toxicol. 35*, 213–218.

102. Yang, D., Tannenbaum, S. R., Büchi, G., and Lee, G. C. (1984) 4-Chloro-6-methoxy-indole is the precursor of a potent mutagen (4-chloro-6-methoxy-2-hydroxy-1-nitroso-indolin-3-one oxime) that forms during nitrosation of the fava bean (*Vicia faba*), *Carcinogenesis 5*, 1219–1224.

103. Ivankovic, S., Seibel, J., Komitowski, D., Spiegelhalder, B., Preussmann, R., and Siddiqi, M. (1998) Caffeine-derived N-nitroso compounds: V. Carcinogenicity of mono-nitrosocaffeidine and dinitrosocaffeidine in bd-ix rats, *Carcinogenesis 19*, 933–937.

104. Lin, J. K., Wu, S. S., and Chen, J. T. (1986) Mutagenicities of nitrosated carboline derivatives, *Proc. Natl. Sci. Counc. Repub. China B 10*, 280–286.

105. Tricker, A. R., Pfundstein, B., Kälble, T., and Preussmann, R. (1992) Secondary amine precursors to nitrosamines in human saliva, gastric juice, blood, urine and faeces, *Carcinogenesis 13*, 563–568.

106. Kumano, T., Yoshioka, T., and Uematsu, T. (1986) Comparative effect of chemical structure of chlorinated *N*-hydroxy-*N*-acyl-aminobiphenyl ethers and their related compounds on rat liver cytosol-catalyzed transacylation, *Drug Metab. Dispos. 14*, 487–493.

107. Gibson, G. E., Blass, J. P., Beal, M. F., and Bunik, V. (2005) The α-ketoglutarate-dehydrogenase complex: a mediator between mitochondria and oxidative stress in neurodegeneration, *Mol. Neurobiol. 31*, 43–63.

108. Ko, L. W., Sheu, K. F.-R., Thaler, H. T., Markesbery, W. R., and Blass, J. P. (2001) Selective loss of KGDHC-enriched neurons in Alzheimer temporal cortex: Does mitochondrial variation contribute to selective vulnerability? *J. Mol. Neurosci. 17*, 361–369.

109. Hefti, F., and Weiner, W. J. (1986) Nerve growth factor and Alzheimer's disease, *Ann. Neurol. 20*, 275–281.

110. Butterworth, R. F., and Leong, D. K. (1996) Thiamine deficiency (Wernicke's) enceph-alopathy: pathophysiologic mechanisms and development of positron emission tomog-raphy (PET) ligands, in *Biochemistry and Physiology of Thiamin Diphosphate Enzymes*, Bisswanger, H., and Schellenberger, A., Eds., Intemann, Prien, Germany, pp. 409–417.

111. Venkatraman, A., Landar, A., Davis, A. J., Chamlee, L., Sanderson, T., Kim, H., Page, G., Pompilius, M., Ballinger, S., Darley-Usmar, V., and Bailey, S. M. (2004) Modification of

the mitochondrial proteome in response to the stress of ethanol-dependent hepatotoxicity, *J. Biol. Chem. 279*, 22092–22101.

112. Kabysheva, M. S., Storozhevykh, T. P., Pinelis V. G., and Bunik V. I. (2009) Synthetic regulators of the 2-oxoglutarate oxidative decarboxylation alleviate the glutamate excitotoxicity in cerebellar granule neurons, *Biochem. Pharmacol. 77*, 1531–1540.

113. Graf, A., Kabysheva, M., Klimuk, E., Trofimova, L., Dunaeva, T., Zundorf, G., Kahlert, S., Reiser, G., Storozhevykh, T., Pinelis, V., et al. (2009) Role of 2-oxoglutarate dehydrogenase in brain pathologies involving glutamate neurotoxicity, *J. Mol. Catal. B61, 60*, 80–87.

114. Kahlert, S., Zündorf, G., and Reiser, G. (2005) Glutamate-mediated influx of extracellular Ca^{2+} is coupled with reactive oxygen species generation in cultured hippocampal neurons but not in astrocytes, *J. Neurosci. Res. 79*, 262–271.

115. Bunik, V. I., Denton, T. T., Xu, H., Thompson, C. M., Cooper, A. J. L., and Gibson, G. E. (2005) Phosphonate analogues of α-ketoglutarate inhibit the activity of the α-ketoglutarate dehydrogenase complex isolated from brain and in cultured cells, *Biochemistry 44*, 10552–10561.

116. Bunik, V., Kaehne, T., Degtyarev, D., Shcherbakova, T., and Reiser, G. (2008) Novel isoform of 2-oxogluratate dehydrogenase is identified in brain, but not in heart, *FEBS J. 275*, 4990–5006.

117. Bunik, V. I., and Degtyarev, D. (2008) Structure-function relationships in the 2-oxo acid dehydrogenase family: substrate-specific signatures and functional predictions for the 2-oxoglutarate dehydrogenase-like proteins, *Proteins 71*, 874–890.

118. McCartney, R. G., Rice, J. E., Sanderson, S., Bunik, V., Lindsay, H., and Lindsay, J. G. (1998) Subunit interactions in the mammalian α-ketoglutarate dehydrogenase complex: evidence for direct association of the α-ketoglutarate dehydrogenase (E1) and dihydrolipoamide dehydrogenase (E3) components, *J. Biol. Chem. 273*, 24158–24164.

119. MacDonald, M. J., Husain, R. D., Hoffmann-Benning, S., and Baker, T. R. (2004) Immunochemical identification of coenzyme Q_0-dihydrolipoamide adducts in the E2 components of the α-ketoglutarate and pyruvate dehydrogenase complexes partially explains the cellular toxicity of coenzyme Q_0, *J. Biol. Chem. 279*, 27278–27285.

120. Elam, J. S., Malek, K., Rodriguez, J. A., Doucette, P. A., Taylor, A. B., Hayward, L. J., Cabelli, D. E., Valentine, J. S., and Hart, P. J. (2003) An alternative mechanism of bicarbonate-mediated peroxidation by copper–zinc superoxide dismutase: rates enhanced via proposed enzyme-associated peroxycarbonate intermediate, *J. Biol. Chem. 278*, 21032–21039.

121. Liochev, S. I., and Fridovich, I. (2004) Carbon dioxide mediates Mn(II)-catalyzed decomposition of hydrogen peroxide and peroxidation reactions, *Proc. Natl. Acad. Sci. USA 101*, 12485–12490.

122. Andrekopoulos, C., Zhang, H., Joseph, J., Kalivendi, S., and Kalyanaraman, B. (2004) Bicarbonate enhances α-synuclein oligomerization and nitration: intermediacy of carbonate radical anion and nitrogen dioxide radical, *Biochem. J. 378*, 435–447.

123. Berlett, B. S., Chock, P. B., Yim, M. B., and Stadtman, E. R. (1990) Manganese(II) catalyzes the bicarbonate-dependent oxidation of amino acids by hydrogen peroxide and

the amino acid–facilitated dismutation of hydrogen peroxide, *Proc. Natl. Acad. Sci. USA* *87*, 389–393.

124. Gomazkova, V. S. (1973) Effect of thiamine pyrophosphate and of the ions of divalent metals on the activity and stability of the α-ketoglutarate decarboxylase from the breast muscle of the pigeon [Russian], *Biokhimiia 38*, 756–762.

125. Markiewicz, J., and Strumilo, S. (1997) The effect of Mn^{2+} on the catalytic function of heart muscle 2-oxoglutarate dehydrogenase complex, *Biochem. Arch. 13*, 127–129.

126. Bunik, V., Westphal, A. H., and de Kok, A. (2000) Kinetic properties of the 2-oxoglutarate dehydrogenase complex from *Azotobacter vinelandii* evidence for the formation of a precatalytic complex with 2-oxoglutarate, *Eur. J. Biochem. 267*, 3583–3591.

AUTHOR INDEX

Advances in Enzymology and Related Areas of Molecular Biology, Volume 77
Edited by Eric J. Toone   Copyright © 2011 John Wiley & Sons, Inc.

SUBJECT INDEX

Advances in Enzymology and Related Areas of Molecular Biology, Volume 77
Edited by Eric J. Toone   Copyright © 2011 John Wiley & Sons, Inc.